A

TEXT-BOOK

OF THE

SCIENCE OF BREWING.

BY

EDWARD RALPH MORITZ,

CONSULTING CHEMIST TO THE COUNTRY BREWERS' SOCIETY,

AND

GEORGE HARRIS MORRIS,

PH.D., F.C.S, F.I.C., ETC.

BASED UPON

A COURSE OF SIX LECTURES

*DELIVERED BY E. R. MORITZ AT THE FINSBURY TECHNICAL COLLEGE
OF THE CITY AND GUILDS OF LONDON INSTITUTE.*

WITH PLATES AND ILLUSTRATIONS.

E. & F. N. SPON, 125, STRAND, LONDON.
NEW YORK: 12, CORTLANDT STREET.
1891.

PREFACE.

———

AT the invitation of Professor Meldola, F.R.S., Professor of Chemistry at the Finsbury Technical College of the City and Guilds of London Institute, I delivered at that College, in the winter of 1889, a course of six lectures on the Science of Brewing. At the conclusion of the course, those who had honoured me by their attendance were so good as to suggest to me the desirability of publishing the lectures in book form. On a consideration of that suggestion, I thought that such a publication might perhaps be of some service ; and it was then my intention to limit myself to a re-perusal of the manuscript, and to its publication as such, after only those alterations had been made which seemed necessary from the literary stand-point. On commencing the work, however, I soon came to the conclusion that to continue on those lines would be unsatis-factory : partly because the scope of the work would thus be very limited ; partly because researches published since the date of the lectures threw an entirely new light upon several important matters, more especially upon the pro-ducts of starch-conversion, and upon the changes occurring during the germination of the cereals. I therefore determined to go over the ground afresh, taking my lectures only as a basis for a more complete and general work than I had at first contemplated. Such a plan, however, involved more leisure than I had at my disposal, and I was obliged to seek assistance. I was so fortunate as to secure the co-operation of my friend, Dr. G. Harris Morris, who, as the colleague of Mr. Horace T. Brown, F.R.S., had been closely associated with those researches which have been referred to as throwing fresh light upon phenomena, only partially explicable anterior to their publication. In the introductory remarks, the scope of this work is briefly sketched. We know it to be very imperfect, but we hope that it will not prove entirely useless to those for whom it is intended.

EDWARD RALPH MORITZ.

72, CHANCERY LANE, LONDON, W.C.

INTRODUCTORY.

THE object of this work is to provide in a convenient and accessible form such knowledge of the processes of brewing and the materials employed in that industry, as is at our disposal; and—so far as we are able—to connect such knowledge with the practice of brewing. We therefore intend it as a text-book in which may be found the results of scientific research together with the practical conclusions which we consider justly deducible from them. We have consequently considered it incumbent upon ourselves to exercise a discretion as to the researches to be included in this work. Apart from those omitted from sheer lack of space, we have purposely excluded those which have been traversed by subsequent investigations, or those which we considered otherwise unreliable. Similarly, we have not hesitated to criticise certain hypotheses still current among practical men when these hypotheses have appeared to us erroneous. In endeavouring to explain practical observations by means of scientific investigations, we have perhaps been guilty of audacity, inasmuch as the explanatory hypotheses we have ventured to put forward may have to be modified as our knowledge extends. But even with this possibility before us, we have thought ourselves justified in doing what we have done. It has seemed to us that the chief value to the practical brewer of such a work as this will lie in the conclusions deducible from scientific researches; and that, if we refrained from drawing such conclusions, he himself would inevitably do so. While carefully differentiating between exact scientific

investigations on the one hand, and the conclusions presumably deducible from them on the other, we have thought
it only right to take upon ourselves the responsibility of in-
terpreting the experimental evidence, rather than place that
responsibility upon our practical readers.

We do not pretend that a perusal of our work will enable
a novice to brew beer; neither will a study of it convert a
purely practical man into a chemist. We do not call this
work a practical book; at any rate, it is not "practical" in
the sense of giving brewers hints or suggestions as to the
routine of their operations, or as to the various forms of
plant in use. Such knowledge is only to be acquired by
experience in the brewhouse, with the assistance of works
like that of the late Mr. E. R. Southby, on the 'Practice
of Brewing,' to which excellent book we consider this a sort of
companion. We do, however, intend this as a practical book
in one sense, in so far that it is meant to lead the brewer to a
better understanding of what we may term the physiology and
pathology of brewing, and, by so doing, put at his disposal a
means for more efficient control over his operations.

We have divided the book into three sections : materials,
processes, and analysis. A brief description of what we consider the most reliable methods for the analysis of materials
and products has appeared to us essential, and for these
reasons :—In the first place, in treating of materials, we have
been forced to explain ourselves by a reference to the results
of analysis; and without giving the modes of obtaining
them the references would be more or less obscure. We
have also thought that the methods of analysis, illustrated by
examples fully worked out, would render the book acceptable
not only to those brewers who have laboratories and some
scientific knowledge, but also to chemical students interested
in technical operations, or such as intend subsequently to
interest themselves in these operations. We have pre-
supposed in the analytical section a knowledge of manipu-

lation, of elementary chemistry, and of ordinary forms of apparatus. We considered that these matters formed no branch of our subject, and that there existed an abundance of text-books adequately dealing with them. We would, however, ask purely practical readers not to entirely ignore the analytical section, for, apart from analytical directions and examples, we have unavoidably had to include in that section much that is of more general interest. Similarly, the first and second section are not very closely demarcated ; we mean that in the first section we have unavoidably had to treat of processes, and in the second section of materials. We have however, endeavoured to meet that difficulty by a system of cross references.

It will be observed by those whose scientific reading is only confined to the trade journals that we have said but little with respect to the albuminoids, or nitrogenous constituents of malts, worts, &c. We are of opinion that the present state of our knowledge of these substances is too restricted to permit of practical deductions ; and on this and other points we have endeavoured to restrain ourselves from putting forth explanatory theories or practical deductions unless sound experimental evidence can be adduced in support of them.

We have freely availed ourselves of all books and journals dealing with the subjects of which we treat, and we have in all cases referred to the original source, not only in common fairness to the author, but for the benefit of those who desire fuller and more complete knowledge on any special points. If we have included work in this book, other than our own, to which we give no reference, we have omitted it unwittingly, and take this opportunity of expressing our regret for that omission. So far as the style is concerned, we are quite aware of its roughness and the entire want of literary polish. We have not, however, sought to correct that defect. We have endeavoured to be lucid, and trust we have succeeded in that limited ambition.

The work being principally intended for practical men, temperatures are in all cases given on the Fahrenheit scale. In the more scientific chapters, however, the temperatures are given in degrees Centigrade as well as Fahrenheit.

We have to thank many friends for samples supplied for our own investigations, and for the information which they have so freely put before us. We also desire to thank our friends Messrs. Matthews and Lott, for the loan of plates which have illustrated their valuable work on the "Microscope in the Brewery and Malt House," which work should be consulted on all points specially connected with that department of this industry.

E. R. M.
G. H. M.

October, 1891.

TABLE OF CONTENTS.

———◦◦◦———

CHAPTER III.

MALT SUBSTITUTES.

CHAPTER IV.

Hops.

SECTION II.—PROCESSES.

CHAPTER V.

Mashing and Sparging.

CHAPTER VI.

THE BOILING OF THE WORT.

CHAPTER VII.

FERMENTATION.

CHAPTER VIII.

THE RACKING AND STORAGE OF BEER.

CHAPTER XI.

The Analysis of Sugars and Hops.

CHAPTER XII.

THE ANALYSIS OF BEER.

APPENDIX.

LIST OF PLATES.

A TEXT-BOOK

OF THE

SCIENCE OF BREWING.

—•—

SECTION I.—MATERIALS.

—•—

CHAPTER I.

BREWING WATERS.

CLASSIFICATION OF BREWING WATERS FROM THE MINERAL STANDPOINT.

THE requirements of waters for brewing purposes are not identical with those for potable purposes. If drinking waters are pure, not excessively hard or salty, and free from injurious metals or by-products, they satisfy all necessary conditions. But in brewing, the dissolved salts are of the very highest importance; different ales requiring different salts and varying proportions of them. Brewing waters must necessarily be free from injurious metals and by-products, and they should be organically pure. The degree of purity necessary in brewing waters is a more intricate question than it is in domestic supplies, complicated as it is by the changes effected in the organic matter during the various stages of the brewing processes. This branch of the subject will be treated subsequently; for the moment the mineral matters dissolved in waters will be considered.

The ordinary terms "hard" and "soft" are hardly sufficiently precise in classifying brewing waters. A hard water—that is, strictly speaking, one capable of decomposing relatively large amounts of soap—may be hard either through dissolved

B

chalk, or gypsum, or other compounds of magnesium and calcium. It is true that hardness, according to this ordinary classification, may be either " permanent " or " temporary,"—that is, destroyed, like dissolved chalk, or not destroyed, like gypsum, by thorough boiling ; still, when a brewer talks of a hard water, he implies its containing a sufficiency of many salts not contemplated in the term " hard," which salts are of distinct importance for brewing purposes. The term " soft " water would be equally misleading for brewing purposes. The ordinary soft water for domestic purposes is one containing minute amounts of salts capable of destroying soap. But a brewer would certainly call those waters soft which contain the sulphates and the carbonates of the alkalies. Yet, these bodies frequently accompany a fair proportion of chalk, and waters of this kind would be customarily classified as hard in the usual phraseology. On these grounds it will be advisable, in classifying brewing waters, to go beyond the terms " hard " and " soft," and to adopt more exact if less terse expressions. It will be convenient to divide brewing waters into the following three arbitrary classes :—

I. Waters containing large proportions of gypsum.

II. Waters containing small proportions of gypsum, or no gypsum.

III. Waters containing no gypsum, but sulphates and carbonates of the alkalies.

Waters answering to these descriptions may be considered typical of the great bulk of English brewing supplies. There are some, however, of rarer occurrence, which possess certain peculiarities, such as, an excess of gypsum, or excess of salt ; these will be subsequently considered. The following analyses of waters, mostly performed by one of us, will exemplify the above mentioned three broad classes. The acids and bases are first stated as found, then given in combination*

---

* In all cases the combined salts are given in the anhydrous form.

according to the scheme for combining them, described in Section III. of this book, under Water Analysis.

Class I.—This class of waters is not widely extended; but as the Burton well waters are typical of the class, and as so much of our best pale ale is brewed in that district, they form a very important section. The more so, as the excellence of these waters for pale ale brewing undoubtedly gave Burton beer its original great pre-eminence; and since it was the recognition of this fact which has led to the artificial treatment of less excellent waters up to the Burton standard, in order that these might give equally good results for pale ales.

The following is an example :—

BURTON (DEEP WELL).

	Grains per gallon.
Silica	0·49
Alumina	0·49
Iron oxide	trace
Lime	36·33
Magnesia	10·15
Soda	7·25
Potash	0·86
Chlorine	2·37
Sulphuric acid	52·29
Nitric acid	1·25

These bases and acids would exist in the water probably combined as under :—

Sodium chloride	3·90
Potassium sulphate	1·59
Sodium nitrate	1·97
,, sulphate	10·21
Calcium sulphate	77·87
,, carbonate	7·62
Magnesium carbonate	21·31
Silica and alumina	0·98
	125·45 *

---

* E. Brown, 'The Geology and the Mineral Waters of Burton-on-Trent,' p. 23.

Class II.—Waters containing insufficient proportions of lime and magnesia salts, form the bulk of the brewing waters of this country. Included in this class are the waters from the chalk, which, being pure, and easily treated with the minerals which they naturally lack, are much prized as brewing waters. Also included in this class are the waters from the rock, which contain mere traces of dissolved salts. When these waters are pure they are readily capable of treatment with the minerals in which they are deficient; but the absence of dissolved chalk is a disadvantage which will be referred to subsequently. Similar to these rock waters are lake water and rain water. Lake water is rarely used, and personally we know of no case where rain water is employed. When pure, however, they could be readily treated and would, as brewing waters, rank with rock waters.

The following are examples of this class of water :—

(a) Water from the chalk (deep well), Reigate, Surrey.

	Grains per gallon.
Silica	1·10
Alumina	0·04
Lime	17·37
Magnesia	0·49
Soda	1·72
Nitric acid	3·51
Sulphuric acid	5·14
Chlorine	1·99

These bases and acids would exist in the water probably combined as under :—

	Grains per gallon.
Sodium chloride	3·28
Calcium nitrate..	5·33
,, sulphate	8·74
,, carbonate	21·34
Magnesium carbonate	1·03
Silica and alumina	1·14
	40·86

(*b*) Water from rock (shallow well), Merthyr Tydfil.

	Grains per gallon.
Silica..	0·42
Alumina	0·60
Lime	3·07
Magnesia	0·21
Soda	1·32
Nitric acid	nil
Sulphuric acid	2·76
Chlorine	0·78

These bases and acids would exist in the water probably combined as under :—

	Grains per gallon.
Sodium chloride	1·29
,, sulphate	1·46
Calcium sulphate	3·30
,, carbonate	3·05
Magnesium carbonate	0·44
Silica and alumina	1·02
	10·56

Class III.—The waters containing no salts of lime and magnesia other than carbonates, but containing sulphates and carbonates of the alkalies, are abundant and extensively used. The deep wells below the London clay yield waters of this class, and similar waters are found between London and Harwich, London and Windsor, and London and Gosport. They are generally pure, and are excellently adapted in their natural state to stout brewing. The success of the London firms in this branch of trade, and the monopoly they enjoyed in it, is in great part attributable to the natural suitability of these waters for black beers. At the same time they are unsuitable for ale brewing in their natural state; it is not surprising, therefore, before the adequate treatment of this class of water was understood, that the London brewers sent to Burton for their pale ales, or erected branch establishments of their own in that district.

This form of water requires something more than the

addition of gypsum and magnesium sulphate. The detrimental effects of sodium carbonate and sodium sulphate are so
marked in ale brewing that it is necessary to decompose these
alkaline compounds, irrespective of the addition to the water
of the minerals naturally present in Burton waters. This is
possible in several ways : either by the addition of calcium
chloride, which converts the sodium sulphate and carbonate
respectively into sodium chloride and calcium sulphate and
carbonate; or by first converting the sodium carbonate into
sodium sulphate, by means of gypsum, then acting upon the
so formed sulphate with calcium chloride ; or by adding
sulphurous acid, or even hydrochloric acid, and so directly
transforming the carbonates into sulphites or chlorides.

We believe the treatment with calcium chloride gives the
most satisfactory results. A moderate excess of it is not only
harmless, but desirable ; and this cannot be said in general
of the addition of hydrochloric or sulphurous acid. Waters
thus treated with calcium chloride and also with the requisite
amounts of calcium and magnesium sulphates, form good
brewing waters for pale ale, although perhaps somewhat inferior to the natural Burton waters, and to waters of the second
class suitably treated. There is, for instance, some difficulty
in obtaining the necessary degree of "cleanness" or delicacy
of flavour in beers brewed with them. The reason of this is
not quite clear, but is probably due to the high proportion of
common salt present in treated waters of this kind. Waters of
this class naturally contain rather high proportions of salt ;
and the natural quantity is increased by that formed through
the double decomposition between calcium chloride and the
sodium sulphate and carbonate. Salt in a moderate quantity
is a good thing, but it is liable, in our experience, to give full,
rather than delicate and clean beers. . Taking the example
which follows, it will be found on calculation that the total
quantity of salt in the water after treatment with calcium
chloride is 31·47 grains ; this is a distinctly large amount.

The following is an example of water of this class :—

LONDON WATER (DEEP WELL).

	Grains per gallon.
Silica	0·38
Alumina	0·01
Lime	3·01
Magnesia	1·82
Soda	16·88
Chlorine	5·18
Nitric acid	0·44
Sulphuric acid	9·38

These bases and acids would exist in the water probably combined as under :—

Sodium chloride	8·54
,, nitrate	0·69
,, sulphate	16·65
,, carbonate	8·25
Magnesium carbonate	3·82
Calcium carbonate	5·37
Silica and alumina	0·39
	43·71

We must now briefly refer to those types of water which are of rarer occurrence, and exhibit distinct peculiarities.

Sub-class A.—Water containing an excess of gypsum. The following water, from a deep well in Leicester, is typical of this class :—

	Grains per gallon.
Silica	1·55
Alumina	0·35
Lime	45·71
Magnesia	14·99
Soda	15·84
Chlorine	4·90
Nitric acid	6·67
Sulphuric acid	79·90

These bases and acids would exist in the water probably combined as under :—

	Grains per gallon.
Sodium chloride	8·07
,, nitrate	10·50
,, sulphate	17·70
Calcium sulphate	111·01
Magnesium sulphate	6·94
,, carbonate	26·63
Silica and alumina	1·90
	182·75

Waters of this class are liable to somewhat reduce the vigour of the fermentation, and to postpone the cask-fermentation, or " condition." The reason for their doing so has hitherto been sought in the precipitation of an excess of the phosphates derived from the malt, which in certain rates are necessary alike for the vigour of the primary and the secondary fermentations. To decide whether this idea were correct or not, we made certain experiments, which were briefly as under.

Two malts were taken, and each mashed under precisely the same conditions of temperature, concentration, &c., with three waters. Water A was distilled water; water B, distilled water containing 50 grains of (anhydrous) calcium sulphate per gallon; water C, distilled water containing 100 grains of (anhydrous) calcium sulphate. The distilled water would represent ordinary soft water; the second water, ordinary gypseous water; the third, excessively gypseous water.

The experimental mashes were in each case made by acting on 50 grams of malt with 270 c.c. of the waters. This rate is approximately 5 barrels to the quarter of malt. After mashing for 1 hour at 150° F. the mash was made up to 515 c.c. with distilled water, the grains filtered off, and the wort boiled down in each case to one half its bulk. The phosphoric acid was now determined in each wort in the usual manner (precipitation by molybdate solution; solution of molybdate precipitate in ammonia; and precipitation with magnesia solution). The results are expressed in grams of P_2O_5 (phosphoric anhydride), in 100 c.c. of the boiled wort. The boiled worts were of sp. gr. 1053·7 to 1054·6; it will thus be seen that they were of average brewing strength.

MALT I.	Distilled water.	50 grains CaSO4	100 grains CaSO4
P_2O_5 grams per 100 cc. wort ..	0·129	0·104	0·089
MALT II.			
P_2O_5 grams per 100 cc. wort ..	0·114	0·091	0·085

. Now although there is a perceptible decrease in the phosphoric acid on increasing the calcium sulphate, we are bound to say that we do not consider that this decrease explains the sluggish primary and secondary fermentations. In fact the proportions of phosphoric acid in the most gypseous of these worts, would suffice for a vigorous fermentation. In saying this, we are guided by a series of phosphoric acid determinations made by Salamon and Mathew,[*] in which worts are quoted containing considerably less phosphoric acid than the above, and which fermented in a normal manner. We are inclined to consider the effect of gypsum as being possibly connected with a checking of the diastatic action (see p. 150), while perhaps the gypsum when in excess might exercise some directly destructive action upon the ferment in a manner similar to that exercised by salts of the heavy metals.

Sub-class B.—Water containing an excess of common salt. The following analysis of a deep well in Essex, in the neighbourhood of the Thames, is typical of an extreme case.

	Grains per gallon.
Silica	0·35
Alumina	0·35
Lime	19·74
Magnesia	11·74
Soda	87·91
Nitric acid	3·56
Sulphuric acid	15·04
Chlorine	108·50

These bases and acids would exist in the water probably combined as under :—

Sodium chloride	165·90
Calcium chloride	12·24
,, sulphate	25·57
,, carbonate	2·14
,, nitrate	5·40
Magnesium carbonate	24·65
Silica and alumina	0·70
	236·60

[*] Journal Society Chemical Industry, iv. (1885) p. 376.

This water was systematically used in the brewery for about a month. The fermentations were, however, so sluggish, the racking gravities so high, and the yeast so weakened, that at the end of this time it had to be discarded. Waters containing a large amount of salt are always liable to give full rather than delicate ales. In this particular case, where the amount of salt was excessive, the pale ales entirely lost their distinctive flavour, and became, to all intents and purposes, full-drinking mild ales.

It is of some interest to know what the permissible limit of salt is. The quantity of salt in a water, the analysis of which follows, taken from a well on the coast of Durham, is probably about as high as would be safe. The water has been in use for some years, and yields beers much liked in the district. As would be supposed, they are full; but there is no tendency to unduly high racking gravities or yeast weakness, as in the previous case mentioned.

	Grains per gallon.
Silica	0·15
Alumina	0·06
Lime	11·77
Magnesia	6·93
Soda	27·87
Chlorine	31·92
Nitric acid	0·98
Sulphuric acid	6·10

These bases and acids would exist in the water probably combined as under :—

Sodium chloride	52·60
Calcium nitrate..	1·47
Calcium sulphate	10·37
Calcium carbonate	12·48
Magnesium carbonate	14·55
Silica and alumina	0·21
	91·68

Most waters contain some quantity of iron. As a rule it amounts to a mere trace. In some cases, however, the quantity may range up to 1 or 2 grains per gallon. If allowed to enter the wort at this rate, it would be prejudicial. Inky compounds would be formed with the hop tannin, and these would affect the colour of the beer. It might also somewhat prejudice its soundness. The hop tannin is required for the precipitation of nitrogenous bodies, and such part of it as would be absorbed by the iron would be lost—and lost, probably, to the detriment of the beer. In spite of these theoretical objections to iron in a water, the risk is not a practical one; for the iron can be readily removed from the water previous to brewing operations. When a water contains a fair amount of chalk in solution, its precipitation during boiling in the hot liquor tank will throw down the iron in an insoluble form; and the only precaution then to be observed is that none of the precipitated sludge shall enter the mash-tun, where, owing to the acidity of the wort, the precipitated iron might be redissolved. When a water contains no chalk, the iron may be precipitated by boiling with a small quantity of crystallised sodium carbonate (soda crystals), preferably at the rate of 5·5 grains per grain of iron (reckoned as Fe_2O_3), and the slight excess of sodium carbonate can then be decomposed by treating with a little calcium chloride solution. The rate of this latter salt need not be specified, for any slight excess of it would be of benefit to the water.

THE EFFECT OF SALINE CONSTITUENTS ON THE CHARACTER OF BEERS, WITH SPECIAL REFERENCE TO THE BREWING OF PALE BITTER ALES.

Having now considered the classes of water which are in general use, it will be well, before describing their treatment in detail, to survey the present state of our knowledge concerning the influence of the saline constituents in relation to the brewing process. It is a somewhat humiliating fact that our knowledge on this important matter is still pretty well limited to the practical teachings of brewhouse experience.

We know for instance, that gypsum in fair quantity is essential to the production of sound pale bitter ales, but why this is the case is not by any means easy to explain. The older writers attributed its beneficial influence to the small quantity of albuminoids extracted by gypseous waters from any given malt. But this idea, like many others, was based on the assumption that the less nitrogenous matter in a wort, the sounder the beer. That this assumption is nonsense, we shall have subsequently occasion to show; but the theory being enunciated and believed, it became desirable to fit the facts into accordance with it. Hence, gypseous waters yielding sounder beers, the gypsum had to be accredited with the property of preventing the extraction of the nitrogenous bodies. It was, however, shown by Southby [*] in 1879 that with any given malt gypseous and non-gypseous waters extracted practically the same amount of nitrogenous matter. This being so, it became necessary to account for the beneficial influence of gypsum in some other way. Accordingly, a fresh theory was put forward: that although the *quantity* of nitrogenous matter extracted by gypseous and non-gypseous waters might be the same, the quality was widely different: gypseous waters extracting a sounder and less putrescible type of nitrogenous matter than non-gypseous. Having regard to our complete ignorance as to what is, or what is not a sound type of nitrogenous matter, and to our complete incompetence to separate one from the other, this theory must be regarded as nothing more nor less than mere verbiage. The only attempt to really test the difference in type of nitrogenous matter extracted from a given malt by water containing different minerals in solution was made by Briant.[†] Briant used for this purpose the method of Ullick, by which the nitrogenous matters extracted by each water are separated into albuminoids, peptones, amides, and unknown nitrogenous bodies. Person-

---

[*] Country Brewer's Gazette, Jan. 22, 1879 ; and ' Practical Brewing,' p. 171.
[†] Brewer's Journal, 1887, pp. 25–26.

ally, we regard this method as valueless. Any attempt
at solving the problem which has to depend upon any such a
method as Ullick's, cannot be regarded with confidence. In
no sense is Mr. Briant's care in carrying out his experiments
questioned ; it is the method which condemns the work in our
eyes, and we base our condemnation upon a very considerable
experience of it.

Leaving the influence of gypsum upon types of nitrogenous
matter extracted, it remains to be seen whether we cannot
discover some surer and sounder ground upon which to
explain its admitted beneficial influence. In the article previously referred to, Southby pointed out that when a given malt
is mashed with a gypseous water and distilled water under
the same conditions, the worts yielded will behave very differently on boiling. The gypseous wort "breaks" well, that
is, the coagulable albuminoids are thrown out in large curdy
flocks, and after a short standing, the wort itself becomes brilliant ; the distilled water wort throws out its albuminoids, not in
curdy flocks, but in the form of a fine powder—and even after
standing, the wort refuses to brighten, owing to the distribution
throughout it of the precipitate, which in this case is of a light
and powdery form. The gypseous wort, then, is brilliant, the
precipitate lying compactly at the bottom of the vessel ; the
distilled water wort is turbid, the precipitate being partly distributed in the liquid, partly resting on the bottom in a powdery
state. Southby gives these observations in full, but apparently
connects them only with the probable superior clarifying
capacity of beers brewed with gypseous waters. We think,
however, that the difference described in great measure
explains the superior soundness of beer brewed with these
gypseous waters, and the matter having appeared to be of
some importance one of us repeated Southby's experiment.
In so doing we used, besides the gypseous water and the
distilled water employed by him (and which would correspond
to Class I. and Class II. of the classification proposed), a third

water, one containing sodium carbonate, and which would correspond to Class III., the effects of which in pale ale brewing are known to be prejudicial in point of stability.   The gypseous water contained 44 grains per gallon of anhydrous calcium sulphate, the sodium carbonate water 31 grains of anhydrous sodium carbonate.   The same malt was used throughout, and, to ensure uniformity of conditions, the worts were made simultaneously and at the same rates   After mashing, the worts were filtered off from the grains, boiled half an hour, allowed to stand ten minutes, and then observed. As regards the gypseous water wort and the distilled water wort, our observations entirely confirm Southby's, and we have nothing to add to his remarks on them which have been quoted above.   The sodium carbonate wort, however, shows all the faults of the distilled water wort, but to a far more marked degree.   In fact, on being boiled, it hardly " breaks " at all ; and after boiling and standing, the wort remains persistently turbid, far more so than with distilled water.   To prove that the non-"breaking" of the sodium carbonate wort was not due to the non-extraction of coagulable albuminoids from the malt, we mashed the same malt again with the gypseous water, and added to the wort just before boiling a sufficient quantity of sodium carbonate to decompose the gypsum, and yet leave about as much sodium carbonate in the wort as was present on the former occasion.   The result was the same : hence it is not to any non-extraction of coagulable albuminoids that the bad "breaking" was due.

The following photograph taken by Mr. Norris of the Finsbury Technical College, shows the beakers containing the three worts after standing for ten minutes, subsequent to the boiling for thirty minutes.

It will be at once observed that the middle wort (gypseous water) was perfectly clear ; the distilled water wort less so, and the coagulated matter of a more powdery description.   In the sodium carbonate wort the liquid is intensely turbid.

Now the relative brilliancy and turbidity of these worts is strikingly in accordance with the respective soundness which we know from experience would result from these three waters respectively—gypseous beer being sounder than those from untreated waters of Class II., and these again

Fig. 1.

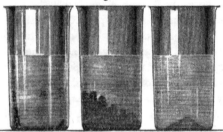

being less unsound than beer made from the alkaline waters (untreated) of Class III. This agreement is probably no mere coincidence, and we certainly think that there is reason to assume some direct connection between the "breaking" of the worts and the soundness of the resulting beer, when we consider what this "breaking" of wort means in practical operations.

After the wort has been boiled in the copper, it is run together with the hops into the hop-back. The hop-leaves form a more or less compact layer through which the wort has to pass, and we may regard this layer of hops as a rough filter-bed. Now it is clear that the efficacy of the filtration process will depend primarily upon the form of the matter which it is required to filter off. When separated in large curdy flocks, the hops will prevent its passing through. When, however, the matter separates in a fine pulverulent form, the rough filtration effected by the hops will not prevent the passage of this powdery precipitate into the coolers and fermenting vessels. The difference between the satisfactory "break" as obtained by the use of a gypseous water, and the unsatisfactory "break" resulting from soft, and more especially from alkaline water, will therefore amount to this : that with the gypseous water we can completely filter off the coagulated albuminoids in the hop-back ; with the other waters, we allow them to pass through and to become an integral part of the worts as they are cooled and fermented. It has been shown by Adolph Mayer [*] that the coagulable non-diffusible albuminoids are incapable of nourishing yeast ; their passage through with the worts cannot therefore be regarded as beneficial in point of yeast nourishment. There is, on the other hand, good reason to suppose that they are not only useless for purposes of yeast nourishment, but that they are distinctly prejudicial to the soundness of the beer. When we come to study those bacteria by the development of which most of the diseases are caused, we shall see that certain of them have the property of rendering assimilable to themselves albuminous matter previously unassimilable. This property, which is known as a peptonising property, therefore puts it into the power of these organisms to manufacture from these coagulable albuminoids, material suited to their development and propagation. Taking our system of brewing as it is, the entire avoidance of these bacteria is impossible ; and our endeavour must be to prepare a wort such as will be unsuited to their development. In the case however mentioned, where through the unsatisfactory "break," and the consequent unsatisfactory removal from the wort of these coagulable albuminoids, we are clearly leaving in our worts those bodies which may become the food for the disease organisms—those very bodies, in fact, which it should be our object to avoid.

The above considerations have led us to suppose that the connection between gypsum in a water and the resulting soundness, depends in some measure upon the removal of the coagulable albuminoids prior to the cooling stage, where the principal infection by bacteria is to be expected ; and we

[*] ' Gärungschemie,' chap. viii.

venture to think that these views are alike in accordance
with the experimental facts which have been quoted and with
practical experience in the brewery. We do not doubt, too,
that the removal of these bodies will induce greater clarifying
capacity as suggested by Southby ; but the equally important
question of stability is probably also intimately connected
with it.

The connection between the removal of coagulable albuminoids and the stability and clarifying capacity of beer, which
we have put forward, has been utilised by Mr. B. W. Valentin, who, within the last few years, has taken out a patent in
connection with this point. Mr. Valentin finds that when the
wort is treated in the ordinary way, the albuminoids, which
coagulate at 180° F., are partially redissolved during the subsequent boiling process in copper. These albuminoids, according
to the patentee, subsequently deteriorate the beer on storage,
inducing instability, cloudiness and frettiness ; and he therefore
advises their removal before ebullition, by filtering all wort,
heated to 180°, over goods, so separating out the coagulable
albuminoids which would otherwise be redissolved during
boiling, and which would therefore become a part and parcel
of the wort as cooled and fermented. It is clear, then, that
the patentee regards the non-elimination of coagulable albuminoids from the wort as prejudicial to the resulting beer ;
and, so far as this goes, his views fit in with those we have
put forward.

Through the kindness of Mr. Valentin we have been furnished
with a sample of the albuminoid coagulated at 180°, and separated from the wort. We thought it would be of some interest to
purify this albuminoid to see whether the saline constituents of
the water exercised any influence in dissolving different quantities of it during boiling. The effect of sodium carbonate is
certainly striking ; a water containing 32 grains per gallon of
anhydrous carbonate dissolving, when saturated with it, more
than ten times as much of it as does distilled water under the

c

same conditions. The numbers obtained (mean of two con-
cordant experiments) were :—

Distilled water	0·010 grams per 100 c.c.
Sodium carbonate water	0·167 ,, ,,

We also experimented with a water containing 44 grains
per gallon of anhydrous calcium sulphate, and with a water
containing 31 grains of common salt. Both the calcium sulphate
water and the common salt water seem to dissolve rather more
than distilled water ; but the differences are too slight to
permit of any conclusions. We are inclined to think, therefore,
that as regards actual quantity redissolved, and putting other
questions out of account, distilled water redissolves just about
as much as gypseous and salty waters, but that alkaline
waters dissolve about sixteen times as much as either.

That the ordinary saline constituents of water exercise
some influence upon diastatic conversion in the mash-tun is
generally admitted. A good deal of work has been done in
this connection, but the results are contradictory, principally
through the mashings being conducted by different operators
under different conditions. Some years ago, one of us
published some work on this matter, which was strongly
opposed by Windisch of the Berlin Brewers' Institute.
It is unnecessary to refer back either to that work or to
Windisch's counter experiments. It may be safely stated that
both sets of experiments were unreliable, and the conclusions
in both cases fallacious, and this because the methods of
analysis used by both operators were inaccurate and faulty.
In the first place, no regard was had to the ready-formed
sugars in malts, and when these are ignored, the percentages
of maltose and dextrin must be inaccurate. In the second
place, the dextrins were determined by the unreliable acid
conversion process ; in the third place, no regard was had to
the maltodextrins. As will be clearly seen from the figures
which follow, any experiments on this subject, carried out

without reference to the amounts and types of those substances, are valueless ; for it is in regard to variations in the amounts and types of them that the salts principally show their influence. In order, therefore, to determine as accurately as possible the precise influence of dissolved salts on the products of diastatic action, the subject has been again taken up by one of us, the resulting worts being analysed with due regard to the points previously neglected.

In the first series of experiments, two malts (I. and II.), were mashed with (*a*) distilled water; (*b*) distilled water containing 30 grains per gallon of common salt ; and (*c*) distilled water containing 37·5 grains per gallon of anhydrous calcium chloride. These waters were then used to mash each malt under the conditions laid down for the analysis of malts (Section III., p. 456), and the worts analysed according to the methods there described. But in this series of experiments no regard was had to the maltodextrins.

The conditions of mashing were throughout precisely the same. The results, calculated on 100 parts of malt, were as follows :—

	MALT I.			MALT II.		
	Distilled.	Salt.	Chloride of Calcium.	Distilled.	Salt.	Chloride of Calcium.
Total Maltose	33·58	33·77	32·36	32·46	34·99	31·65
Total Dextrin	14·38	13·78	15·26	15·26	15·32	16·25
Ready-formed Sugars ..	16·19	16·09	16·81	16·81	15·58	15·76

We consider that the above figures prove that the influence of the dissolved salts on the total maltose and dextrin yielded is practically nil, and the above results are partly quoted by us with the object of showing that without reference to the maltodextrins the real influence of the salts remains obscure.

A further series of experiments was now made on

C 2

Malt III., which was mashed in precisely the same manner as above, with the same waters as above described, and also with a fourth water containing 50 grains per gallon of anhydrous calcium sulphate (gypsum). The resultant worts were then analysed for the maltodextrins contained in them, and the results obtained were as under. The method of determining these substances was that stated on p. 477, but no correction was made for the unfermentable residue since this would obviously be constant throughout the series. The results are here calculated on 100 parts of dry wort-solids :—

MALT III.

—	Distilled.	Gypsum.	Salt.	Calcium Chloride.
Combined Maltose	6·29	6·80	5·25	9·01
Combined Dextrin	6·88	9·12	6·27	6·35
Total Maltodextrin ..	13·17	15·92	11·52	15·36

Experiments, both in the first and second series, were made with a water containing sodium carbonate (30 grains per gallon of the anhydrous salt); but owing to the strongly deterrent influence of the salt on diastatic activity, the worts contained so much starch, and so much starch was left behind in the grains after mashing, that no reliable analysis of the worts could be carried out.

From the above results certain interesting conclusions may be drawn. As will be subsequently shown, the maltodextrins obtained during mashing are not perceptibly fermentable during the primary fermentation, and consequently the more of them we have in our worts, the greater will be the palate fulness of the racked beer. It will also be shown that upon them we depend for the "condition," or cask fermentation, and in great measure for the keeping properties or stability of ales ; and within certain limits it may be said that the larger the percentage of the maltodextrins, the more persistent the cask fermentation, and the more pronounced the keeping properties. But apart from percentage, the type of the malto-dextrin plays a very important part, and by type is meant the relative proportions of maltose and dextrin in the compound ; and it will be shown later that for any given percentage of maltodextrin the higher the type, i.e., the more dextrin there is in it, the more pronounced will be the stability, and the more persistent the condition ; while the palate fulness imparted by such a type is rather of the dry, delicate kind than of the full, sweet kind. A lower type, on the other hand, as we shall afterwards see, will give a more rapid condition, although it is less persistent ; the keeping properties will be less pronounced, and the palate fulness will be rather of the sweet, full kind than of the dry and delicate. These considerations are fully dealt with on pp. 232–239, but a brief *résumé* in this place is necessary for a proper understanding of the figures before us.

It will be clear then that for pale stock ales, in which the essential requirements are pronounced stability and clean dry flavour, we shall require a more dextrinous maltodextrin than for mild running ales, where the essential requirements are early condition and sweet fulness.

On turning now to the figures obtained, it will be observed that by the employment of the gypseous water we do obtain distinctly more combined dextrin than by any other saline compound used. It is not surprising, therefore, that practical men find gypseous waters pre-eminently suited to the production of pale stock ales. If now we turn to the calcium chloride we find the combined maltose much increased, and the influence of this salt, therefore, must be of decided benefit in the production of sweet mild ales. It is true that the salt is sometimes used in pale ale brewing : but the rate employed is always small compared to that of the gypsum, and, consequently, its influence would be more than covered by that of the latter substance. We should certainly have expected that sodium chloride (salt) would have shown a greater proportion of combined maltose, for ales brewed with salty waters certainly show the sweet fulness associated with a low type of malto-dextrin. We think, however, that the explanation lies in the fact that the salt, when exceeding say 50 grains per gallon, exercises a direct deterrent influence upon the attenuative vigour of the yeast in the primary fermentation ; and, conse-quently, that the beer will rack with a relatively large amount of those abnormally low maltodextrins, which under ordinary circumstances are fermented out in the fermenting vessel. In the analysis of the worts no trace of such substances is dis-coverable, since in the laboratory fermentation they will inevitably be completely fermented out. This matter is more fully explained on p. 385.

It is clear from the abortive experiments with the car-bonate of soda that this substance would be unsuited alike for pale and mild ales. Putting aside a possible loss of extract, the presence in the worts (and therefore beers) of starch, or of the very high maltodextrins allied to it, would greatly, if not entirely, impede clarification ; and it therefore follows that this substance is only permissible in the production of stouts and porters, in which clarification need not be thought of.

We consider that our experiments fully bear out Southby's general statement as to the effect of saline constituents, viz., that the sulphates give delicacy and cleanness, and are there-fore wanted for pale ales, while the chlorides give the sweet fulness requisite in mild ales.

When we approach the questions of colour, and of the flavour of the hop-extract in beers brewed with different types of water, there is at any rate complete agreement among chemists. Briefly, it may be said that gypseous waters give pale beers, and prevent the extraction of the rank, coarse flavour of the hop ; sodium carbonate water, on the other hand, gives deeply tinted beers, extracting much of the coarse and rank hop flavouring. Soft waters come intermediate between these extremes. Sodium sulphate gives a thin, harsh, cold flavour, but a pale colour. Common salt and calcium chloride also yield pale beers, and the hop flavouring is neither rank nor cold. Magnesium salts are not used apart from gypsum, and their effects alone are therefore unascertained ; but they seem, when in due proportion, to co-operate with gypsum in giving the best results for pale ales. It is more than possible that the salts exercise other influences upon flavour than those dealt with ; but since there is no definite knowledge on the point, we omit further reference to it.

Having regard, then, to the previous considerations for pale bitter ales, we require waters either of the first class, or water of the second or third classes so treated as to be made similar to them.

The treatment of waters of Class II. amounts to an endeavour to dissolve in them the salts naturally contained in the Burton waters. When the added salts are introduced in the proportions demanded by the natural deficiencies ascertained by analysis, these waters (putting purity out of the question) are as good as the natural waters imitated. As an example, the first water of this class would be brought up to about the Burton standard, by—

Gypsum (powdered) .. .. .. 6 oz. per barrel.
Magnesium sulphate .. .. .. 1 oz. „ „
Calcium chloride .. .. .. 1 oz. ($\frac{1}{10}$ pint saturated solution) per barrel.

This treatment should apply to the whole of the liquor, both mashing and sparging, the treatment preferably taking place overnight. Southby advises the addition of the gypsum and magnesium sulphate in the form of a thin whitewash, which can be readily prepared in a tub with some of the water previously abstracted for the purpose. The water should either be briskly boiling when the mixture is added, or else it should be thoroughly agitated. Southby also advises that the calcium chloride should, on account of its deliquescing, be

added in the form of a saturated solution.   The correct strength
is represented by a specific gravity of 1·38, and $\frac{1}{80}$th pint of
this solution added to 1 barrel of water means the solution
in the water of about 10 grains per gallon of anhydrous
chloride.   The requisite volume of the calcium chloride
solution may be added to the whitewash of the other
minerals.

In treating waters of Class III., which contain sodium
carbonate and sodium sulphate, it is necessary to decompose
these bodies, as well as to add the other minerals in which
these waters are deficient.   To do so we add a larger proportion of calcium chloride, using enough to completely
decompose the sodium sulphate and sodium carbonate, and
to leave from 5 to 10 grains per gallon in excess of this
amount.   The gypsum may be slightly reduced ; for some
of it is formed by the double decomposition of the sodium
sulphate and calcium chloride.

Taking the first example of this class of water the additions should be

    Gypsum (powdered) ..   ..   ..   5 oz. per barrel.
    Magnesium sulphate ..   ..   ..   1 oz. ,,   ,,
    Calcium chloride   ..   ..   ..   2 oz. ($\frac{1}{40}$ pint saturated solution) per barrel.

It is unnecessary to say that the above suggestions are
very general, and that when waters are naturally richer in
certain salts than those quoted they will require less of these
added, and *vice versâ ;* again, where a special result is desired, for instance, fulness, a greater quantity of chlorides
should be used.   Such variations must, however, be left to
the discretion of the brewer.

The separation of iron has been alluded to.   The separation
of salt is not possible by any practicable means, while in the
case of an excess of gypsum (Sub-class A) the only satisfactory way of dealing with these waters is to dilute them down
with suitable proportions of non-gypseous waters.

THE REQUIREMENTS OF WATER FOR MILD ALE BREWING.

In this class of trade the requirements are somewhat different to those of pale ales previously considered. The consumption is far more rapid, and the principal points to aim at besides a moderate stability, are fulness, early condition, and early clarifying capacity.

On these grounds high proportions of gypsum are not necessary; indeed, inasmuch as the after-fermentation, or condition, would be retarded through the high dextrin rate in the maltodextrin, they would be undesirable. The fulness, however, imparted by the chlorides is most desirable, and waters for running ale should always contain from 10–20 grains of them. Running ales can be successfully brewed from waters of Class II. in their natural condition. They can be also successfully brewed from waters of Class III. untreated, i.e. those containing sodium carbonate and sulphate; but better results follow the previous decomposition of these alkaline salts, and the addition of chlorides to both classes.

Taking the example given for the waters of Class II., the additions should be—

Gypsum (powdered) .. .. 1 oz. per barrel.
Kainit .. .. 2½ oz. ,, ,,

For a water of Class III. the above, together with $\frac{1}{20}$ pint of the saturated solution of calcium chloride, would give good results.

The mineral Kainit contains in itself the substances most desirable for mild ale brewing and it is a convenient form in which to add them in the desired proportion. The following[*] analysis shows its average composition :—

	Per cent.
Potassium sulphate	23.0
Magnesium sulphate	15·6
,, chloride	13·0
Sodium chloride	34·8
Water	13·6
	100·0

[*] Hake, Journal Society Chemical Industry, ii. (1883) p. 47.

THE REQUIREMENTS OF WATERS FOR STOUT AND PORTER BREWING.

In this class of beers the requirements are very different to those previously considered; we want as much colour as possible, and as much fulness as possible; delicacy of hop flavour is a minor consideration, and brilliancy need not be regarded. Stability should result from the limited diastatic conversion resulting from the employment of high dried and caramelised malt.

The waters used in the London and Dublin breweries are of course those most suitable for the production of black beers. The analysis of the water used by the London firms for stout brewing has been already stated (Class III.); the following is an example of a soft water of this class used for stout brewing by the Dublin breweries.

	Grains per gallon.
Carbonates of lime and magnesia precipitated on boiling	11·0
Lime not precipitated on boiling	0·9
Magnesia not precipitated on boiling	0·9
Sulphuric acid (SO_3)	0·5
Chlorine	1·2

—SOUTHBY.

Probably equally good results would be obtainable from either class of water. The addition of sodium carbonate to waters of Class II., to bring them up to the London standard, is in our experience unnecessary. Although sodium carbonate and sodium sulphate are perhaps of service in black beer brewing, their addition does not improve a water which is sufficiently free from gypsum, &c.

When a water of Class II. contains calcium carbonate, the water should be well boiled before mashing, so as to precipitate the greater part of it, and beyond this no treatment is necessary. Similarly, when the water belongs to Class III. no treatment other than boiling is required. In the case of waters of Class I., and still more so of Sub-class A, stouts such as are ordinarily demanded, cannot be successfully brewed until the gypsum and magnesium sulphate are decomposed by the addition of sodium or potassium carbonate. In such cases sufficient alkaline carbonate should be added to destroy the calcium and magnesium sulphates, leaving only a slight excess over; every grain of gypsum (anhydrous) requiring 2·10 grains of crystallised sodium carbonate or 1·28 grains of crystallised potassium carbonate; and every grain of magnesium sulphate (anhydrous) requiring 2·38 grains of the sodium, or 1·45 grains of the potassium carbonate. The carbonate should be added in the form of a concentrated solution to the water, which should then be well boiled for 45 minutes.

Before leaving the question of the effects of the saline constituents of a water in brewing, we must lay stress on this : that although of very high value—and this specially applies to gypsum in pale ales—yet successful brewing depends upon factors more important than the right quality and proportions of mineral constituents. The excellence of Burton ales, and London and Dublin stouts is by no means entirely due to the natural suitability of the waters found in these districts for the classes of beer for which they are respectively famous ; and no brewer should form the idea that the mere addition of gypsum to his water entitles him to expect perfect pale ales. The mineral constitution of a water is one (but only one) of the very many important points, all of which must be properly carried out to ensure success.

It is perhaps unnecessary to say that analysis does not determine the salts as they exist in water; all that it is possible for chemists to do is to determine the quantities of the bases and acids. We think there can be little question that for practical purposes, these bases and acids after statement as such, should be stated in combination in the manner which our experience of their usual affinities leads us to regard as most probably correct. It is true that these combinations are not indisputable : yet having regard to the fact that the treatment of a water must necessarily be based upon the amounts of compounds presumably present, and not upon the amounts of separate bases and acids actually so, and to the fact that there are certain broad lines of admitted correctness upon which the combinations can be based ; we consider that the analyst does not go beyond his province in stating the forms of combination in which the acids and bases probably exist. This statement of combination should however, be preceded by a statement of the acids and bases as they are actually found.

The lines recommended for the combination of the acids and bases will be found in the third section of this book, and they follow the methods for the analysis of water.   In the analyses quoted for the exemplification of the various classes of water, the combinations are given in the anhydrous form. To work out and insert the water of crystallisation of each salt, enormously complicates the calculations, and serves no useful purpose.

Brewing Waters considered in regard to their Organic Purity or Impurity.

In considering the effects of organic impurity in brewing waters, the enquiry resolves itself into two quite distinct and separate questions ; firstly, as to the effect of the micro-organisms ; secondly, as to the effect of unorganised organic matter.

(a) Micro-organisms in Brewing Waters.

In regard to the effects of the micro-organisms in brewing waters, much will obviously depend upon whether these organisms survive the brewing processes and pass into the beer ; or whether, on the other hand, they succumb to the conditions of the brewing processes. If they survive, their presence is directly dangerous; if they succumb they can at most be but indirectly responsible for damage to the beer.

In a later chapter (p. 256) we shall treat fully of the destruction of all organisms during the boiling of the wort in the copper, and when doing so shall quote the experiments of one of us * on this subject, which show beyond the shadow of a doubt that the conditions under which brewers' worts are boiled are such as to entirely annihilate any organisms which may exist in them prior to the boiling process. These researches deal rather with the destruction of the organisms adherent to the surface of the malt than with those possibly introduced with the water. But since the water-borne organisms are less resistant to adverse conditions than the dry organisms adherent to the malt, and since the latter are indubitably killed during the boiling in copper, it will follow that the water-borne organisms are similarly, and in fact the more easily, annihilated during that process. Previous to the publication of the experiments in question, it was customary for chemists to regard water-borne organisms as surviving the boiling process, and subsequently developing in the beer, to its detriment. This view arose probably from the admitted instability of beers brewed with a water containing any decided excess of organisms. We have no wish in any way to gainsay the evil effects of such a water, but the reasons for these ill-effects we now know to be quite other than the survival of the contained organisms.

It is clear that if a water contains a large number of organisms, such organisms must find in the water substances capable of nourishing them; and if we brew with such a water, the wort as it runs from the mash-tun will contain these substances. During the boiling of the wort, the organisms are killed; but the wort, as it runs from the copper at the conclusion of boiling, will still contain the nutrient substances. The process in the practice of brewing which follows the

* G. H. Morris, Transactions Laboratory Club, iii. (1890) p. 24.

boiling process is the cooling of the wort, which is effected by exposing the wort to the atmosphere on the coolers, and again exposing it as it passes over the refrigerators. The atmosphere contains in abundance the organisms which, under conditions favourable to their propagation, will produce the diseases of beer; consequently, under practical conditions of brewing, we must of necessity infect our worts, at a stage subsequent to the boiling stage, with disease-organisms. If the worts thus infected contain the substances upon which they can readily propagate, such worts will be the more liable to yield unstable beers. Now since a water containing these organisms must necessarily impart these nutrient substances to the wort, it will follow that such a wort will contain the matters upon which organisms infecting the brewings subsequent to the boiling process can develop; and their development must tend to the deterioration of the beer. In employing polluted water containing organisms, it is not the organisms themselves, therefore, which constitute the real danger; it is the substances which accompany them which are dangerous; substances upon which they developed in the water, and upon which air-borne organisms will develop at a later stage.

The assertion of the sterilization of wort in the copper, until finally substantiated by the experiments of Morris above referred to (and which will be given later), was, when first put forward on purely logical grounds, vigorously opposed. The opponents to that view based their contentions upon experiments by Tyndall, Prazmowski, and von Huth; but these experiments, when critically examined, in no way really weaken the arguments in favour of sterilisation. It is true that Tyndall [*] conclusively proved that the dry spores of the hay bacillus survived continuous boiling for some time. But this we consider no relevant argument, since boiling in water (as in Tyndall's experiments) and boiling in the presence of acid and hop-extract (as in a brewer's copper), are two entirely

* Tyndall, ' Floating Matters of the Air,' pp. 205 *et seq.*

different things; and although the spores referred to can admittedly resist the one, they cannot resist the other. Prazmowski's experiments on the resisting power of *Bacillus subtilis* are also relied upon. We find, however, that Grove [*] states that the spores of the organism are destroyed by boiling in water for two hours, while Prazmowski [†] himself asserts that they are similarly killed in five minutes. Assuming even that the longer period be really necessary for their destruction, it is clear that they must inevitably succumb during an ordinary copper boiling; for boiling for ten hours in water is far less adverse to the retention of vitality by these organisms than boiling for ten minutes in a slightly acid and hopped wort. Von Huth claims to have traced *Sarcina* in a beer to the water used in its preparation. At first sight this might seem to indicate their survival during boiling. It has, however, been recently shown by Lindner,[‡] who has subjected this organism to the closest investigation, that it forms no resting spores, and that in all circumstances it is killed by eight minutes' exposure to water at 140° F. It follows, therefore, that the *Sarcina* found by von Huth in the beer could not have been due to the water; or, if due to that source, the water must have gained accidental access to the wort or beer after the boiling process.

There is nothing, therefore, in these statements to weaken the experiments in question, and we may therefore take it as a well-substantiated fact that all organisms are annihilated during copper boiling, and that the danger of a water containing these organisms lies in their being accompanied by substances which will promote the development of the air-borne organisms which infect the worts after boiling is completed, and principally during cooling and refrigeration.

But there is a further point to take into account. The

[*] Grove, 'Bacteria and Yeast Fungi,' p. 28.
[†] Klein, 'Micro-organisms and Disease,' p. 73.
[‡] P. Lindner, 'Report of the Berlin Brewers' Institute,' 1888, p. 35.

organisms killed in the boiling process, will themselves yield to the wort matters capable of nourishing organisms subsequently introduced; and there is every reason to suppose that these matters will be peculiarly assimilable by them, and therefore, peculiarly adapted to promote their development. Every living creature is most effectively nourished on such matters as compose the body of itself and of its fellow-creatures. Obvious reasons prevent our feeding on our fellow-men, yet there is every reason to suppose that a cannibal diet would be peculiarly nutrient. If, however, we refrain from cannibalism, we choose, when requiring particularly assimilable sustenance, such foods as are nearest in chemical composition to the flesh of our own bodies; the raw beef selected by athletes in training is sufficiently significant. Going lower in the scale, we know that yeast is, in respect of nourishment, best fed upon " yeast-water," that is, the substance which yeast yields to boiling water. Recent research has indeed shown, for instance, that a certain carbo-hydrate,* galactose, is unfermentable by yeast unless that yeast is fed with yeast-water. There is, therefore, good reason to assume that bacteria killed during boiling will yield to the wort substances peculiarly suited to the development of bacteria subsequently introduced, and capable of nourishing, at the very least, an equal number of them.

The exact position of micro-organisms in a water then is, firstly, as indicating the presence of highly putrescible substances which are dangerous to the stability of the beer, and in contributing—if, perhaps, slightly, yet distinctly—to the amount of these substances when destroyed. They therefore deserve to be ranked, from the first standpoint, as indicators of dangerous impurity; while, from the second standpoint, they exist as potential putrescible matter.

The position now assigned to micro-organisms in a brewer's supply puts the so-called bacteriological analysis of

* Brewing Trade Review, 1889, p. 263 (abstract).

waters on a different footing to that attached to it by some chemists. Assuming that the organisms in a water survive the boiling in copper, it is clearly of importance to estimate the number of these water-borne bacteria, and ascertain their species; and, in that case, no chemical analysis unsupplemented by a bacteriological examination could be regarded as sufficient for the proper formation of an opinion concerning its suitability as a supply for brewers' purposes. But having regard to the ascertained destruction of the bacteria during subsequent operations the bacteriological examination may be ignored. In any case these bacteriological examinations cannot be regarded as very precise or reliable. Experience as to the precise connection between the number of bacteria and their effects is wanting; while the nourishing medium selected by the analyst for their artificial growth plays an important part in showing up either large or small numbers of them. These bacteriological examinations, too, are a little sensational. It is so striking to be informed that a water contains so and so many bacteria in such and such a bulk, that the practical man is apt to regard this apparently precise (but really far from reliable) information with more confidence than it deserves, and with more confidence than he regards the more sober and less taking results of ordinary organic analysis. This being so, it is not unfortunate that, recent researches having proved the destruction of water-borne bacteria, all need for these bacteriological examinations is removed. It is true that bacteria indicate impurity, and become impurity, but on these grounds they do not require to be numbered. The chemical analysis of the water, and its microscopical examination, will tell us, for all practical purposes, what we wish to know. ·

It has been suggested that bacteriological analysis is of use for all water employed for cleansing purposes, in so far as although the brewing water will be robbed of its bacteria during boiling, that used in cleansing the plant will not be

D

subjected to the same process.   We are not inclined, however, to attach very much weight to this suggestion.   The
vessels to be cleansed are freely exposed to the air, which is
so rich in organisms that any few introduced with ordinary
water can assuredly be of only small moment.   Of course,
where the water is swarming with bacteria, the matter
stands differently ; but even here, the chemical and microscopical examination will yield all the information required.

There is one purpose, however, to which water is sometimes put in a brewery, where the bacteriological analysis will
serve a distinct purpose.   Water used for mixing with wort
subsequent to the boiling stage is here referred to.   This practice is, however, uncommon, and it is by no means a commendable one.   On the Continent, and in a few English breweries,
it is customary to wash the yeast with water.   Water used for
this purpose comes under the same consideration as that
employed for admixture with wort after boiling.

The present will be a convenient opportunity for briefly
referring to the microscopical examination of water sediments.
The sediment collected and examined as suggested in the
analytical section (Chap. IX. p. 424), is capable of serving a
distinct purpose ; for, from the nature of the organisms and
their relative amount, we gain information supplementary to
that yielded by the chemical analysis, as to the intensity and
nature of the organic contamination, if any be present ; and
we are thus able to surmise to what extent bacteria introduced
subsequent to the boiling process will be nourished on such
organic matter.

Comparatively few waters are entirely free from suspended
particles ; and comparatively few sediments are quite free
from some forms of organised matter.   Even in pure waters
we find some form of vegetable matter, which, in the process
of decaying, discharges the cell-sap, endochrome, &c., which
after a time loses its distinctive green colour, becoming
yellow to brownish in tinge.   The decaying vegetable matter

is frequently surrounded by the lower forms of vegetable life, the existence of which is supported by the original decaying vegetable matter. When the amount of decaying vegetable matter is excessive, the infusoria make their appearance. Infusoria are also found largely in waters polluted by sewage, in which case they are accompanied by swarms of bacteria. It is to the bacteria, therefore, that in microscopic examinations we look for guidance as to whether or no a water is polluted with organic matter of animal origin. The ordinary bacterium of putrefaction (*B. termo*) is the form most common in waters of this class, though others, such as *Spirillum*, *Spirochæta*, &c., are occasionally found.

Swarms of bacteria accompanied by infusoria generally point to sewage contamination, and the deduction thus drawn will as a rule be abundantly confirmed by the results of the chemical analysis. In judging waters from new wells, which are generally very turbid when first used, a certain allowance must be made. Bacteria are usually then present, even in the case of waters which subsequently become free from them on continued use.

The bacteria in waters are generally found in the zooglœic stage, imbedded in a gelatinous mass of their own preparation. Desmids, confervoids, diatoms, &c., are frequently found, more especially in waters from rivers, lakes, and ponds. They may also indicate unclean water-tanks, pipes, and pumps.

Hard waters frequently throw out on standing a portion of the lime and magnesia salts they contain, and the sediments of some waters present a beautifully crystalline appearance. Minute particles of amorphous mineral matter, as well as of unorganised organic matter are frequently found in water sediments. As they possess a tremulous movement (the Brownian movement) they are likely to be mistaken for bacteria or other low types of organised matter. The directions given in the analytical section for discriminating between them and the organised matter should therefore be rigidly followed.

Plate I. will sufficiently explain the types of organised matter most frequently found in water deposits. It has been drawn from the plates in Parke's 'Hygiene,' and represents organisms found in well-water, and in river-water.

(b.) *Unorganised Organic Matter in Brewing Waters.*

The effect of unorganised organic matter in brewing waters has been already implied. The substances of which it is composed (mostly nitrogenous) are such as tend to nourish bacteria introduced subsequent to the boiling stage, and which, thus nourished, may proceed to deteriorate the yeast and beer. In other words, a water containing them in excess will, by their becoming an integral part of the resulting wort, tend to lower the resisting power of that wort. At the same time, but few waters are absolutely free from them, and it becomes necessary to see what amounts of them are permissible, and what amounts constitute a dangerous excess. We must therefore turn to the chemical tests, on which we rely alike for their detection and their approximate measurement.

The three principal tests are :—

1. The determination of carbon and nitrogen (Frankland's process).

2. The determination of the albuminoid and free ammonia (Wanklyn's process).

3. The determination of oxygen absorbed from permanganate of potash (Forschammer's process).

The four subsidiary tests are:—

(i.) The determination of the fermentative powers of the water on solutions of cane sugar (Heisch's process).

(ii.) The determination of nitrous and nitric acids.

(iii.) The determination of chlorine.

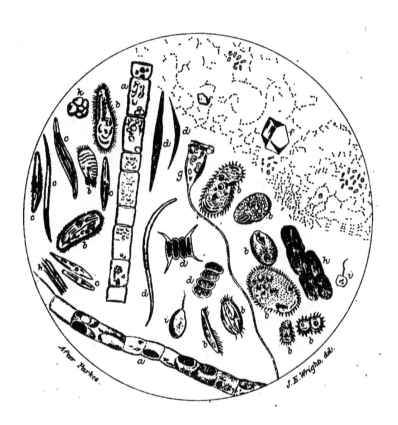

After Parkes

J.E. Wright, del.

a . Conferva .	f . Crystalline particles. (probably quartz.)
b . Infusoria.	g . Vorticella.
c . Diatoms .	h . Vegetable matter.
d . Desmids .	i . Monads.
e . Bacteria .	k . Foraminifera.

E. & F. N. Spon, London & New York.   Tho⁸ Kell & Son, Lith

(iv.) The determination of phosphates.

(v.) The behaviour of the solid residue of a water on ignition.

1. All these tests are necessary for the satisfactory analysis of a brewing water, except perhaps the first main test (determination of carbon and nitrogen), which may be regarded as optional. It is expensive and cumbrous, and although, for hygienic purposes unquestionably the most accurate and conclusive, it is in no sense superior (if not actually inferior) for brewers' purposes to the second main test (Wanklyn's process). Brewers do not require to know so much the *quantity* of the organic matter as its quality, i.e. its relative putrescibility or assimilability by bacteria. In our opinion, Wanklyn's process gives us more information on this point than Frankland's. On this ground the Frankland process has not been described in the third section of this work, though all the other tests (principal and subsidiary) are given in sufficient detail. Those interested, however, in the matter, will find full particulars of the Frankland process, with the interpretations to be put upon results, in the work published by its distinguished originator.[*]

2. The interpretations put upon the results of the Wanklyn process are based upon the quantities (absolute and relative) of the free and albuminoid ammonia. When, for instance, the free ammonia is high and the albuminoid ammonia also so, we conclude we have much organic matter, and of a putrescible character. The quantity of free ammonia roughly measures the activity of bacteria in much the same way as the quantity of alcohol produced by a yeast will measure that yeast's activity. The high albuminoid ammonia indicates a considerable amount of matter left over for the bacteria, subsequently introduced, to feed upon. Waters of this class are to be condemned. When, however, the high albuminoid ammonia is unaccompanied by high free ammonia, we have evidence that although

[*] E. Frankland, 'Water Analysis,' Van Voorst, 1880.

contaminated with organic matter, the organic matter is such
as is not readily putrescible. On this ground it is the less
prejudicial, and within certain limits waters of this class may
be passed.   Waters contaminated with matter of vegetable
origin come under this head ; waters defiled by sewage come
into the class previously mentioned.   In the ordinary way
0·10 parts per million of free or of albuminoid ammonia is
the outside limit permissible. But in this and in all other
constructions placed upon determinations of organic matter, no
very hard or fast lines can be drawn. For instance, the waters
of Class III. must be considered distinct exceptions to the
limit laid down for the free ammonia. These waters are as a
rule pure and yet contain large amounts of free ammonia (from
0·88 to 1·90 parts per million). It is true, however, that their
purity is indicated by the low albuminoid ammonia (0·02 to
0·06 parts per million). The high free ammonia in this type
of water is due to a destruction of albuminoid matter (and
perhaps to a reduction of nitrates) incident to the percolation
of the water through sandy and chalky strata.

It should be added, perhaps, that free ammonia has but
little direct action in brewing; it may slightly stimulate
the yeast, but if so, only to a small extent. Albuminoid
ammonia is an expression. It means the quantity of ammonia
evolved on treating the albuminoid matter with boiling alkaline
permanganate, the amount evolved being a more or less constant fraction of the contained albuminoid impurity. This
ammonia simply measures the albumin; it does not exist as
such in the water.

3. The estimation of the oxygen necessary for the complete
oxidation of the organic matter in a given amount of water,
and in a given amount of time, is admittedly a rough test; it
is so simple, however, that it should always be performed,
especially as it will show up contamination with fresh urine,
which escapes detection by the ammonia test. It is interfered
with by ferrous salts and nitrites; but a correction may be

made by making a double determination; one at the end of 25 minutes, the other at the end of 4 hours. The first will give the oxygen absorbed by ferrous salts and nitrites (which are acted on in this time, while organic matter is not so to any appreciable extent), and the second the oxygen absorbed by the organic matter as well as by the ferrous salts and nitrites. The difference between them gives the oxygen absorbed by the organic matter. This amount should not as a rule exceed 0·10 parts of oxygen absorbed by 100,000 parts of water.

(i.) The determination of the fermentative power of a water on sugar solutions is more useful than most chemists seem inclined to admit. To regard the fermentative action as a direct indication of sewage contamination is of course absurd; the manipulation of the test, as ordinarily performed, is not such as to preclude the possibility of fermentation due to contamination by air-borne bacteria; again, the assistance to the fermentative change given by phosphates is so marked that, as pointed out by Frankland, the test may be considered rather as an indication of phosphates in a water than of sewage defilement. Yet the sugar test is one capable of yielding very useful information; and such information as it yields is of special use in the case of waters required for brewing purposes. If a water shows a marked fermentative action, we get evidence that it contains bodies (whether phosphates only, or phosphates and putrescible organic matter) which will tend to nourish the bacteria introduced after the boiling stage; and on this ground it is well worth performing.

We consider it useful to supplement the ordinary test, as originally proposed by Heisch, by taking the fermentative action of the water after boiling as well as before boiling. For it is always possible to boil a brewing water before mashing, and it is desirable to see whether the nutrient matters are altered during boiling. The fermentative action of a water is nearly always very considerably less after than before boiling, and this is probably in great part ascribable to the deposition of the nutrient matters together with the calcium carbonate generally separated by boiling. Most waters in this country contain some dissolved carbonate of lime, and this constituent is therefore desirable; since, in depositing when the water is boiled, it brings with it much of the contained organic matter. In those few waters which contain no chalk, boiling does not seem to appreciably lessen the fermentative capacity of the water.

The activity of the fermentative change is reckoned by the time which elapses before the solutions commence to decompose. The period of observation should extend to 72 hours. Much stress need not be laid on a fermentative change before boiling, if up to the end of 72 days no such change is observable in the boiled sample. When, however, the boiled sample shows signs of decomposition within 72 hours, the water should be regarded with suspicion. Iron seems to exercise a deterrent influence on the fermentative change; and many impure waters containing it do not show the fermentative change otherwise anticipated. Other substances, however, as pointed out by Lott,* seem inoperative.

(ii.) The determination of nitric and nitrous acids is an important one, for we seldom find waters polluted which do not contain considerable amounts of them. They are formed from the decomposition of the organic matter in waters, or from the free ammonia which is nitrified by specific organisms, upon which Warington has done excellent work.† For brewers' purposes we have to consider them from two standpoints: as indices to pollution and as saline constituents when combined with the bases. From the first standpoint, no hard or fast rules are possible. Although we seldom find a decidedly polluted water without a very large proportion of nitrates or

---

* Journal Society Chemical Industry, vi. 1887, p. 495.

† Warington, Journal Chemical Society, 1878, p. 44; 1879, p. 429; 1884, p. 637; 1885, p. 758; 1887, p. 118; 1888, p. 727.

nitrites (unless perhaps the defilement is quite recent), yet many unquestionably pure waters (especially the pure chalk and gypseous waters) contain them in almost equal abundance. It is therefore perhaps wise to regard them as undesirable, unless the water is proved to be unquestionably pure by other tests ; and to only consider them as constituting some special danger when the water is proved to be of only moderate purity or of decided impurity. A great deal of prejudice exists against nitrates and nitrites as such, regarded as saline constituents, and apart from their bearing on organic purity. Southby, for instance, condemns a water simply on the ground of it containing over 5 grains per gallon of combined nitrates and nitrites. This rather sweeping condemnation is not, we think, warranted by facts. Pure waters (and they are hard waters) are known to us containing up to 10 grains per gallon of nitric and nitrous acids which yield excellent results, and our experience is that when a water is pure, *and especially when it is naturally hard*, a large amount of nitrates and nitrites for all practical purposes is harmless. When, however, it is impure (and especially when the water is soft), they do constitute a special danger, possibly as affording food upon which bacteria can subsequently develop. The influence of the saline constituents in regard to harmfulness or the reverse of nitrated waters, is not altogether plain. Warington [*] has found that nitrification proceeds more rapidly in the presence of chalk and gypsum than in their absence. It may therefore be that a given amount of nitric and nitrous acid in the presence of these mineral substances may mean organic matter completely oxidised, while, in their absence, this destruction by oxidation would be only partial. Digestion of the water with a sulphite, or sulphurous acid, has been frequently recommended in the brewing journals for the removal of nitrates and nitrites. In the dilute state in which the nitrates and nitrites exist, and in which the sulphite would operate,

[*] Journal Chemical Society, 1885, p. 758.

no such destruction occurs.   There is, in fact, no practicable method by which nitrates and nitrites in a brewing water can be removed.

(iii.) The determination of chlorine has some bearing upon the organic matter. Sewage contains sodium chloride, and contamination with it must raise the amount of chlorine. When a water contains more chlorine than undefiled supplies from the same formation in the district, suspicion is cast upon the supply.

(iv.) The position of phosphates is pretty much the same as that of nitrates.   The matters constituting the more dangerous forms of contamination contain phosphorus; its presence in any amount therefore raises suspicion; again, the phosphates would certainly assist the development of micro-organisms. However, many waters admirably suited for brewing contain an appreciable amount of phosphates, and it is probably safest to attach little importance to them when the water is proved pure by other tests, but to attach considerable importance to them when it is found impure.

(v.) To the practised eye the behaviour of the solid matters, when ignited, is suggestive. Roughly speaking, the more intense and persistent the blackening, the more probable is it that the pollution is considerable, and of animal origin. The fumes evolved during ignition give by their odour some idea of the intensity and nature of pollution. The blackening of the solids during ignition, and the fumes evolved, are, however, interfered with by nitrates and nitrites, when these are present in any quantity; for the nitrates and nitrites will, if present in sufficient amounts, oxidise the organic matter, preventing the blackening and the evolution of organic fumes, they themselves being decomposed and expelled as red fumes.

THE PURIFICATION OF IMPURE BREWING WATERS.

When a water is only slightly contaminated, boiling for about an hour before mashing does all that is necessary. This boiling is especially efficacious when the water contains dissolved chalk (calcium carbonate). For, as the chalk is precipitated, and subsides, it takes down with it the greater part of the organic matter, and the greater part of the micro-organisms.

When a water is more impure, it should be boiled and treated with some antiseptic. In the case of chalky waters, bisulphite of lime is undesirable, for this will redissolve a portion of the precipitated chalk, and so disturb the compact mass formed by the subsidence of the chalk and the organic matter. Some soluble salt of sulphurous acid may be added when the water is boiling, preferably a solution of potassium sulphite. In the case of non-chalky waters bisulphite of lime is the best preservative. It is best added after the water has boiled, and when it has cooled to about 180°; the valuable sulphurous acid is thus in great part retained.

In the many cases where a sufficient degree of purification results from the subsidence of the chalk and organic matter, there are one or two simple precautions which require observance. As long a period as possible should be given after the boiling of the water, and before its use, in order that the chalk, &c., may settle down into as compact a mass as possible. It should therefore be treated overnight. It is necessary, too, to have a good space between the outlet pipe and the bottom of the vessel, otherwise the deposited matter may be sucked down into the mash-tun when the supply is being drawn upon for mashing purposes. Indeed, an intermediate vessel is desirable, into which the water may be slowly and carefully run after the subsidence of the precipitate, and from which the requisite mashing liquor may be drawn. When the mashing liquor is drawn direct from the vessel covered with deposit, there is some risk of the lighter and less compact particles on the top of the deposit being sucked down into the mash-tun, owing to the rapid rate at which the mashing liquor is required, and to the inevitable agitation of the contents of the liquor back.

Filtration for brewers' purposes we regard as only really useful when a water is very distinctly polluted; in this case the water may be rendered moderately pure. We know of no instance, however, in which a water has become perfectly purified by filtration, even when the original amount of impurity was not very large. We are talking, of course, of the usual filters used in this country. Pasteur-Chamberland filters should give perfectly sterile waters, but their practical utility has yet to be proved. The filters the action of which is known are, animal charcoal, silico-carbon, and spongy iron ones. These filters are of use in transforming a water so impure as to be unusable into a moderately pure one; but they will not completely purify either moderately impure water, or very impure water. Having regard to this fact, and to the danger that always exists of not renewing the filtering media with the frequency demanded by the impurity of the water, brewers should, as a rule, be chary of resorting to filters.

That the filtering media must be periodically changed is now recognised as urgently necessary, but there is one other necessary precaution which is not so well known; i. e., the necessity for using filtered water at once, or raising it at once to a temperature prejudicial to the development of bacteria. It has been shown that no filter such as is ordinarily used completely sterilises the water; and in face of the equally incomplete removal of organic matter, and of the mineral substances favouring bacterial development, it is clear that if we allow the filtered water to lie about at ordinary temperatures, the remaining organisms will rapidly develop. At the rate at which they are able to propagate they would thus soon equal, if not exceed, those originally present in the water prior to filtration. We consider that it is better to have the unorganised matter present as such, than existing as the plasma of organisms. In the latter state it will, in our opinion, have been rendered more readily assimilable to the organisms to be subsequently introduced during cooling and refrigeration.

The reduction of micro-organisms by the filtration of contaminated waters through various filtering media has been exhaustively studied by P. F. Frankland, Salamon and Mathew, Bischof, and others ; but it is only recently that attention has been drawn to the reduction in the dissolved matter which results on passing waters through certain forms of filters.* Snyders gives, among others, the following experimental results :—

—	Solid Residue at 230 F.	Potassium permanganate reduced.	Organic Matter.	Chlorine.	Ammonia.	Lime and Magnesia.
Experiment I.—Pasteur-Chamberland filter filled with animal charcoal.						
	Grns. per gal.	Grns. per gal.	Grns. per gal.	Grns. per gal.	Grns. per gal.	Grns. per gal.
Unfiltered	93·13	1·02	5·07	9·10	·0063	21·70
Filtered	46·90	0·24	1·22	8·90	·0021	8·40
Experiment II.—Maignen's Filtre Rapide.						
Unfiltered	88·55	0·98	4·90	8·75	·0063	21·70
Filtered	32·20	0·05	3·15	8·75	·0021	4·55

Seeing that brewers frequently use their brewing waters for the steeping of their barleys, and for boiler-feed purposes, it will be desirable to briefly consider waters from these standpoints.

* Brewing Trade Review (abstract), 1888, p. 335, (Berichte der Deutschen Chemischen Gesellschaft, xxi. (1883), p. 1683.)

MALTING WATERS.

As in the case of brewing waters, it will be convenient to consider malting waters, first from the standpoint of their dissolved salts, and secondly from the standpoint of their organic purity. It must be confessed that, on the whole, such work as has been done with respect to the former consideration is of an unsatisfactory character. It is limited to researches by Lintner,[*] and Mills and Pettigrew.[†] Lintner's experiments go to show that salt impedes the germination of barley, and reduces the elimination of albuminous matter. His results are as under :—

—	Hours in Steep.	Days on floor.	Extract of malt per cent.	Albuminoids in malt.
Ordinary water	97	6	69·9	8·85
Water with 21 grains of salt per gallon	125	7	66·3	11·91
Water with 25 grains of salt per gallon	132	9	63·5	12·65

Lintner's statement as to the checking of germination by salt is in accordance with what would be anticipated from the ascertained action of salt on vegetation generally ; the value of the above results, indicating a lessened extraction of albuminoids, is reduced for practical purposes, seeing that the steeping periods are far in excess of those adopted in this country.

Apart from this we are quite in the dark as to whether a great or a small extraction of albuminoids gives the best practical results in the brewery. No great weight can be attached to these experiments, until we have ascertained the connection between certain proportions of albuminoids in our malts, and the quality of the resultant beers.

[*] Lintner, ' Lehrbuch der Bierbrauerei,' p. 137.
[†] Journal Chemical Society, xli. (1882), p. 38.

The researches of Mills and Pettigrew are more elaborate, but of no appreciable value at present for practical purposes. They find that waters containing much lime compounds impede the extraction of nitrogenous matter, but not to any very marked extent. They also find that the soft (and somewhat impure) Lichfield water, used by the Burton firms for steeping purposes, extracts more nitrogen and as much non-nitrogenous organic matter, as waters artificially prepared, and containing (1) chalk, (2) gypsum, (3) salt. It is pretty well shown by them that waters containing gypsum (waters of Class I.) extract less phosphates and nitrogenous matter than waters deficient in gypsum, &c. But this information, in the present state of our knowledge, is not very useful. We are in complete ignorance as to whether it is desirable to extract little or much phosphates and nitrogenous matter, and until we know which is right, the value of the experiments before us must rest in abeyance. What we require is a series of practical maltings, made on the same material, and under otherwise identical conditions, but with waters of different types. The resulting malts will then require the most searching analysis, and should preferably be practically brewed with under the same conditions, and the resultant beers examined as exhaustively as the malts. By this means a connection would be established between the steeping water and the malt, and the beer. Until that connection is made, it will be safest to make no statement as to the salts desirable or undesirable in malting operations.

Our present system of steeping induces apparently a large extraction. This may be right or wrong ; at any rate, it has not been proved wrong, and it gives good practical results. It would seem, therefore, that the use of gypseous waters is adverse to the spirit of our present system ; and the experience of the Burton brewers, who prefer the soft (and not very pure) Lichfield water to their own hard and pure water, cannot be ignored. It certainly *suggests* that non-gypseous

are preferable to gypseous waters, and that is about all that
can at present be said.

With respect to the organic standpoint, we believe great
organic purity is unnecessary. Southby,[*] although indifferent
to the saline constituents of a malting water, insists on its
purity. It is difficult to see, however, what effect impurities
in the water would have, considering the highly putrescible
matters extracted from the barley, and the myriads of organisms by which they are covered.

Hansen [†] says, "I have not paid especial attention to the
(bacteriological) examination of malting water, and for the
reason that I consider it superfluous. The barley grains have
on their surface an abundance of bacteria and other organisms
before they are brought into contact with the steep water;
therefore the greater or less number which ordinary water may
introduce can in this instance have no importance. The
principal evils to be feared in malting are the mould-fungi,
and these are as rare in water as they are frequent in aerial
dust."

This view of Hansen must strike everyone as reasonable.
Not only so, but it is hard to believe that any unorganised
organic matter in water can be of any material moment when
we are extracting matters from barley which, as every maltster
knows, putrefy most rapidly. The addition of bisulphite of
lime to the steep liquor is of some little value, for it will help
to check the only too ready putrefaction of the substances
extracted. We distinctly hold that the organic impurity of
a water, and its saline constituents, are of far less practical
importance in malting than the temperature of the steeping,
and the sufficient frequency with which the water is renewed.
We do not of course mean that a water polluted with sewage
would be permissible; but any ordinary pollution would
probably not matter.

* Southby, ' Practical Brewing,' p. 201.
† Brewing Trade Review, ii. (1888), p. 69.

BOILER FEED WATERS.

The nature and degree of incrustation produced depends upon the saline character of the water used. It is unnecessary to point out the loss due to scaled plates; fuel will be clearly lost through the resistance offered to the transmission of heat by the film of baked saline compounds; while the boiler plates themselves are, when incrusted, liable to be overheated.

The most objectionable salts in waters in this respect are calcium and magnesium carbonates and sulphates. Of these, however, the calcium sulphate is by far the most pernicious. Neither of the carbonates is in itself so detrimental, especially if certain precautions be adopted; they are precipitated, not as hard scale, but in a light powdery form; and if the flues are allowed to cool down before the boiler is blown off, little difficulty is experienced in removing any that may adhere to the surface of the plates. On the other hand, if this precaution is not taken, the heat of the flues, after the water has been withdrawn, will bake this substance into a hard scale. Further, if any considerable quantity of calcium sulphate is introduced into the feed liquor with the carbonates, and is allowed to remain unchanged in the boiler, the whole of the deposits (including the carbonates) will then assume the form of a hard tenacious coating. Again, if any tallow is allowed to find its way into the boiler, through the exhaust from engine cylinders, it forms (with the precipitated lime and magnesia) a soapy mass which is liable to become the source of considerable danger owing to its low conductivity. When engine exhaust is utilised directly for heating boiler feed, this lubricant should be discarded.

There are two principal systems of treating boiler feed. The water can either be chemically treated before it enters the boiler, or it may be subjected to treatment within the

E

boiler itself. The most perfect system is the preliminary process. This process, however, is rarely adopted in breweries, as it necessitates the employment of rather cumbrous apparatus required for the ready deposition and removal of the precipitated bodies. It will not, however, be wise to ignore this process, for it may become more general than at present. The precipitants selected must be efficacious in the cold, and thus differ somewhat from those added to the boiler itself, where the reactions are aided by the temperatures there obtaining.

There is this further difference between preliminary treatment and treatment in the boiler, that, while in the former we wish to separate the calcium and magnesium salts by precipitation, the precipitate being removed by subsidence, in the latter we wish to render the calcium and magnesium salts soluble.

As has been said, our object in treating a water outside the boiler is so to alter the hardening salts of the water that they are eliminated, and that those entering the boiler shall be of a more soluble nature. When this is effected, concentration may take place without fear of scaly deposit. The simplest form of this class of treatment is the addition of lime. This removes the carbonates only, but is quite without action upon the sulphates and other salts of calcium and magnesium. The lime and caustic soda method removes practically all the hardness ; it does so without the aid of heat, and gives a water containing only soluble sodium salts, and no excess of alkali. In order to precipitate salts other than carbonates, it is necessary to have sodium carbonate. This is provided by the caustic soda and the free carbonic acid usually present in the water. Only enough caustic soda is employed to decompose the sulphates, chlorides, &c., any excess of free carbonic acid being removed by lime. To precipitate magnesium salts, caustic soda, and not the carbonate is to be preferred. The following equations exemplify the reactions :—

$$CaCO_3,CO_2 + CaO = 2CaCO_3$$
$$CaCO_3,CO_2 + 2NaHO = CaCO_3 + Na_2CO_3 + H_2O$$
$$CaSO_4 + Na_2CO_3 = CaCO_3 + Na_2SO_4$$
$$MgSO_4 + 2NaHO = MgH_2O_2 + Na_2SO_4.*$$

Where the permanent hardness due to calcium salts is excessive, or the quantity of free carbonic acid too small, it becomes necessary to supplement the caustic soda with sodium carbonate, in order to make up the total quantity of carbonate required. The caustic soda and sodium carbonate, when added in regulated quantities, are practically wholly converted into sulphates or chlorides, which are both soluble and harmless. It is, however, impossible to remove the whole of the calcium and magnesium by this process, owing to the slight solubility of calcium carbonate and magnesium hydrate. There is some variation, too, according to the impurities present, and the time allowed for the reaction. In this process the magnesium salts are either precipitated as carbonate or hydrate, according as sodium carbonate or caustic soda is used as the precipitant. Although the whole of the calcium and magnesium salts cannot be removed, water so treated is immensely improved, and, if proper precautions are observed, boilers that are fed with it need never develop a scale. There are two points worthy of special notice with regard to the influence of magnesium salts in boiler feed water. It is a curious fact that magnesium carbonate, although precipitated as such, is eventually converted into the hydrate after prolonged exposure to the high temperature within the boiler. This magnesium hydrate is much more readily baked into a hard scale than is the calcium carbonate. It is therefore of importance that when magnesium salts are present in the feed-water, greater care than usual should be taken to allow the flues to cool down before the water is withdrawn from the boiler. When any magnesium exists as chloride, it

* Readers interested in the softening of water supplies outside the boiler should consult an interesting paper by Messrs. Macnab and Beckett, *Journal Society Chemical Industry*, v. (1886), p. 267, which has been freely used.

is of special importance to ensure its conversion into the carbonate or hydrate. The chloride is rapidly decomposed at a high temperature, such as is maintained within a boiler, hydrochloric acid being set free. This of course might prove most injurious in time unless a sufficiency of alkali were present to neutralise it.

Sometimes a soluble salt of aluminium is resorted to, in order to facilitate the deposition of the calcium and magnesium precipitate. Enough lime and soda must in this case be added to precipitate the aluminium as hydrate, as well as to remove the hardness of the water. The usefulness of this precipitated alumina is, however, chiefly felt in the case of waters which are organically impure. With these, precipitation is generally sluggish, and without the alumina there is a tendency to turbidity a long while after the lime and soda have been added. If, too, the original amount of hardness is small, the whole precipitate exists in such a fine state of suspension that subsidence is very slow. Alumina here is of great service.

As a rule the boiler feed waters in a brewery are treated in the boiler itself. Of the substances used for this purpose the most general are free alkali (soda), boiler compositions containing sodium tannate, and neutral bodies which, by their mechanical agitation in the boiler, to some extent prevent the hard silting on the boiler plates, and allow of their being more readily cleaned. Malt-combes are frequently added for this latter purpose. The sodium tannate has, on the other hand, a specific chemical action. The tannic acid combines with the calcium and magnesium, with the production of a soluble calcium or magnesium tannate, and the acid of the previous calcium and magnesium combines with the sodium, these sodium compounds being all soluble. The action of free alkali in decomposing calcium and magnesium salts has been already referred to.

It is a common fallacy to add large quantities of the soda,

or boiler composition, all at one time after the boiler has been cleaned. This is a mistake in many ways. There is always a chance, if the boiler is regularly and efficiently blown off, of the material being entirely used up previous to the addition of the next quantity. Again, the large excess of alkalinity at first is very apt to induce priming. It is preferable in every way to add definite quantities daily, depending upon the amount of water evaporated, and the amount of insoluble matter it is desired to render soluble.

Very good results may be obtained by the daily injection of sodium carbonate. It has no influence upon the calcium and magnesium carbonates; but these are, as has been stated, comparatively harmless if unaccompanied by the sulphates, which the sodium carbonate would remove. To carry out the system, it is necessary to arrive at an estimate of the average quantity of water daily injected into the boiler. The next step is to ascertain the quantity of calcium sulphate contained in the liquor supply, as well as any other calcium and magnesium salts, besides carbonates, which, might use up the alkali, and for which due allowance must be made. Brewers are acquainted, as a rule, with the quantities of the saline constituents in their waters, and only a simple calculation is necessary ; the following figures may, however, be of use.

		Sodium carbonate crystals.
10 grains of calcium sulphate (anhydrous)	require	$21 \cdot 03$ $Na_4Co_3 + 10H_2O$.
10 grains of magnesium sulphate (anhydrous)	,,	$23 \cdot 83$,,
10 grains of calcium chloride (anhydrous)	,,	$25 \cdot 77$,,
10 grains of magnesium chloride (anhydrous)	,,	$30 \cdot 10$,,

CHAPTER II.

BARLEY AND MALT.

INTRODUCTORY.

IN barley we have to do with the most important of the brewing materials, since it must always form the basis of all brewing operations, and, beyond a certain point, it can never be replaced by any other material.

Barley, like wheat, oats, and rye, belongs to the cereals, and is placed by botanists in the Natural Order Gramineæ, or Grasses. It is a member of the genus *Hordeum*, the species which is usually met with being *Hordeum vulgare*. This species embraces many varieties, and is divided, according to the arrangement and position of the grain upon the spikelet, &c., into four groups, each of which includes a large number of sub-species and varieties. Körnicke * has classified these groups as under :—

     I. Six-rowed barley (*Hordeum hexastichum*).
     II. Four-rowed barley (*Hordeum tetrastichum*).
     III. Intermediate barley (*Hordeum intermedium*).
     IV. Two-rowed barley (*Hordeum distichum*).

It is only the members of the last group which are of any importance in brewing, and of these it is the many varieties of the long two-rowed barley that are generally employed. The best known and most widely distributed of these varieties is, perhaps, the Chevalier, which is largely cultivated in all barley-growing countries, and which probably combines in itself all the best qualities of every variety of barley.

* Zeitschrift f. d. gesammte Brauwesen, 1882, p. 114, *et seq.*

VALUATION AND EXAMINATION OF BARLEY.

The desirable characteristics of a sample of barley may be classed under two heads, essential and non-essential. Of the essential qualifications, good "condition" is the first. This comprises all those qualities upon which perfect germination depends: *maturity, dryness, freedom from damaged and from heated and discoloured corns.* In addition to condition, mealiness is, as a rule, a desirable characteristic. Actual age is important, and also odour, as an indication of condition. Dryness is of particular importance, because, as is well known, damp barley invariably fails to germinate satisfactorily, and if stored it is certain to become mouldy. Its vitality, instead of improving, as is the case with dry barley up to a certain limit, is often very seriously reduced. Maturity, or ripeness, is also important, since neither under- nor over-ripe grain germinates satisfactorily. By mealiness, is meant the condition of the interior of the corn, and this to a considerable extent is dependent upon ripeness, and, to a less extent, upon the climatic conditions under which the ripening has taken place. If, when bitten, the exposed section of the grain is regular and white, the germination will probably be vigorous; if it is of a bluish-grey or brownish tint, and of a hard and brittle nature, there is less likelihood of a perfect modification on the growing-floor. This peculiarity is probably due to the presence of small and densely-packed cells, with highly resistant cellulosic walls, and points to unsatisfactory conditions of growth and harvesting. Those corns which are situated at the extremity of the ear are more likely to be vitreous or steely than the lower ones, hence, perhaps, the presence of steely corns to a certain extent in all samples of grain. Freedom from damaged corns is highly important. Broken or half corns are easily seen at a glance; but very careful scrutiny is needed to detect

slight abrasions of the husk. If the corns have been broken
off too close to the ear, much damage may also be done. On
the whole, samples in which the awns are incompletely removed are preferable to those that have been too closely
dressed. The last mentioned imperfection affords excessive
encouragement to germs of mould and bacteria. Freedom
from heated and discoloured corns is also essential for
regularity of germination. These corns are the result of the
heating of the grain in stack or store, following bad conditions
of harvest. It is seldom that heated corns germinate at all, if
the evil is very pronounced ; but dark and stained corns of
which the defects are due to bad weather alone, grow quite
readily, although it is questionable if the quality of the malt
from such grain is really satisfactory. Those corns which have
started germinating in the field are easily detected by the
appearance of the germ, and the corresponding softness of
that end of the corn. It is almost certain that with such
corns germination will never be resumed. In addition to the
above there is another class of discoloured corns which presents
a greenish-yellow appearance, and is indicative of incomplete
ripeness. A slight admixture of these is often found in
otherwise good samples of barley, and is probably due to
portions of the crop having been grown under hedgerows, or
in situations where they were deprived of the direct rays of
the sun. Some corns of this class grow apparently well,
but still they must be considered decidedly objectionable.
Unevenness in maturity is often due to an admixture of the
seed-barley in the first instance, and it is then the direct result
of agricultural incompetence.

The above are the essential qualifications of barley for
malting purposes, which all lead up to the main consideration, that of vitality. Although much may be determined,
as regards the vitality of a sample, by an experienced
and critical examination, yet the only reliable test is to
subject it to such conditions as favour germination, whereby

the degree and regularity of growth can be ascertained. This may be done by sowing in moist flannel, or sand, but probably the most accurate and certainly the quickest results are obtained with the Coldewe and Schoenjahn apparatus.

If the grain is in sound condition to start with, vitality increases for a time, thus, samples that early in the season contain many idle corns, often grow much better after a few months' storage. Barley, however, does not grow readily after the second year.

Corn which has been harvested under adverse circumstances and stacked and stored before it is dry, generally contracts a disagreeable smell indicative of fungoid growths.

The following may be termed non-essential qualifications : *size, weight, uniformity, and cleanliness.* Size and weight are naturally closely connected, and were, under the old fiscal system, considered of prime importance. Heaviness generally implies fulness and maturity with a consequent smaller proportion of husk. In some cases it is evidence of careful cultivation. Outside these considerations mere weight is of little importance. Uniformity applies both to regularity of size, and also to the general homogeneity of the sample. In the latter respect it may of course have an important bearing upon vitality, since mixtures of dissimilar barleys cannot possibly grow regularly. In the absence of grading machinery, irregularity of size may also have an important influence upon satisfactory germination. Small corns are more quickly saturated with water than larger ones, and therefore either start more quickly, or, if the steeping is prolonged, never start at all, having had their vitality destroyed by the prolonged immersion. The latter is, of course, an extreme supposition, but it helps us to realise the disadvantage of attempting to work an uneven sample.

By cleanliness is meant the absence of other seeds, which are often particularly noticeable in foreign barleys. Unless these are previously removed, the flavour and aroma of the beer are liable to be affected. Cleanliness also denotes the absence of stones, dirt, and other rubbish, and is on the whole an important consideration in determining the value of a sample. No reference has yet been made to general colour, which varies considerably with the district and season. Depth of colour, if accompanied by brightness, does not necessarily imply inferiority, but the original colour of the barley can hardly fail to have some influence upon the tint of the beer, irrespective of the curing process. We therefore find that for pale ale brewing a light coloured barley is preferred.

THE MORPHOLOGY AND EMBRYOLOGY OF THE BARLEY-GRAIN.*

When a grain of ripe barley is examined, it is at once seen to be spindle-shaped, and with the point at one end much more regular and sharper than at the other. The sharper end is the lower end of the corn and often exhibits the scar which marks the point of the former attachment of the grain to the spikelet of the ear.

One side of the grain is much more convex, both laterally and longitudinally, than the other. The former is hence known as the *ventral* side (Plate II., Fig. 1, *b*), and it is the side which faces inwards whilst the grain is still attached to the ear. Throughout the entire length of the ventral side, there runs a furrow (Plate II., Fig. 1, *c*), which, at the lower end, contains a fine *bristle*, known as the " basal bristle " (Plate II., Fig. 1, *d*), and furnished with very fine lateral hairs. This bristle, which on closer examination is found to be in intimate connection

---

* For more detailed information on this subject, see :—

Holzner and Lermer, ' Beiträge zur Kentniss der Gerste,' München, 1888.

W. Johannsen, ' Développement et constitution de l'endosperme de l'orge.' Comp. rend. de Carlsberg, 1884, p. 60.

H. T. Brown, ' A Grain of Barley.' Trans. Burton-on-Trent Nat. Hist. and Arch. Soc., 1888, p. 110.

Brown and Morris, ' Researches on the Germination of the Gramineæ.' Journ. Chem. Soc., lvii. (1890), p. 459.

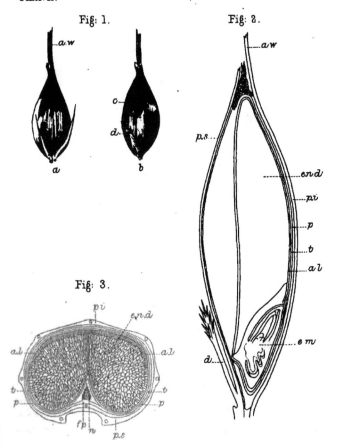

Plate. II.

Fig: 1.   Fig: 2.

a.w   a.w

o   p.s

d   en.d

a   b   pi

p

t

Fig: 3.   a.l

pi   en.d

e.m

a.l   a.l

d

t   t

p   p

f.p   n   p.s

Fig: 1. Grain of Barley. a. Dorsal view. b. Ventral view.
o. Ventral furrow. d. basal bristle. (After Holgner.)

Fig: 2. Diagrammatic sketch of longitudinal section of a grain of
barley. p s. Palea superior. p.i. Palea inferior. p. Pericarp.
t. Testa. a l. Aleurone layer. en.d. Endosperm. e m. Embryo.
d. Basal bristle. a w. Awn. (After Holgner.)

Fig: 3. Transverse section through Barley Grain. n. Remnant of
nucleus. f p. Section of "pigment string". Other parts as
in Fig: 2. (After Holgner)

E & F. N. Spon, London & New York.   That Roffe Son, Lith.

with two scale-like processes, the so-called *lodicules*, occurring under the paleæ and enwrapping the embryo, was formerly supposed to absorb and conduct water to the embryo and thus hasten its growth. Experiment has, however, shown this view to be incorrect, and it is now generally accepted that these lodicules are the remains of organs which played their part in the earlier life of the grain; they correspond, in fact, to the calyx and corolla of other flowering plants.

The outer and less convex side of the grain is termed the *dorsal* side (Plate II., Fig. 1, *a*). The evenness of its surface is broken by five slight ribs which run the entire length of the grain, and mark the position of the same number of vascular bundles. These vessels serve to nourish the outer coating, and are well seen in a transverse section of the grain, to the dorsal side of which they give an angular appearance (Plate II., Fig. 3).

In nearly all varieties of barley, the grain is covered by a thick outer coating, consisting of the two paleæ. Strictly speaking these paleæ do not belong to the grain at all, since they are the remains of the protecting envelopes of the flower. The paleæ are closely adherent to the dorsal and ventral sides respectively of the grain. That adhering to the latter is the smaller and innermost, and is known as the *palea superior* (Plate II., Fig. 2, *p.s.*), whilst that adherent to the dorsal side is known as the *palea inferior* (Plate II., Fig. 2, *p.i.*), it is the larger and outermost of the two, slightly overlapping the other, and its upward prolongation forms the *beard* or *awn* of the grain (Plate II., Figs. 1 and 2, *aw*). In the majority of cereals, such as wheat, rye, &c., the paleæ are non-adherent, and readily separate from the ripe grain in the form of chaff.

Underneath the paleæ, we find the true integuments of the grain; these consist of the *pericarp*, and the *testa*, the former being the outermost. Of these coatings, the testa (Plate II., Figs. 2 and 3, *t*) is the survival of the wall of the embryo-sac, and forms the integument of the seed, whilst the pericarp

(Plate II., Figs. 2 and 3, *p*) is derived from the walls of the ovary, and corresponds to the covering of the fruit in other plants. Both these coatings can be differentiated into several layers of cells. The outermost layers of the cells of the pericarp are arranged longitudinally, and give rise to the striated appearance which a barleycorn shows when the paleæ are removed. Both the pericarp and the testa have their point of origin in the so-called "pigment-string" (Plate II., Fig. 3, *f.p.*), a line of tissue, which runs the whole length of the grain, and has a distinct reddish-yellow colour ; it is the remnant of the line of attachment of the ovule to the walls of the ovary.

A close comparative study of the development of the barleycorn shows that the ripe grain includes the united products of ovule and ovary, that is to say it is a combination of both seed and fruit ; the fruit or ovary is, however, represented only by the pericarp, referred to above.

The two essential parts of a barleycorn are the embryo and the endosperm. The former, which constitutes about $\frac{1}{80}$th of the corn, is situated at the base of the dorsal side of the grain (Plate II., Fig. 2, *em.*) ; the latter (Plate II., Figs. 2 and 3, *end.*), constitutes the remaining portion of the grain.

On the side adjacent to the endosperm, the *embryo* has a shield-like expansion, the *scutellum* (Plate III., Fig. 4, *scut.*), which, lying in close contact with the endosperm, serves as a special organ of absorption, through which, during germination, the nutritive matter stored in the endosperm must pass on its way to the growing portions of the embryo. The actual line of separation between the embryo and endosperm is marked by a line of cells forming the *epithelium* (Plate III., Fig. 4, *ab. ep.*), which covers the whole of that portion of the scutellum in contact with the endosperm. The cells of the epithelium are columnar in shape and are arranged with their longer dimensions normal to the surface of the scutellum. The

Plate III.

a.c. Nutrition dis-ap dinorizas
gelatican of sentidum f_1, f_2, f_3,
hilum shionule plum, sh Plumule sheath
rad Radicle, r cap Root cap, r sh
Root sheath.

s.c. Sarch containing cells of endosperm
c.e Emptied and consumed cells of
endosperm al Eurone cells t Testa.
p Pericarp p.i. Fulca inferior P.s
Fulca superior, d Basal bristle f_1
Funiculus, or "pigment string"
(After Holzner)

bases of the cells are closely united with the subjacent tissue of the scutellum, and their free ends are in the very closest contact with the deplated layer of cells (Plate III., Fig. 4, *c. c.*) of the endosperm, but are not in any way organically united with this tissue; in fact it has been shown by Brown and Morris that the relation of the embryo to the endosperm is purely parasitic, or more correctly, saprophytic.[*]

The cell-walls of the epithelium are very thin, and not in the least cuticularised. The cell-contents, before germination, are very finely granular, with a large and distinct nucleus. The epithelium has long been regarded as being especially concerned in the transference of nutritive matter from the endosperm to the growing parts, but Brown and Morris have shown that the function of absorption is one which the epithelium shares with the deeper-seated cells of the scutellum and the principal organs of the axis, whilst the properties which markedly differentiate it physiologically from the other tissues are its special secretory functions.

The cells of the scutellum underlying the epithelium are polyhedral, and have well-marked laminated walls and conspicuous intercellular spaces. The cell-contents of these cells consist principally of minute spherules of aleurone, embedded in a fine protoplasmic network; there are also present, mingled with the aleurone-grains, minute globules of fat. The aleurone-grains are completely soluble in a 10 per cent. solution of sodium chloride, and appear to consist of at least two compounds, one belonging to the class of *albumoses*, the other to that of the vegetable *globulins*.

Immediately below the scutellum, and in intimate organic connection with it, are the main organs of the axis, the *plumule* and the *radicle*. The former consists of four rudimentary leaves (Plate III., Fig. 4, f_1, f_2, f_3, f_4), inclosed in the

---

[*] The distinction between these two terms is that whilst a *parasite* obtains its nourishment from another living plant or animal, a *saprophyte* flourishes on its host when dead and inert.

plumule-sheath (*plum. sh.*), whilst the primary radicle (*rad.*), with its root-cap (*r. cap.*), is completely embedded in the root-sheath (Plate III., Fig. 4, *r. sh.*).

Returning to the endosperm, we find that this portion of the grain of barley is principally filled with a mass of thin-walled cells closely packed with starch-granules embedded in a very fine network of proteid material (Plate III., Fig. 4, *s. c.*), and arranged with their larger dimensions parallel with the axis of the grain. These starch-containing cells are surrounded by a triple layer of thick-walled rectangular cells (Plate III., Fig. 4, *al.*), the so-called *aleurone-cells*, the cell-contents of which consist mainly of closely packed aleurone-grains and fat, embedded in a protoplasmic matrix with well-defined nucleus. The aleurone-layer, where it is in contact with the starch-containing cells of the endosperm, usually consists of a triple or quadruple row of cells, but in the portion surrounding the embryo it dwindles to a single row of cells, which gradually dies away towards the base of the grain. This aleurone-layer does not, however, enter the embryo at any point, but is only in close contact with it.

Lying between the starch-containing portions of the endosperm and the embryo is a comparatively thick layer of emptied and compressed cells (Plate III., Fig. 4, *c. c.*), belonging to the endosperm. These cells are emptied of their contents during the later stage of the growth of the grain, when the development of the embryo goes on at the expense of some portion of the starch-containing cells of the endosperm. As these latter are emptied of their contents, they are gradually compressed by the growing embryo, and are found in the mature grain in the form described.

Having now described in brief the chief points in the anatomy of the barleycorn, we are in a position to examine the morphological changes which take place during germination. The results of these changes are of a threefold nature, there being in the first place, the dissolution of the cell-walls of the endosperm by a special enzyme; in the second place, the dissolution by diastase of the starch-granules thus set free, and the transference of the soluble products of the action to the growing points of the embryo; and finally, the changes which render the reserve nitrogenous material of the endosperm-cells available for the food of the young plant.

We know very little of the changes which the protoplasm of the endosperm undergoes, or of the agent which brings about the modification of this complex and non-diffusible nitrogenous material and renders the products of its conversion capable of diffusing through the cell-walls of the scutellum. That some action of this kind takes place there can be no doubt, since H. T. Brown has stated that he found, during eleven days' germination of barley, 40 per cent of the total reserve nitrogen originally present in the endosperm had passed through the scutellum to the young plant. And from the work of Green and others, on the germination of certain leguminous and other plants, there can be little doubt that this transference is brought about by the agency of a proteo-hydrolytic enzyme somewhat analogous to the pancreatic ferment of the animal body. Experiment has shown that the latter enzyme is capable of converting the coagulable albuminoids of cold-water barley-extract into readily diffusible and crystallisable substances, and indeed, cold-water malt-extract itself exerts a similar, but very much feebler action. Here we undoubtedly have a breaking down of the albuminoids into the simple amido-bodies, leucin, tyrosin, asparagin, &c., and, as before mentioned, experiments with other seeds have shown that the same action takes place during germination. In those cases in which a ferment of this nature plays a part, it has been shown to be present in the form of a *zymogen*, a substance which requires the presence of a small quantity of acid in order to develop its full activity.

Our knowledge of the other changes which take place in the germinating grain is much more extended and precise.

It has long been known* that a portion of the cell-walls of the endosperm disappears during germination, but it is only quite recently that the true nature and full extent of this disappearance of cellulose has been understood.† Brown and Morris have shown that within from 24 to 36 hours of the commencement of germination the cellulose of the depleted layer of cells between the embryo and endosperm becomes softened, and finally almost completely disintegrated and dissolved ; whilst, as the germination proceeds, this action is extended to the starch-containing cells of the endosperm, and proceeds progressively throughout the length of the corn. This action is always antecedent to any action upon the starch-granules ; in fact, Brown and Morris show that these are never attacked as long as the walls of the cells containing them are intact. The first indications of this action are seen in the swelling up of the cell-wall, the stratification of which becomes much more apparent, owing to a partial separation of its constituent lamellæ. As the action proceeds, these are gradually disintegrated, and break down into minute fragments, which finally disappear, leaving no visible sign of separation between the contents of contiguous cells.

When this action has taken place throughout the whole of the endosperm, the grain is readily friable or "mealy" throughout, and without doubt the process of malting consists in bringing about this action to the fullest possible extent.

The dissolution of the cell-wall takes place in certain definite directions in the germinating grain. Commencing, as stated above, in the depleted layer of cells in contact with the epithelium of the scutellum, it progresses most rapidly along the grain immediately under the advancing plumule. This action, which is shown diagrammatically in Plate IV., is not due to any direct action of the growing plumule, but to the

* Brown and Heron, Journ. Chem. Soc., 1879, p. 622.
† Brown and Morris, Journ. Chem. Soc., 1890, p. 468.

Plate.IV.

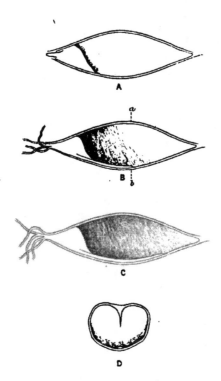

Diagram shewing the progress of the dissolution of the
cell-wall of the endosperm of barley at different periods
of growth. The advancing disintegration of the
contents of the grain is indicated by shading.

A at 3 days
B , 6 ,
C , 10 ,

D Transverse section of B at a.b.

fact that the cells in this portion of the grain are the youngest in point of development, and their cell-walls are consequently less resistant to the ferment which brings about the dissolution of the cellulose. The soluble ferment, under whose influence the cell-wall is dissolved, has not yet been isolated in a pure state, but Brown and Morris have conclusively shown the existence of such an enzyme in germinating barley.

We have already stated that the dissolution of the cell-wall always precedes the attack upon the starch-granules; the former action also proceeds much more rapidly than the latter, but both follow the same course in the grain, and Plate IV., which represents the progress of the former action, also represents the lines of attack of the diastase upon the starch-granules.

The first indications of the action of diastase upon the starch are seen in minute pittings on the surface of the granules. These pittings increase in number and depth, and finally radial clefts appear, and the concentric laminations show a more or less complete separation from each other. This action takes place under the influence of the diastase of the grain, and consists in the conversion of the starch into maltose, which then serves for the nutriment of the young plant. The diastase increases very much, and also alters greatly in nature during germination. We shall have occasion to refer more particularly to these changes in a later section of this chapter (see p. 70.)

The aleurone-cells of the endosperm undergo no visible alteration in appearance during a limited period of growth, such as we are considering here.

Coincident with the above-mentioned changes in the reserve material of the endosperm, starch begins to appear in the tissues of the embryo, which in the mature and resting stage is free from starch. These starch granules are not, of course, translated as such, but are elaborated from the cane-sugar, which, as we shall see later, is derived from the starch

F

of the endosperm through the agency of the epithelium of the scutellum.   The first appearance of growth is seen in the elongation of the primary radicle, and immediately this begins, starch-granules are deposited in the tissues of the radicle, and to a lesser extent in those of the plumule.   As germination progresses, the secretion of starch-granules spreads through the tissues adjacent to the radicle and plumule, and is especially abundant in the tissue immediately surrounding the young vascular bundles, which themselves, however, contain no trace of starch.   The cells of the scutellum at the same time begin to show indications of small starch-granules, which appear first of all in the cells of the scutellum immediately underlying the absorptive epithelium.   The appearance of starch in these tissues is coincident with the first signs of action upon the cellulose of the depleted cells of the endosperm, and affords the first indication of the passage of reserve material from the endosperm to the growing embryo.

We have already referred to the fact that the epithelium plays a most important part as a secretory tissue, and we have mentioned that the changes which occur in the endosperm during germination, all take their rise in the parts immediately under this tissue.   The epithelial cells themselves undergo certain alterations during germination.   The cell-contents become much coarser in structure, and the granulations increase in size and number until the nucleus is quite invisible.   The free ends of the cells also become club-shaped, and form a sort of velvety pile which projects into the endosperm.

It was mentioned above that Brown and Morris have shown the absence of any organic connection between the embryo and endosperm of the cereals, and have also proved the parasitic nature of the former, and the entire absence of vitality in the starch-containing cells of the latter.   The recognition of these most important facts have enabled these workers to throw much light on several questions connected with the embryology of the barleycorn, since the changes which take place in the endosperm during germination must be regarded as the result of influences communicated through the embryo.

The method adopted for the study of these questions consisted in dissecting out the embryos from barley, after steeping for a few hours in water, and then cultivating the excised embryos upon artifical endosperms of definite and known composition. In this way it was possible to follow the changes brought about by the growing embryos, and to determine their nature. The results obtained by this method of experiment will be more particularly described in later sections.

NATURE OF CHANGES DURING GERMINATION.

The preceding review of the anatomy and morphology of the barleycorn, will enable the changes which occur in malting to be the better understood, and will assist in elucidating the objects for which the grain is subjected to this preliminary process.

In classifying the principal objects of malting we must undoubtedly give pre-eminence (1) to the dissolution of the cellulosic walls of the starch-containing cells, and (2) to the development of the diastase. We have already referred to the former action and mentioned that the friability* or mealiness of the malted grain depends entirely upon the thoroughness with which the action of the soluble cyto-hydrolytic† ferment has spread through the grain. It has been found that the different varieties of barley vary very greatly in the rapidity and ease with which this action

* We are not referring here to the friability of malt as determined by the conditions of curing.

† Derived from κίτος, a cell, and analogous to amylo-hydrolytic.

proceeds through the grain. This appears to depend greatly upon the influence of soil and climate, but as a rule, those varieties in which the cell-walls give way with the greatest facility possess the greatest value for malting purposes. The product of the action of the dissolution of the cellulose has not yet been identified, but it undoubtedly serves as a food for the embryo during the early stages of germination.

The development of the diastase during germination is also of the very highest importance. As soon as germination commences, the embryo begins to secrete diastase, which passes over to, and diffuses into, the endosperm, and rapidly attacks the cellulose and starch-granules immediately under the scutellum. This was proved by Brown and Morris, by cultivating the excised embryos of barley upon artificial endosperms, consisting either of porous tiles moistened with water, or of gelatin alone, or of "starch-gelatin." The latter was prepared by adding finely divided barley-starch to a slightly warmed solution of gelatin (5 per cent.), and stirring until the moisture was on the point of solidifying. The excised embryos were placed upon the surface of the "starch-gelatin" when this was on the point of setting, a gentle pressure being used in order to bring every part of the convex surface of the scutellum in contact with the cultivation medium. In this manner it was possible, by the examination of small pieces of the starch-gelatin cut from beneath the growing embryo, to follow the gradual progression of the solvent action upon the starch-granules.

The action was seen to arise, first of all, at the limiting surfaces of the scutellum and starch-gelatin, and to be independent of the presence of micro-organisms. The action of the embryo is also dependent upon its growth, or rather vital activity. When the embryos were exposed to chloroform before being placed on the gelatin, it was found that they were incapable of growth, and no action took place on the starch.

Tested by the method we have just described, it was found that the excised embryos are entirely without diastatic power, but that after four days' culture on water and 5 per cent. gelatin respectively, both the embryos and the media on which they were grown showed distinct evidence of the formation of an amylo-hydrolytic enzyme. The appearance of the enzyme outside the embryo, which it will be remembered was free from diastatic power, points to its formation from some material stored in the embryo, and modified in its passage outwards; showing that the formation of diastase is a secretory and not an excretory process.

We have already stated that it is the epithelium which is concerned in this secretory function, and we may mention that this was clearly shown by cultivating, side by side, ordinary embryos and embryos from which the epithelium had been removed; in the former case the action proceeded as usual, whilst in the latter case the embryos had entirely lost their power of attacking the starch-granules.

The secretion of diastase by the epithelium during germination is increased by the presence of a little acid, but it is in no way stimulated by the presence of either starch-granules or soluble-starch. The presence, however, of readily assimilable carbohydrates completely checks the secretion of diastase. This was shown by cultivating embryos upon starch-gelatin containing a little cane-sugar or other assimilable carbohydrate; under these conditions the epithelium does not secrete any diastase whatever during germination. It was further shown that when barley itself is germinated in such a manner that a free supply of cane-sugar is available for the nourishment of the growing embryo, no diastase is secreted, and the grain remains as tough and resistant as before germination. This is a very important fact, and shows us that the power of secretion is so adapted to the requirements of the young plant, as to be only exercised when the supply of soluble nutriment begins to fail. In fact, the secretion of diastase by the epithelium must be regarded as one of those starvation phenomena of which several are known to botanists.

In practice, the diastase secreted by the epithelium passes into and permeates the endosperm, and acts, as we have already seen, upon the starch-granules liberated by the removal of the cell-walls by the cellulose-dissolving enzyme, which has been previously secreted by the epithelium and passed into the endosperm. A steady accumulation of diastase thus goes on during the germinating period.

As we shall have occasion to see later, diastatic activity is of two kinds, namely, saccharifying and liquefying, and all the evidence seems to point to these two functions being due to two distinct ferments. The diastase of barley is almost entirely without any liquefying action, that is to say, a cold-water extract of barley is practically without solvent action upon starch-paste ; its saccharifying action is, however, very considerable, and when barley-extract is allowed to act upon a solution of soluble-starch, it very rapidly and completely saccharifies the starch. During germination, the liquefying power is developed, and the saccharifying power is also greatly increased. This is well shown in the following experiment quoted by Brown and Morris :—

Diastase in 50 corns of *steeped* barley   = 2·325 grams CuO.

,,   ,,   green-malt from}
                 same barley   } = 13·726 grams CuO. *

---

* Throughout the paper previously referred to, Brown and Morris have expressed the diastatic power in terms of grams of CuO per 50 individual corns, embryos, or parts of embryos, &c. This was arrived at by a modification of Lintner's method (see p. 452), and consisted in taking a certain number of the corns or parts of corns, and making an extract of these with a definite volume of water. A certain definite volume of this extract was then allowed to act upon a solution of Lintner's soluble-starch, in such proportions that the reducing power of the transformed solution fell within the "law of proportionality." The action was allowed to proceed for exactly an hour, at 30° C. (86° F.), and the cupric-reducing power of the solution determined gravimetrically. After due correction, the amount of copper oxide due to the transforming action of the diastase contained in 50 corns or parts of corns could be easily calculated, and formed a convenient basis for comparison.

Much of the diastatic power of green malt is lost during kilning, and it has been shown by Bungener and Fries that the saccharifying power of certain green malts was in some cases reduced to as little as one-ninth of its original value by kiln-drying. This is confirmed by the every-day experience of one of us.

The aim upon the malting floors is therefore not only to develop this most essential liquefying function, but also to stimulate the saccharifying energy of the grain in order to withstand the crippling which it will sustain upon the kiln. A striking example of the potency of the former function is afforded by the custom of adding a little malt flour to the meal, when brewing with unmalted grain. The facility with which the starchy matter is prevented from setting, even when the temperature has exceeded the saccharifying limit (190–200° F.), is strong proof of the liquefying influence of the malt being an entirely distinct function.

Closely connected with this increase of diastatic power in the germinated grain, is the increased solubility of the nitrogenous compounds after germination. We have every reason to believe that there is an intimate connection between the soluble and coagulable albuminoids of malt, and the diastatic power; indeed, Brown and Heron showed that in a malt-extract, loss of diastatic power proceeded *pari passu* with the removal of the coagulable albuminoids by heat, and that the complete removal of the latter by passage through a porcelain diaphragm, caused a total loss of diastatic power. But, although there is this connection between the soluble and coagulable albuminoids and the diastatic power, there is none whatever between the total nitrogen of a barley, and the latter function. It is quite incorrect to infer that a barley rich in nitrogen will necessarily yield a highly diastatic malt: the production of diastase evidently depending not upon the total proteid matter, but upon the presence of certain varieties of nitrogenous substances.

Many attempts have been made to determine the nature of the changes which take place in the nitrogenous substances during germination, but the analytical methods employed have not been sufficiently exact to allow any definite conclusions to be drawn from the results.   Quite recently, however, Hilger and van der Becke[*] have adopted methods which promise more hope of success.   They show that the soluble nitrogen increases during germination as follows :—

	Calculated on total nitrogen.
Barley	6·74 per cent. soluble nitrogen.
Steeped barley	3·75 ,, ,,
Green-malt	21·96 ,, ,,
Kilned-malt..	24·44 ,, ,,

Their results on the degradation of the soluble nitrogenous substances are also far in advance of any previous workers, and show clearly the large increase in peptones, amido-acids, and amides (pp. 138–141), which takes place during germination.   The following table gives a summary of these results, calculated on the dry matter of the barley and malt:—

	Barley.	Steeped barley.	Green-malt.	Kilned-malt.
Nitrogen as albumen	0·0600	0·0354	0·1671	0·1194
,, as peptones	0·0046	0·0009	0·0058	0·0233
,, as ammonium salts..	0·0169	none	0·0290	0·0057
,, as amido-acids.. ..	0·0417	0·0294	0·1417	0·2257
,, as amides	none	none	0·0505	0·0029

Another, and most important change, which goes on during germination, is the disappearance of starch and the consequent increase of ready-formed sugars in the grain. The " pitting " which the starch-granules undergo during the latter portion of the germinating period has been already referred to, and this " pitting " is of course indicative of the dissolution of the starch which is taking place.   Stein places

* Archiv für Hygiene, x. (1890), p. 477.

the loss of starch during germination at 5·2 per cent., and Märcker gives 4 per cent. as the amount, but recent numbers given by Brown and Morris point to a much greater loss than either of the above. The figures they give, are—

	Starch in 1000 corns of barley before germination.	Starch in 1000 corns after 6 days' germination.	Loss of Starch.
	Grams.	Grams.	Grams.
Experiment 1 ..	20·0552	15·4398	4·6154
,, 2· ..	19·9158	15·3636	4·5522

A portion of this starch is entirely lost in the form of gas, having been evolved as carbonic anhydride and aqueous vapour, as shown by T. C. Day, during the respiration of the growing grain; a second portion has been used up in the formation of new tissue in the young plant; but the third and largest portion is retained within the grain in the form of cane and other sugars. The quantity of ready-formed sugars pre-existent in barley is thus considerably increased during germination. Kjeldahl[*] first pointed out that the amount of cane-sugar increased from 1·5 per cent. in barley to 4·7 per cent. in green-malt, and O'Sullivan[†] later published the following complete analysis of the sugars in barley before and after malting:—

Sugars.	No. 1 Barley.		No. 2 Barley.	
	Before germination.	After germination.	Before germination.	After germination.
Cane-sugar ..	0·9	4·5	1·39	4·5
Maltose.. ..	} 1·1	{ 1·2	} 0·62	{ 1·98
Dextrose ..		3·1		1·57
Levulose ..		0·2		0·71

[*] Résumé du compte-rendu des travaux du Laboratoire de Carlsberg, 1881, p. 189.
[†] Journal of the Chemical Society, xlix. (1886), p. 58.

Recently, Brown and Morris have more closely examined the distribution of these sugars through the grain, and for this purpose carefully separated the embryos and endosperms of a large number of corns, and determined the sugars in each portion. The following table expresses their results, calculated on the dry weights of embryos and endosperms respectively :—

Sugars.	Barley after steeping in water for 48 hours.		Barley after germination for 10 days.	
	Embryos.	Endosperms.	Embryos.	Endosperms.
Cane-sugar ..	Per cent. 5·4	Per cent. 0·3	Per cent. 24·2	Per cent. 2·2
Invert-sugar ..	1·8	0·2	1·2	2·2
Maltose	—	—	—	4·5
TOTAL.. ..	7·2	0·5	25·4	8·9

These results are interesting, as showing that the maltose is confined to the endosperm, the portion of the grain in which it originates; whilst the cane-sugar, and the product of its decomposition, invert-sugar, are almost entirely confined to the growing embryo. This points most forcibly to the conclusion, that the maltose is absorbed, as such, from the endosperm by the epithelium of the embryo, and at once converted by the living cells of the latter into cane-sugar. That such an action actually does take place, Brown and Morris proved by cultivating excised embryos upon solutions of maltose and dextrose; in the former case, cane-sugar in considerable amount was found in the embryos after six days' culture, whilst those embryos grown on dextrose solution contained no trace of cane-sugar.

The amount of fat in barley is also much diminished during germination. According to John, quickly-grown malt loses 20·7 per cent., and slowly-grown malt 30·2 per cent. of the original amount.

Among the minor changes which take place during malting are—(1) the decrease of α- and β-amylan, which undoubtedly yield products which go for the nourishment of the embryo ; and (2) the decrease in the amount of ash. The following are some numbers showing this decrease of mineral constituents :—

Barley contains 2·95 per cent. of ash, calculated on the dry weight.
Malt ,,   2·61   ,,   ,,   ,,

This ash consists, in the respective cases, of—

	Barley.	Malt.
	per cent.	per cent.
Potash	16·4	14·4
Soda	6·3	4·9
Lime	4·5	5·0
Magnesia	7·7	8·3
Ferric oxide	0·9	1·4
Phosphoric acid	36·9	31·2
Sulphuric ,, ...	1·5	1·3
Soluble silicic acid	23·2	23·4
Insoluble ,, ,, ...	8·4	9·3
Chlorine	1·2	0·8

A marked increase of acidity takes place during germination, and during the kilning process. Bĕlohoubek gives the numbers for this increase :—

Barley .. .. 0·338 per cent. as lactic acid.
Green-malt .. .. .. 0·590   ,,   ,,
Kilned-malt .. .. .. 0·942   ,,   ,,

The extent to which the above-mentioned changes take place in any particular sample of barley is largely influenced by the rapidity of germination, and by other variable conditions, such as temperature and moisture. In normal germination there can be no doubt that the extent of the change is indicated by the development of the plumule ; but if the growth is forced, the plumule may extend the full length of the corn in half the time required under more moderate conditions, and, naturally, the grain in the former instance would be imperfectly malted.

We have already seen that the principal objects in malting are threefold, namely (1) the development of the diastase or starch-transforming agent, (2) the preparation of the starch of the grain for transformation by the dissolution of the cell-walls, and (3) the development of the agent which brings about the modification of the nitrogenous substances. But in carrying out these objects it is most essential that we should guard against undue loss from respiration and rootlet development, as well as against the ravages of mould, acidity, and putrefaction.

In order to attain the desirable conditions, without encouraging the evils just alluded to, the necessity for a prolonged, rather than a rapid germination, and for cool temperatures, cannot be too strongly insisted upon. The waste of material due to respiration and rootlet growth is subject to very considerable variation, but in some extreme cases the loss by respiration has been found to reach nearly 10 per cent. of the weight of the grain. This loss is, of course, at the expense of the reserve materials of the grain, chiefly the starch, therefore it is of great practical importance. The losses incurred through respiration and rootlet development are intimately connected: high temperatures and rapid germination being conducive to both, but whilst the latter is very evident, the gaseous evaporation proceeds invisibly.

John gives the following numbers as the loss on slowly and rapidly germinated barley :—

	I. Rapidly germinated.	II. Slowly germinated.
Respired	8·36	6·38
Malt (exclusive of rootlet and acrospire)	83·09	85·88
Acrospire	3·56	3·09
Rootlet	4·99	4·65
	100·00	100·00

Thus the dry-substance of the malt (without rootlets) forms 88·97 per cent. with the slowly germinated barley, and 86·65 with the rapidly germinated grain.

With regard to rootlet waste, moreover, not only is their quantity affected, but their composition is also influenced by the rapidity of the growth. There is evidence to show that the greater the rapidity of growth the greater the proportion of starch abstracted by the rootlet.

Now if by rapid germination we were able to improve the quality of the malt, there might be some excuse for extra loss from these two causes ; but as the reverse is actually the case, it is of double importance to adopt all means of reducing the loss to a minimum.

In addition to the above changes, others, with which we are at present unacquainted, no doubt take place ; indeed the fact, first proved by T. C. Day,* that more oxygen is absorbed during germination than is required for the oxidation of the carbon, evolved as carbonic anhydride, points to some further changes taking place.

SCREENING, KILN-DRYING, AND STEEPING.

The remarks already made respecting cleanliness and uniformity are sufficient to prove the necessity for a thorough grading and screening of the barley prior to steeping. There cannot be the slightest doubt that the adoption of the best possible machinery is advisable, and that, although expensive, it quickly repays the outlay of capital by the superior quality of the malt produced.

It is very desirable to remove all broken corns and foreign seeds, and to thoroughly cleanse the grain from adhering dirt and dust. The removal of the latter is of very distinct importance ; for knowing, as we do, that the dust is largely composed of various micro-organisms,—bacteria, wild-yeasts,

* Journal of the Chemical Society, xxxvii. (1880), p. 645.

mould-spores, &c.—it must be apparent that the dissemination of this dust among the growing grain can hardly fail to have an injurious influence upon the product. As the barley is shot into the cistern, there is often a dense cloud of dust, especially if the grain is let in before the water. It would always be much better to isolate the cistern so that this dusty cloud could not contaminate the floors ; or, when this could not be conveniently arranged, it would be a good plan to let the grain into the cistern through a simple form of automatic mashing machine, so that the barley should be settled before it has time to make any dust. In any case, the better plan is to erect proper cleansing machinery, and to collect all the dust in water. This collection of the dust is particularly important, because in most maltings there is ample opportunity for some of it to find its way to the germinating floors. If by no other means, it may readily gain access through the windows of the working floors, supposing these and those of the screening room to be left open together, and the wind to be favourable. There is every reason to believe that the marked prevalence of mould noticeable in some malthouses is in part attributable to the ready access of this dust.

With regard to the kiln-drying of barley, there can be no doubt that it is but seldom the grain fails to be improved by this process, although it is frequently not an actual necessity. In good seasons the improvement is not always sufficient to warrant the extra labour and expense ; but whenever the condition of the barley is in the least degree doubtful, we consider that this process is essential for the production of a really satisfactory malt. The influence of the process, beyond reducing the moisture percentage to a uniform rate, and retarding mouldiness and deterioration, is very marked in connection with the vitality of the grain. It is evidently akin to the natural maturing of the barley in the stack, a process which is now unfortunately far too hurried, if not neglected altogether.

This effect is an interesting one, and has but recently received an explanation. Brown and Morris have shown that the epithelium of mature embryos is incapable of secreting any soluble-ferment until the ripening of the grain has advanced to partial desiccation. In other words, the embryo is unable to exercise its secretory functions until the necessary amount of water has been expelled, and when this does not take place naturally, it is necessary to bring it about artificially by kiln-drying.

Natural sweating is decidedly preferable when it takes place in the stack or rick, for the grain is here more readily accessible to the air than after threshing, when it lies in closer bulk. Unless grain is extremely dry it is very dangerous to store it for any lengthened period in dense bulk, owing to the risk of its becoming fusty and heated, whereby the vitality of the germ is endangered. This of course may occur in stack if the corn has been got up in bad condition ; in bad seasons, therefore, it should be artificially dried as soon after the harvesting as possible, in order that it may be stored in thoroughly sound condition.

Unfortunately this is not always possible, on account of limited kiln and storage accommodation, and the British farmer is essentially backward in adopting mechanical and scientific improvements.

In kiln-drying barley, great care must be taken not to allow the temperature to rise sufficiently high to endanger its vitality, and a temperature of 110° F. must be regarded as the maximum. The abstraction of the moisture must depend more upon the volume of air than its temperature. When dried, the barley should be stored for at least a fortnight before it is steeped. This is important, as it allows the grain to mellow, and to undergo those ill-understood changes upon which its vital energy depends—in fact, the value of drying is in great measure negatived if the grain is immediately wetted.

The grain is now steeped, i. e. moistened with water in a suitable tank. We have already referred to the influence of the dissolved salts in steeping waters, and to the influence of organic purity, and as our present system is one which appears to aim at a maximum extraction, we are apparently doing right in using those waters which extract the most, that is to say, soft waters. We may, however, be entirely wrong; in which case hard waters would give the best results; as a matter of fact we do not yet know what we want.

Some barleys lose much more than others during steeping, and in the same sample the smallest corns lose the most. The total loss, however, seldom exceeds 1·5 per cent., and is frequently much less. The mineral matter extracted is generally in excess of the organic, but different barleys vary very much in this respect. Temperature and duration are powerful factors, and have a marked influence on the vitality of the germ. Excessively low temperatures naturally retard germination; excessively high ones, on the other hand, encourage mould, acidity, and putrefaction, all of which are adverse to healthy growth upon the floors. The most desirable period of steeping depends upon the temperature of the weather and the nature of the barley, and is a question for practical experience to decide; as a rule, barley is steeped for from 45 to 72 hours. Frequent changes of the steep water are very essential, and we consider that the frequency of change is of far more importance than the nature of the water. If the barley is unscreened, it is clear that the water must swarm with myriads of germs; and however perfect the screening, there still remains an infinite number of these organisms on the grain. Not only have we to deal with the germs themselves, but the water extracts the very matters on which they will freely develop; we have here, therefore, every opportunity for the propagation of mould and the bringing about of putrefactive changes which can only be kept in check by frequently renewing the steep waters, which should be done, at least every twenty-four hours. The grain should also be aërated : this is done when water is drawn from below and supplied by a sparge arm from above.

After being steeped for the requisite time the grain is usually couched, and then put to germinate in thin layers on the floors; sometimes the couch is entirely dispensed with.

CONDITIONS OF HEALTHY GERMINATION AND DRYING.

In order that germination may proceed in a healthy and vigorous manner, there are three principal conditions which must be carefully attended to. These conditions are, moisture, temperature, and ventilation. Moisture is of prime importance, since there must be sufficient water present in the grain to effect the solution of the food which supplies the germ. It is often necessary, on account of the evaporation that takes place on the floor, to supplement the water absorbed by the grain during steeping; this is done by sprinkling the grain some five or six days after it has left the cistern. At the same time it is very injudicious to continue sprinkling operations at the end of the growing period ; when this is done it is impossible to avoid loading the grain excessively wet. Maltsters are careful to keep their vacant flooring damp, in order that the air of the house may be moist. It is of great importance that the temperature of the growing malt should be kept well under control, and as a rule it should not exceed 60° F. at any stage ; high temperatures not only promote the growth of moulds and putrefactive ferments, but they lead to unnecessary waste by the excessive development of the rootlets, and by increased respiration. Apart from the waste, the quality of the malt is deteriorated ; malts grown at excessive temperatures never yielding sound beers. It is only by slow growth, at moderately low temperatures, that we arrive at a really satisfactory and perfect internal modification. Ventilation is an absolute necessity ; not only are relays of fresh

G

air needed throughout the house, but the layers of grain must be frequently turned, or moved, so that the carbonic acid gas, together with any products of putrefaction, may be thoroughly dispersed, and made to give place to fresh volumes of air. A part of the oxygen absorbed by the grain is evidently taken into combination, as has been shown by T. C. Day,[*] for more of it is absorbed than is exhaled as carbonic acid. This oxygen seems to be chemically fixed in the same manner as is the oxygen taken up by hot worts on the coolers. There is, however, no fixed ratio between the oxygen which is retained and that which is exhaled. In addition to the carbonic acid exhaled as a decomposition product, water also appears to be liberated, and the ratio between the two may be expressed by the following equation, which represents the oxidation of a glucose :—

$$C_6H_{12}O_6 + 6O_2 = 6CO_2 + 6H_2O.$$

Before the green malt is loaded on the kiln it is usually allowed a period of withering, which consists in checking the active growth by keeping the floors cool. Much of the moisture of the grain is dissipated in this way, and the rootlets fade. The withering process is really a preliminary stage of drying, and when it takes place satisfactorily there is less work to be done upon the kiln, and a shorter period for the detrimental influence of moist heat. With perfect kilns, however, it is perhaps questionable if the process of withering is of much practical importance, but with the ordinary kilns no point is of greater moment than loading the corn fairly dry, that is, with less than 45 per cent. of water.

When malt is loaded wet, the influence of the long-protracted moist heat is similar to that exercised by a forced growth on the floors, but to a more marked degree. We have found almost invariably that there is under these conditions a tendency to the formation of an excess of ready-formed sugars, usually accompanied by the development of a sweet

[*] Journal of the Chemical Society, xxxvii. (1880), p. 645.

and sickly flavour, which is permanent to the malt. The mere increase in these ready-formed sugars is in itself hardly sufficient to account for the invariable unsoundness of beers produced from malts containing such an excess; but it is probable that the breaking down of an excess of starch into these substances, under these conditions, is accompanied by a similar degradation of the nitrogenous substances; and it may well be that it is to this latter change that the instability of beers brewed from such malts is due. In dealing with excessively wet corn on the kilns, some maltsters attempt to avoid the detrimental change induced by a long-protracted moist heat, by raising the heats from the start, and thus reducing the water percentage in a minimum of time. But this is no real remedy; for the malt will be scorched and rendered steely, and there will be a tendency to produce "split-ended" corns; such corns will naturally be very liable to become affected with mould. The only safe course is to load the malt fairly dry.

One of the many difficulties of practical malting is caused by the tendency of the malt to become mouldy and putrid. This deterioration starts on the damaged and bruised corns, which naturally afford a ready field for the reproduction of the mould and other germs. But if the disease becomes pronounced, then the contamination of these damaged corns will more or less affect the bulk of those that are intact. High growing temperatures are also especially provocative of this evil, and with barley of doubtful character it is more than ever essential to prevent the slightest excess of heat; insufficient ventilation also stimulates these unhealthy growths. Antiseptics, such as bisulphite of lime or salicylic acid, added to the sprinkling water, are of little or no service, because in order to be thoroughly effective they must be employed in quantity sufficient to have an injurious influence upon germination.

The ill effects of mould and putrefaction are twofold: on

the one hand, the wort becomes impregnated with hostile germs, which, when killed in the wort-copper, afford future nourishment for the organisms taken up during cooling; and on the other hand, the most injurious results can be traced to the degrading influence of these growths upon the starchy and nitrogenous bodies in the malt. Through their agency the nitrogen compounds are degraded, and there is reason to believe that a portion of the carbohydrates is decomposed by the micro-organisms which accompany or follow the mould.

The corn having germinated to a sufficient degree, a point indicated by the acrospire extending to about three-fourths of the length of the corn, by the entire mealiness of the grain, and by other practical considerations, it is now put on the kilns, where it undergoes two distinct processes, namely, the drying and the curing stages.

It may be safely stated that the more quickly the moisture can be withdrawn from the green malt the better. But there is one very important qualification: it is most essential that the malt should not be exposed to excessive heat whilst still moist. There are two methods of drying, the one by the agency of little air and much heat; the other by more air and less heat. It is well known that the higher the temperature of a volume of air, the greater is its capacity for absorbing moisture. This fact has been relied upon far too much in the drying of malt. The perfection of drying consists of the passage of large volumes of air, at a relatively low temperature, through the layers of malt.

The necessity for guarding against a high preliminary temperature, is due to the susceptibility of the diastase of the malt to moist heat and to the danger of gelatinising the starch. Whilst the malt is damp the diastase is seriously crippled by a temperature which, after the malt is dry, would be harmless. Indeed there is far more danger in a temperature of 150° F. whilst the malt is still damp, than from 220° F.

after it has been thoroughly dried. The principal objection to slow drying is the encouragement of mould and of putrefactive organisms and the risk of a prolongation of the growing process under the most unwholesome conditions ; and if a malt is exposed for long upon a slow kiln, at a temperature of 90° or 100° F., there is danger of its grave deterioration in this respect. Therefore, upon badly constructed kilns, it appears preferable to run the risks attached to a somewhat higher heat at the commencement, rather than to expose the malt to the above-mentioned dangers. But, given a quick draught, relatively low commencing heats are best. Speaking generally, it is not advisable ever to allow the temperature of the malt to exceed 125° F. before it is hand-dry.

It is most essential that, whilst drying, the malt should be kept frequently lightened, in order that there may be the least possible obstruction for the air. Malt should not be actually turned at this stage, because it is better to leave the layer of grain, originally in contact with the tiles, to act as a protection to the upper layer. There is nearly always some danger of scorching, and it is better that the deterioration should be limited to as small a number of corns as possible. Whilst damp, the malt is much more amenable to the caramelising influence of heat, hence a further reason for a relatively low commencing temperature.

CURING.

It has been said that the kilning of malt consists essentially of two stages, the drying and the curing stage. We have already briefly glanced at the drying process, and have seen that, so long as certain important conditions are fulfilled, the more quickly the malt is dried the better. With regard to curing, however, duration is a distinctly important feature.

In the ordinary manner the malt is cured by the direct heat

of the furnace, and is thus exposed to the products of combustion.   In many Continental kilns, and in a few English, the
heat is applied indirectly, and the products of combustion are
thereby excluded.   It is supposed that the products of combustion increase the soundness of the beers brewed from malts
exposed to them.   This may be so, but we think their importance is over-rated in this direction.   In point of flavour,
however, there is no doubt that these combustion products
certainly have some influence ; whether that influence is
restricted to the malt only, or whether it is permanent
throughout the brewing process, and passes into the beer, is at
present not known.

In the ordinary way we should prefer malts cured in
contact with the combustion products, but we are bound to say
that we have known malts cured without them which have
yielded very excellent beers.

The plan adopted by certain maltsters of enormously
diluting the combustion products, during drying, with vast
quantities of air from the outside, and, when drying is complete, of cutting off these outside air supplies, and then subjecting the malt to the influence of the products of combustion,
appears to be the best.   Although the influence of the latter is
probably over-rated, yet they cannot do harm when the fuel is
well selected, and they may do more good than we are in a
position to say is or is not the case.

Whatever system be adopted, it must be admitted, that
although there are kilns which can be made to yield satisfactory results upon the ordinary system of heating, yet the great
general defect of English kilns is the difficulty of ensuring
regularity, and of maintaining a sufficiently high temperature,
without the risk of scorching.   There is an impression that
the soundness of high-dried malts is partly due to the influence of the caramelised bodies they contain.   This is an evident
mistake, because caramel in itself has but little, if any, antiseptic property.   The deep colour of a malt is some indication of

complete curing, and any benefit derived from the use of these malts is due to the reduction in diastatic activity and other changes connoted by the increase of colour.

As soon as the malt is dried, the temperature may be gradually raised to the final point, the speed with which this can be safely attained depending on the construction of the kiln. If allowed to rise too rapidly there is danger of scorching, owing to the heat of the lowest layer of grain becoming too high before the thermometers register the maximum heat. The temperature of the tiles is always considerably in advance of the main bulk of the grain, and the thermometers usually employed do not register the actual heat of the tiles at all. When the maximum temperature has been attained the heat should be kept up at that point for at least six hours, in order that it may penetrate to the heart of the grain. This is a most important point, and is one frequently neglected, since many maltsters, as soon as the maximum is arrived at, fancy they have done all that is required. The interior of the grain of malt is a bad conductor of heat, and consequently, penetration is slow, the heat passing more readily between the corns with the upward current of air. For this reason it is desirable to curtail the draughts during the curing stage, and, upon the ordinary system, to confine it entirely to that which passes through the fire. Whilst considering the slow penetration of heat, we may mention the desirable practice of lumping-up the malt upon the kiln whilst it is still hot. When lying in bulk, there is of course practically no draught through the grain at all, and the effect of this is very striking. The grain absorbs heat much faster, and a lower temperature under these conditions is more effective than a considerably higher one when the malt is spread out on the tiles. This is seen by the deepening of colour which takes place whilst the malt is lying in heap, although the temperature may have been allowed to drop some twenty degrees prior to lumping-up.

It has been mentioned that during germination not only is the liquefying function of diastase brought into existence, but that the total saccharifying power of the grain is greatly increased. Quite apart from all other considerations, the employment in brewing of the enormously diastatic green-malt would be fatal : for, as we shall subsequently see, malt, to yield sound beer, must not possess the diastatic capacity of green-malt, or anything approaching it. Indeed much of the brewery trouble incident to the employment of defective malt is in no small number of cases due to the excessive diastatic capacity of the malt. Now by kiln-drying we reduce the diastatic capacity ; and it is indeed in great part on account of this diastatic reduction (and all it means) that so much importance attaches to the kiln-drying process. It has been said that the heat must not be applied while the grain is wet because diastase, being more sensitive to wet than to dry heat, would under these circumstances be unduly crippled. This is certainly the case. A certain amount of diastatic capacity is obviously necessary, and this would stand in danger of de-struction were the heat applied to the wet corn. But once dry, and other things being equal, the higher our kiln heats—i. e. the more the diastase is checked—the better for the soundness of the beer. A limit is placed upon our kiln heats by con-siderations of attenuation, flavour, brilliancy, and colour of beers ; and even apart from colour we obtain a better flavoured pale ale by curing relatively lightly than by curing very thoroughly. But even in the case of pale ale malts, although curing is to be relatively light, yet it should be sufficient, and as a rule an exposure to a final heat of 180° F. for some hours should be insisted upon ; for mild ale malts, the final tempera-ture should be not less than 200° F., and the malt should be exposed to that heat for some hours.

But in addition to reducing the diastase, kilning produces other changes not less necessary. The excess of water is expelled, and the malt is fitted for relatively lengthy storage ;

the plumule growth is arrested ; the development of the mould and other germs is checked, the malt is rendered friable, while it develops the pleasant empyreumatic taste so necessary for the production of well-flavoured beers. Beyond these changes, there is obviously some change in the nitrogenous matters. It is difficult to say exactly how curing affects these substances, what new compounds are produced, how they are produced, and which of these are produced to the detriment or to the advantage of the brewer, but within reasonable limits the changes effected in this connection by a thorough curing seem to be desirable.

The chemical valuation of malts, by analysis, is discussed in the last section of this chapter (p. 163).

THE STORAGE OF MALT.

It is a well-known fact that when malt is stored without due regard to the exclusion of moisture it absorbs water and becomes "slack," as the brewer terms it. Slack malt is a most unsatisfactory material, yielding thin, unstable and frequently fretting beers. It is difficult at first sight to see why this should be so, for the absorption of say 3 or 4 per cent. of water fails to explain why these troubles should result. We think, however, that the following facts may throw some light on the matter. We have found that as malt absorbs water, its diastatic capacity (as determined by Lintner's process) materially increases, and in many experiments on this point we have always found a considerable increase, such increase roughly varying in proportion to the duration of storage. Not only does the diastase increase, but the worts made under the same conditions from these malts show a dissimilar percentage of maltose and malto-dextrin. So far as we can see there is no reason to believe that an actual increase of the diastase takes place, and we are inclined to regard it as apparent rather than real. We think, in fact, that the diastase remains constant, but that the absorption of water by the malt leads to a softening of the tissues, and that when the malt is infused or mashed, it more readily yields the contained diastase than when dry. This would serve to explain both the apparent increase of diastase during slackening, and the lower amount of maltodextrin produced under standard conditions.   We have no reason to doubt that the complete extraction of diastase, which we have noticed in the laboratory from material of this kind, would not apply similarly in the brewery ; so that when mashing malt of this kind on the practical scale on ordinary lines, the worts would be poorer in normal maltodextrins and richer in free maltose and the lower maltodextrins than would similar malts which had not "slackened."   That this alteration in the constitution of the wort would produce thinness, instability, and a tendency to frettiness, will be clear on referring to the later chapters of the work where these defects are fully treated of.

In practice it is usual to re-dry slack malts, and by this process (which consists of kilning at about 160° for 24 hours) the diastase will be reduced and the moisture in the tissues reduced.   The benefits thus attainable are clearly in unison with the explanation we have advanced.

We think that we have observed an increase in the ready-formed sugars in slack malts.   But we hesitate to say that such is the case since the analytical methods employed for their estimation were too crude to give absolutely reliable results.   It is not impossible, however, that, in the presence of the absorbed water, the diastase might break down a portion of the starch as it does on the malting floors, and this would serve as the explanation for any increase in these substances.   But we expressly wish to state that, until we have confirmed our suspicion by more accurate analysis, we do not in any way commit ourselves to this view.

———————

THE CHEMICAL CONSTITUENTS OF BARLEY AND MALT.

Having now discussed the anatomy and physiology of the barleycorn, and the conditions necessary for its conversion into malt, we pass on to the consideration of the constituents of the dry substance of barley, which amounts to about 85 per cent. of the total weight.

These constituents may be divided into three classes.—

    I. Non-nitrogenous organic compounds.

    II. Nitrogenous organic compounds.

    III. Mineral salts (ash).

I. The Non-Nitrogenous Organic Compounds.

The chief members of this group of compounds belong to the class of substances known as *carbohydrates;* these substances are made up of the three elements, carbon, hydrogen, and oxygen, and the latter two always occur in the atomic proportion of $2:1$; that is to say, in the same ratio as they exist in water (H_2O), hence the name carbohydrates. The members of this class which we have to consider are cellulose, starch, the sugars, and the gummy matters, including the amylans. In addition to the carbohydrates, the fats and the so-called extractive matters are members of this group.

Cellulose.

Cellulose is very widely distributed in the vegetable kingdom, and occurs in many forms. There are, however, only three forms which possess any interest for the brewer; these are (1), the lignified cellulose of the paleæ and other parts of the grain; (2) the parenchymatous cellulose of the endosperm; and (3) the so-called starch-cellulose.

1. The lignified cellulose of the paleæ plays no part in the germination of the grain, but remains unaltered even when the growth of the plantlet is pushed to its extreme limit. It may be separated and purified, and is then obtained in the form of white shreds, that retain the shape of the scalariform vessels from which they are derived. This form of cellulose amounts to about 8 per cent. of the grain.

2. We have already referred at some length to the cellulose forming the cell-walls of the endosperm, and its disappearance during germination by the solvent action of an enzyme. This cellulose contributes to the food-supply of the young plant, and its complete dissolution is a matter of the utmost importance in malting. It is very much more readily soluble in all reagents than the lignified cellulose of the other parts of the grain.

3. The third form of cellulose is the so-called starch-cellulose, which is generally supposed to envelop the starch-granules, and which is obtained when starch-paste is acted upon by malt-extract in the cold;[*] the starch is quickly liquefied and saccharified, and the cellulose remains as a flocculent precipitate, which can be separated by filtration and washed with cold water. When the starch-conversion is made at an elevated temperature, the so-called starch-cellulose does not separate out, but undergoes a conversion similar to that of the starch.

Pure cellulose is a white, odourless and tasteless substance ; it is insoluble in water, alcohol, ether, benzol, &c., but dissolves in ammonio-cupric hydroxide, from which solution it is reprecipitated by the addition of acids. It is not coloured by iodine solution alone, but after treatment with strong sulphuric acid or zinc chloride (which hydrolyses it) it gives a blue colouration with iodine solution. Its formula is $(C_6H_{10}O_5)n$.

[*] See Brown and Heron, Journ. Chem. Soc., 1879, p. 612.

Starch and its Transformations.

The most important constitutent, for the brewer, of the dry substance of barley and malt is the starch, which in barley forms about 60–66 per cent. of the total weight. Starch is very widely distributed throughout the vegetable kingdom, and in the cereals and grasses especially it constitutes the bulk of the reserve material. Starch also is a substance of very considerable importance for many purposes, and the manufacture of starch in a state of more or less purity is an industry which has attained to very large dimensions. The chief sources of starch for manufacturing purposes are the potato, barley, wheat, maize, rice, and sago.

When purified starches from different sources are examined under the microscope, it is seen that the granules vary much in shape and size, being in most cases distinctly characteristic of the source from which they are derived. Starch-granules are built up of distinctly stratified layers, and the growth proceeds from outwards by the successive deposition of layers of starch-substance. Of the starches above mentioned, the granules of potato-starch are the largest and most characteristic; those of barley, wheat and rye are of nearly the same size and appearance, whilst those of maize and rice are very distinctive.

Pure starch has the same percentage composition as cellulose, and its formula may be written $(C_6H_{10}O_5)n$. The molecule is a very large one indeed, Brown and Morris having arrived at the conclusion that the molecule of soluble-starch (see p. 123) cannot be less than 200 times the group $(C_6H_{10}O_5)$.

Starch-granules are commonly supposed to be made up of two substances—*granulose and starch-cellulose.* This division is that of C. Nägeli, who states that when intact starch-granules are acted upon at a slightly elevated temperature (45–50° C.) by saliva, pepsin, dilute acids, or diastase, the former is dissolved, leaving the latter as skeletons of the granules. Granulose is coloured an intense blue by iodine, and starch-cellulose is stained yellow by the same reagent. According to Brown and Heron starch-cellulose consists of a mixture of two substances, one readily acted upon by boiling water and converted into soluble-starch, the other unchanged by boiling with water; the latter is, however, converted into soluble-starch by boiling with potash. Starch-cellulose, the amylo-cellulose of some authors, has been much confused in the past with the amylodextrin of W. Nägeli, which Brown and Morris have recently shown to belong to the series of amyloïns (p. 131).

Starch is insoluble in all the ordinary solvents; when heated with water, the granules first of all swell up enormously, owing to the absorption of water, and finally the granules become ruptured and form the very viscid mass which is known as "starch-paste." The viscosity of starch-paste varies with the temperature, the higher the temperature the less viscid the paste. The temperature at which the starch has been dried before gelatinisation also influences the viscosity of the resulting paste; Brown and Heron showed that the viscosity of a paste prepared from starch dried at 100° C. (212° F.) was nearly four times as great as that made by gelatinising starch dried at 30° C. (86° F.).

Considerable differences of opinion have always existed as to whether granulose is or is not in a true state of solution in starch-paste. When a thin starch-paste is filtered, the filtrate is clear, and gives a deep blue colouration with iodine, but filtration of this liquid through a thin animal or vegetable membrane, or through porous earthenware, completely removes the starch; from this fact Maercker and others are of opinion that the granulose is not in a true state of solution, but only distributed through the water in a very finely divided condition. Brown and Heron incline to the opposite view, and from determinations of the specific gravity of starch in the form of starch-paste, and in the dry state, conclude that it is extremely probable that granulose exists in starch-paste in a state of true solution. We are inclined to think, however, that when starch is gelatinised and boiled, the granulose is converted into soluble-starch, which goes into solution and gives the characteristic reactions. This substance being highly colloidal, like all colloids, will not pass through a membrane, or septum of earthenware.

The different varieties of starch gelatinise at very different temperatures; the following table, recently published by Lintner,[*] gives the temperatures for the commoner starches.—

	Deg. C.	Deg. F.
Potato starch	65	149
Barley ,, ...	80	176
Green-malt starch	85	185
Kilned-malt ,, ...	80	176
Oat starch	85	185
Rye ,, ...	80	176
Wheat ,, ...	80	176
Rice ,, ...	80	176
Maize ,, ...	75	167

The gelatinisation of starch is a matter of very considerable practical importance, since the ready and complete conversion of starch by diastase depends upon the thoroughness with which it has been previously gelatinised, and thus brought into a condition suitable for the action of the soluble-ferment. The above table gives the temperatures at which the various starches gelatinise in a pure state; when they are in a crude condition, and more or less enclosed in the cell-walls of the starch-containing cells, a much higher temperature is necessary to rupture the cell-walls and liberate the starch-granules, in order that they may be acted upon by water and heat.

Both starch-paste and soluble-starch are precipitated by tannic acid.[†] Compounds are also formed with potassium, sodium, barium, calcium, copper, and lead hydroxides. The first two are soluble, and the remainder insoluble in water. The potassium compounds have a rotatory power considerably less

[*] Brauer and Mälzer Kalendar, 1889.
[†] Griessmayer, Der bayerische Bierbrauer, 6.

than that of starch, or soluble-starch ; they are easily decomposed by dilute acids or carbonic anhydride, yielding the original starch.[*]

The characteristic reaction for starch in all its forms is the production of a deep-blue colouration with solution of iodine. Intact starch-granules are stained a deep-blue, whilst starch-paste and soluble-starch give a deep indigo-blue colouration. The delicacy of the reaction varies with the temperature, Fresenius stating that a solution containing $\frac{1}{538000}$ part of iodine as potassium iodide will, when added to a thin starch-paste at 0° C. (32° F.), produce a blue colouration in the presence of acid, but at higher temperatures (13°, 20°, and 30° C.; 55·4°, 68°, and 86° F.) larger quantities of iodine are necessary. Formerly the consensus of opinion was that this blue compound was not a chemical compound, but a mixture of iodine and starch, probably a solution of the former in the latter. Recently, however, Mylius[†] has shown that the blue compound is a hydriodide of starch with the formula $(C_{24}H_{40}O_{20}I)_4HI$. It is only formed in the presence of hydriodic acid, or an iodide ; it forms compounds in which the hydrogen atom is replaced by metals.

The blue colouration is discharged when the compound is heated, but it returns again on cooling, if the heat has not been applied for too long a time. For this reason it is always necessary to cool a wort before testing it for starch with solution of iodine. It is also most important in testing worts for starch that the iodine should be added in excess, as the dextrin and maltodextrins form more or less colourless compounds with iodine, and the affinity of these compounds for the latter is much stronger than that of starch ; therefore, supposing starch to be present, together with these other substances, no blue colouration will be produced until sufficient iodine has been added to saturate the whole of the dextrin and malto-dextrins. It may be mentioned here that the so-called erythro-

---

[*] Brown and Heron, *loc. cit.*   [†] Berichte, xx., p. 688.

dextrin (see p. 127), gives a red colouration with iodine, and the production of this colour always precedes that given by starch, when the two substances are present together.

When starch is acted upon by dilute acids in the cold, it rapidly loses its power of gelatinisation, and is converted into soluble-starch. This change is effected without any alteration taking place in the appearance of the starch-granules. If the action of the acid is allowed to proceed for some time, the granules gradually become disintegrated, a portion of their substance going into solution in the acid, and being found there as dextrose, whilst the insoluble residue consists of crude *amylodextrin* (see p. 129); during this change the iodine reaction of the residue changes from blue to reddish-brown.

When starch is acted upon, however, at the boiling-point with dilute mineral acids, it is very quickly converted into dextrose, the change proceeding the more quickly the stronger the acid. With very dilute acid, or by stopping the action at various stages, it is found that a series of intermediate products between starch and glucose may be obtained. At a very early stage soluble-starch is almost entirely present; later the products consist of maltose, dextrin, and a series of maltodextrins; later still dextrose makes its appearance and the conversion then rapidly contains this substance only. When a temperature much above 100° C. (212° F.) is employed for the conversion, dextrose makes its appearance very early in the reaction, being formed apparently at the expense of the maltose, the amount of which, present at any one time, is reduced to a minimum.

The most important point we have to consider in connection with starch is the action of soluble-ferments upon it. The most familiar and at the same time the most important soluble-ferment is the diastase contained in malt. This, as is well known, has the power of converting starch into dextrin and maltose, a power which is shared by other soluble-

II

ferments, namely, saliva (ptyaline), pepsin, the pancreatic juice, &c.

Starch-granules from some sources, such as barley, are soluble in malt-extract in the cold, yielding maltose; whilst other starches, such as potato-starch, are quite unacted upon by cold-water malt-extract until the granules have been broken up by trituration with powdered glass or some similar material. In this case the products which go into solution consist of maltose and dextrin, with varying amounts of an inactive substance which Brown and Heron consider to be starch-cellulose.

Starch-granules from all sources, however, dissolve more or less in malt-extract at temperatures considerably below the gelatinisation point. The following table gives the results of a series of experiments recently made by C. L. Lintner[*] to determine the quantities dissolved at different temperatures. In each case the digestion was continued for four hours, although at the higher temperatures the action was generally complete in about twenty minutes :—

	50° C.	55° C.	60° C.	65° C.
	per cent.	per cent.	per cent.	per cent.
Potato starch	0·13	5·03	52·68	90·34
Rice ,, ...	6·58	9·68	19·68	31·14
Barley ,, ...	12·13	53·30	92·81	96·24
Green-malt starch	29·70	58·56	92·13	96·26
Kilned-malt ,, ...	13·07	56·02	91·70	93·62
Wheat starch	—	62·23	91·08	94·58
Maize ,, ...	2·70	—	18·50	54·60
Rye ,, ...	25·20	—	39·70	94·50
Oat ,, ...	9·40	48·5	92·50	93·40

When starch-paste is acted upon by diastase, either in the cold or at temperatures below 70° C. (158° F.), the viscosity of

[*] Brauer and Mälzer Kalendar, 1890; and Wochenschrift für Brauerei, 1890, p. 310.

the paste at once disappears ; it becomes limpid, and after a time acquires a sweet taste. Treated with iodine solution at intervals during this reaction, it is seen that the blue colour gives place to a violet colouration, which changes to a red, and finally iodine fails to give any colouration at all. It is thus evident that the diastase has effected an entire change in the starch-paste. The nature of this change has been the subject of much speculation and research, the results of which we must now briefly examine.

In 1811, a French chemist named Vauquelin discovered that when starch is somewhat strongly heated it is converted into a substance which completely dissolves in water, and possesses many of the properties of gum arabic ; a year later Vogel found that a similar gummy substance is formed by the action of a hot dilute acid upon starch. About the same time, Kirchoff announced that starch when boiled with dilute sulphuric acid yields a crystalline sugar ; and two years later, the same worker showed that when starch-paste is acted upon by the soluble albuminoids of grain, the same substance is produced. He also observed that the action of the albuminoids is much intensified by subjecting the grain to the malting process.

The gummy substance referred to above was more fully described and examined by Biot and Persoz in 1833. These chemists found that the substance had the power of rotating a ray of polarised light strongly to the right, and hence they gave it the name of *dextrin*. Biot and Persoz regarded dextrin as consisting of the contents of the starch-granule freed from the outer cellulosic coating by the action of the acid ; and this view of the formation of dextrin caused Payen and Persoz to give the name *diastase* to the particular trans-forming agent present in malt-extract. The latter workers were the first to prepare a dextrin from starch by the action of diastase ; they describe it as a substance soluble in cold water and in weak alcohol, and not coloured by iodine—the colourations which starch and some of the products of its transformation give with iodine being discovered in 1813 by Stromeyer. A year or two later, Payen showed that the rotatory power of this dextrin was equal to that of starch, and that its elementary composition was expressed by the formula $C_6H_{10}O_5$. He also arrived at the conclusion that the dextrin formed by the action of dilute acid, diastase, and heat respectively, were simply physical modifications of the same substance.

From these workers we pass to comparatively recent times, when Musculus, in 1860, threw an entirely different light on the action of acids and diastase on starch. Until this time it had been universally supposed that the crystalline sugar, then supposed to be dextrose, was the product of the hydration of the dextrin ; in other words, that the starch was first converted into dextrin, with which it was isomeric, and this dextrin was then converted into the sugar. Musculus now brought forward strong experimental evidence to show that this view was incorrect, and that the dextrin and sugar were produced simultaneously by the breaking up of the starch-molecule, accompanied by the fixation of water. This theory was strongly opposed by Payen and Schwarzer, but it has outlived opposition, and is the one which, with some modifications, is now universally held by all who have worked on the subject.

About this time, the question of the proportions in which dextrin and sugar were formed came under discussion, Musculus, in the paper above mentioned, asserting that when diastase acts upon starch at temperatures of 70° to 75° C., (158–167° F.) or when dilute sulphuric acid acts under certain conditions, one molecule of sugar and two molecules of dextrin are formed, and that no further action takes place. Payen and Schwarzer, on the contrary, asserted that at least 50 per cent. of the transformation products consist of sugar, and the latter stated that the temperature at which the reaction takes place exerts a great influence upon the ratio of the products. Schwarzer also strenuously opposed the theory of the simultaneous formation of dextrin and sugar, and considered dextrin to be first formed and then sugar.

The behaviour of the dextrin as regards optical power, cupric-reducing power, and colouration with iodine, was also the subject of much controversy, resulting, as we shall see later, from the view then entertained that but one dextrin was formed, and that the sugar consisted of dextrose.

But it now began to be recognised among chemists who were occupied with the study of this question, that there were at least two dextrins produced by the action of diastase or acid upon starch, the one giving a reddish-brown colouration with iodine, and the other giving no colouration at all with this reagent. Griessmayer in 1871, described these two substances as dextrin (I.), and dextrin (II.) ; and in the following year O'Sullivan distinguished them as a- and β-dextrin. They are now more generally referred to as erythrodextrin and achroodextrin, terms expressive of their behaviour with iodine solution, and first proposed by Brücke.

The recognition of these different dextrins, and the rediscovery of maltose by O'Sullivan in 1872,[*] explained much that was formerly obscure in the work of earlier investigators, and prepared the way for a critical examination of the question at issue.

In this paper O'Sullivan gave the results of the examination of the dextrins prepared by the action of diastase and of acid upon starch. These results led him to the conclusion that the dextrins from both sources have the same specific rotatory power, and an elementary composition indicated by the formula $C_6H_{10}O_5$. Although he never succeeded in obtaining any of the dextrins absolutely free from a reducing action on Fehling's solution, yet he brings forward good evidence to prove that in a state of complete purity they would be without any cupric-reducing power. The experiments led him to the conclusion that the three substances—

> Soluble-starch, coloured blue by iodine ;
> a-dextrin, coloured reddish-brown by iodine ; and
> β-dextrin, not coloured by iodine,

have each a specific rotatory power of $[a]_j = + 213°$, and are therefore alike in their action on polarised light.*

* In describing these and other researches we shall have frequently to employ three expressions which require some explanation (see also chapter on the Analysis of Malt and Worts, p. 452). The three expressions are :—

> (1) The solid matter per 100 cc. ;
> (2) The cupric-oxide reducing power ; and
> (3) The specific rotatory power.

Solid Matter per 100 c.c.—In all quantitative work it is, of course, absolutely necessary that we should know, or have some means of knowing, the exact amount of any substance which we may have in solution. This knowledge is usually obtained by evaporating a given volume of the solution, and carefully drying and weighing the residue. In working with the carbohydrates, this method is, however, attended with so much difficulty, and is also so unreliable, that it became necessary to adopt some other means of making this determination. It was evident that if the carbohydrates were prepared in a pure state, and the specific gravity of solutions of different known strengths carefully determined, it would then be possible to deduce from the specific gravity the strength of any solution of the carbohydrate. For instance, in the paper mentioned above, O'Sullivan states that a solution containing 10 grams of either pure maltose or pure dextrin in 100 c.c. of solution has a specific gravity of 1038·5, water being taken as 1000 ; and a solution of 1 gram in 100 c.c., a specific gravity 1003·85 ; and solutions containing intermediate quantities have proportionally intermediate gravities. Hence the number of grams of solid in 100 c.c. of a solution can be found by dividing the specific gravity, less 1000, by 3·85. Thus a solution with a specific gravity of 1060·0 would contain 15·584 grams of solid matter on the 3·85 divisor, for—

$$\frac{1060 - 1000}{3·85} = 15·854.$$

This divisor is used by O'Sullivan throughout the paper mentioned above. In 1879, Brown and Heron pointed out that the divisors for dextrin and maltose were higher than this, and we may conclude that O'Sullivan has found this statement to be correct, since in his paper on the estimation of starch in cereals, published in 1884, he gives the solution constant or divisor for the starch products as 3·95.

The divisor which is more generally used is, however, neither of these, but the one proposed by Brown and Heron †—namely, 3·86. This number is admittedly not correct for the starch-transformation products, it being the constant for a

† Journal of the Chemical Society, 1879, p. 605.

O'Sullivan also shows that under the influence of malt-extract (diastase), the reducing power of a solution of dextrin slowly increases, and becomes constant when the cupric-

solution of cane-sugar of the specific gravity of 1050·0; but it is extremely convenient that all cupric-oxide reducing powers and specific rotatory powers should be referred to one constant and unvarying divisor for the determination of the solid matter, and it is usual to append this number to the symbols denoting the above factors, thus $\kappa_{3\cdot86}$ and $[a]_{j_3\cdot86}$. In order to convert numbers calculated on this divisor into absolute values, it is only necessary to increase or diminish the numbers in the proportion of 3·86 to the absolute divisor. So long as the same solution-constant is used, the values of $[a]_j$ and κ always bear a constant relation to each other, and the percentages of substance calculated from these values also always bear the same ratio to each other; the factors for the different carbohydrates can also be always compared *inter se*; it is thus far preferable in work of this nature to adopt one divisor to which all cupric-reducing and specific rotatory powers can be referred, than to use a different divisor for each carbohydrate.

The cupric-oxide reducing power, or, as it is now more usually termed, the cupric-reducing power of a substance, is defined by O'Sullivan to be the amount of cupric oxide, calculated as dextrose, which 100 parts reduce. Dextrose was taken as the standard of comparison, since it is the type of the reducing sugars, and it was also the first substance for which the amount of cupric oxide reduced by a known weight of the sugar was determined. The weight of cupric oxide reduced by 1 gram of dextrose being known, the amount of cupric oxide reduced by one gram of any substance, calculated upon the number for dextrose as a percentage, will give the cupric-reducing power of the substance. This value is denoted by O'Sullivan by the symbol K,[*] and by Brown and Heron by the symbol κ.[†]

In his 1876 paper, O'Sullivan gives 65 as the cupric-reducing power of maltose (calculated on the 3·85 divisor); but in 1879 Brown and Heron pointed out that this number was too high, and gave the value of $\kappa_{3\cdot86}$ as 61·0. In 1884, in the paper mentioned above, O'Sullivan gives the value of K (calculated on the 3·95 divisor) as 62·5, a number which when converted to the 3·86 divisor, gives 61·07—almost exactly the same as that determined by Brown and Heron.

The specific rotatory power of any substance is the angle through which a ray of plain polarised light of definite refrangibility is deflected on passing through a layer of the solution of the substance 1000 mm. in thickness, and containing 10 grams of substance per 100 c.c. of solution. There are several instruments employed for measuring this angle, but they may be divided into two classes—namely, those with which the neutral tint, corresponding with the medium yellow ray of the solar spectrum, is employed, and of which the Soleil-Ventzke-Scheibler is the best type; and those with which the light emitted by incandescent

[*] Journal of the Chemical Society, 1879, p. 775.
[†] ,, ,, ,, ,, 1879, p. 606.

reducing power equals 66 per cent. of dextrose calculated on the dextrin employed, whilst at the same time the optical activity decreases, until the matter in solution has $[a]_j =$ $+ 150°$. Following this up, O'Sullivan proved that the sugar which was formed by the action of diastase on starch is not, as was generally supposed, dextrose, but a sugar isomeric with cane-sugar and milk-sugar, and having a considerably higher optical activity than dextrose, and a much less cupric-reducing power.

Strange to say, this sugar had been isolated as far back as 1819 by De Saussure, who correctly described its crystalline habit; and in 1847 it was again prepared by Dubrunfaut, who recognised it as a distinct sugar, and stated that its optical activity was three times that of dextrose. He named it "maltose"; but his discovery passed without notice at the time, and appears to have been completely forgotten, until O'Sullivan rediscovered the sugar in 1872, and recognised it

---

sodium vapour is used, and which is represented by the Laurent instruments. With the former class of instrument the specific rotatory power is denoted by the symbol $[a]_j$, whilst the expression $[a]_D$ represents the latter. In this determination also it is necessary to use the solution-constant in calculating the values for the carbohydrates, and, as with the cupric-reducing power, the employment of different divisors has resulted in apparent discrepancies in the values given by various observers. Much confusion is also caused in this case by the neglect of some workers to specify the nature of the ray employed by them, and the consequent inability of subsequent observers to ascertain whether the readings given are for $[a]_j$, or for $[a]_D$, which for the carbohydrates stand in the ratio of approximately 24 to 21·54.

An example of the apparent discrepancies alluded to is afforded by the numbers given for the transformation products of starch. O'Sullivan in his earlier papers gives the values for maltose and dextrin as $[a]_j$ 150° and 214° respectively; these numbers are based on the 3·85 divisor; in his later papers, where he uses the 3·95 divisor we get maltose = $[a]_j$ 154°·0, and dextrin = $[a]_j$ 222°·0; whilst Brown and Heron give, for the 3·86 divisor, the values maltose 150°·0, and dextrin 216°·0. When the numbers of O'Sullivan are calculated on the last-mentioned divisor we get almost identical figures to those of Brown and Heron, thus :—

	Maltose.	Dextrin.	
$[a]_{j\,3\cdot86}$ (observed)	150°·0	216°·0	Brown and Heron.
$[a]_{j\,3\cdot86}$ (calculated)	150°·5	216°·9	O'Sullivan.

as the end product of the action of malt-extract upon starch-paste. He followed Dubrunfaut in giving the name maltose to this sugar.

Some three years after O'Sullivan had satisfactorily demonstrated the existence and properties of maltose, Bondonneau, described three dextrins, which he styles a-, β-, and γ-dextrin respectively. Throughout the paper he considers the sugar formed to be dextrose, and he endeavours to show that the theory of Musculus is correct ; and that by the action of diastase or acid upon starch, two dextrins are formed successively, the one—a-dextrin—being coloured red by iodine, the other—β-dextrin—not coloured by iodine. After the precipitation of these dextrins by alcohol, the solution, according to Bondonneau, contains dextrose, together with from 24·6 to 29·8 per cent. of a non-reducing body, which he names γ-dextrin, and which he considers to have a specific rotatory power of $[a]_D = 164°$. He states that both a- and β-dextrin are attacked by diastase, but the former much more readily than the latter.

There can be no doubt, from its behaviour with yeast and other reagents, that the γ-dextrin of Bondonneau is comparatively pure maltose, and when we remember that he imagined the cupric-reducing substance to be dextrose, it was natural that he should suspect the presence of another body with a higher optical activity than dextrose, but with no reducing power. This is the view taken by O'Sullivan in a second paper on maltose,[*] in which he gives the results of a more complete examination of this sugar, and confirms the analytical values given in the former paper.

The properties of dextrin and maltose having been thus definitely established, and these two substances having been shown to be the only products obtained by the action of diastase upon starch-paste under ordinary conditions, workers turned their attention to the influence of temperature, time,

* Journ. Chem. Society, L [2], 1876, p. 479.

and concentration upon the reaction by which these substances are produced.

The first experiments in this direction were those of O'Sullivan, who divided the action of malt-extract on starch-paste at elevated temperatures into three stages—namely, at temperatures below 63° C. (145° F.), at temperatures between 64–70° C. (147–158° F.), and at temperatures between 70° C. (158° F.), and the point at which the activity of diastase is destroyed, viz. 76° C. (169° F.); at each of these stages a very different proportion of maltose and dextrin is obtained.[*]

The three stages are thus formulated—

A. When starch is dissolved by malt-extract at any temperature below 63° C. (145° F.), if the solution be immediately (5 to 10 minutes) cooled and filtered, the product invariably contains maltose and dextrin in proportions agreeing closely with 67·85 per cent. of the former, and 32·15 per cent. of the latter, the cupric-oxide reducing power being equal to 44·1, and the specific rotatory power $[a]_{j\,2\text{-}85} = 170\text{·}6$.

Under these conditions O'Sullivan considers that a definite reaction takes place, which is represented by the equation—

$$\text{(A.)} \quad \underset{\text{Starch.}\dagger}{C_{12}H_{120}O_{60}} + 4H_2O = \underset{\text{Maltose.}}{4C_{12}H_{22}O_{11}} + \underset{\beta\text{-dextrin iii.,}}{C_{24}H_{40}O_{20}}$$

which corresponds to maltose 67·85 per cent. and dextrin 32·15 per cent. He states that if the malt-extract is not in considerable excess very little change takes place in the proportions, even if the solution is retained for four or five hours at the temperature at which the decomposition was made; but if the malt-extract is in excess, or contains more acid than usual, the maltose is found to increase at the expense of the dextrin, and to undergo a slight hydration itself. It was also invariably found that, in the conversions conducted under these conditions, the rotatory power calculated from the cupric-

---

[*] Journ. Chem. Society, ii. [2], 1876, pp. 125–144.

† This is the formula of the starch molecule adopted by O'Sullivan (Journ. Chem. Soc. 1879, p. 770).

oxide reducing power was greater than that observed. This, O'Sullivan concludes, is due to the slow and gradual conversion of the maltose into dextrose.[*]

B. When starch is dissolved by malt-extract at any temperature between 64° and 68–70° C. (147–158° F.), if the solution be immediately cooled and filtered, the product invariably contains maltose and dextrin in proportions agreeing closely with 34·54 per cent. of the former and 65·46 per cent. of the latter; the cupric-oxide reducing power being equal to 22·4 and the specific rotatory $[a]_{j\,s\bullet}$ 191·8°.

This change is expressed by the equation—

$$\text{(B.)} \quad C_{12}H_{100}O_{66} + 2H_2O = 2C_{12}H_{92}O_{11} + C_{66}H_{66}O_{66}$$
$$\text{Starch} \qquad\qquad \text{Maltose} \qquad \beta\text{-dextrin i.,}$$

which requires maltose 34·54 per cent. and dextrin 65·46 per cent. Here, again, O'Sullivan states that prolonged digestion, with an excess of malt-extract, results in the conversion of a portion of the maltose into dextrose. The third stage is—

C. When starch is dissolved by malt-extract at temperatures from 68–70° C. (147–158° F.) to the point at which the activity of the transforming agent is destroyed, if the solution be cooled and filtered at the end of five to ten minutes, the product contains maltose and dextrin in proportions agreeing closely with 17·4 per cent. of the former and 82·6 per cent. of the latter, the specific rotatory power of the mixture being $[a]_{j2.6} = 202·8°$ and the cupric-oxide reducing power equal to 11·3.

These numbers are required by the equation—

$$\text{(C.)} \quad C_{72}H_{100}O_{66} + H_2O = C_{12}H_{22}O_{11} + C_{60}H_{100}O_{60}.$$
$$\text{Starch.} \qquad\qquad \text{Maltose.} \qquad a\text{-dextrin.}$$

[*] There is no doubt that maltose is never hydrated into dextrose, however abundant the diastase and however favourable the conditions of conversion. The results upon which O'Sullivan based this view were doubtless attributable to the changes which take place in the constituents of malt-extract during digestion. These changes were first pointed out by Brown and Heron, who showed the necessity in all transformations of digesting the malt-extract side by side with the conversion.

To these three equations O'Sullivan added a fourth in the paper published in 1879.[*] This he calls the B' equation. It occupies a position between the A and B equations, and is expressed thus—

$$\text{(B'.)} \quad C_{12}H_{120}O_{60} \; + \; 3H_2O \; = \; 3C_{12}H_{22}O_{11} \; + \; C_{36}H_{44}O_{30}$$
$$\qquad \text{Starch} \qquad\qquad\qquad \text{Maltose} \qquad \beta\text{-dextrin ii. ;}$$

but we are not told at what temperature the conversion takes place according to this equation.

The dextrins corresponding to the four equations were isolated in a more or less pure state by repeated precipitations with alcohol ; they are distinguished as—

a-dextrin, coloured brownish-red by iodine.
β-dextrin i. ⎫
β-dextrin ii. ⎬ not coloured by iodine,
β-dextrin iii. ⎭

All four have a specific rotatory power of $[a]_{13\cdot08} = 222°$, and in the pure state are quite free from reducing power.

Shortly before the publication of O'Sullivan's second paper, Musculus and Gruber published the results of certain researches on the influence of temperature on the action of diastase upon starch ; they prepared three so-called achroodextrins, with the following optical and reducing properties :—

Achroodextrin I. [a] 210° Reducing power, 12.
      ,,   II. [a] 199°   ,,   12.
      ,,   III. [a] 192°   ,,   12.

When these dextrins were isolated, and acted upon with fresh diastase, their properties were modified as follows :—

Achroodextrin I. [a] 159° Reducing power, 36.
      ,,   II. [a] 168°   ,,   0.
      ,,   III. Unchanged.

In order to explain these results, they brought forward a theory according to which starch is a polysaccharide with the formula $n(C_{12}H_{20}O_{10})$, and undergoes, under the influence of diastase and acids, successive hydrations and decompositions.

[*] Journ. Chem. Soc., 1879, p. 770.

A most important contribution to the study of this question was made by Brown and Heron, in 1879[*], who very completely examined the influence of time and temperature on starch conversions.

In order to investigate the action of malt-extract upon starch-paste at elevated temperatures, Brown and Heron found it necessary to arrest, in a portion of the liquid, any further transformation at any desired point of the reaction. For this purpose they employed a small quantity of salicylic acid, 0·05 gram of acid per 100 c.c. being found sufficient to completely arrest all action. Having found this means of stopping the action of malt-extract, they were able, by withdrawing portions of the transformation at stated intervals, and analysing them, to express the progress of the reaction in a graphic form by means of a curve.

The previous treatment of the malt-extract was found to have a most important influence on the nature of the resulting transformation, provided the treatment had been at the same or a higher temperature than that at which the transformation was made. Thus, when the results were expressed graphically, the transformations conducted at 60° (140° F.) and 66° (151° F.), with malt-extract heated respectively to these temperatures, were represented by two totally distinct forms of curve ; but when transformations were made at 60° (140° F.) and 66° (151° F.), with malt-extract previously heated to 66° (151° F.), both curves were found to be identical, and to correspond to that of the higher temperature. This holds good for all temperatures above 50° C. (122° F.).

Brown and Heron found that conversions made at 40° C. (104° F.) and 50° C. (122° F.), yield a product closely approximating to 81·3 per cent. of maltose, and 18·7 per cent. of dextrin. This mixture requires—

$$[\alpha]_{j\,5\cdot66} = 162\cdot3$$
$$\kappa_{5\cdot66} = 49\cdot7$$

* Journ. Chem. Soc., 1879, pp. 596–655.

this composition remaining unaltered for some time even if a further quantity of malt-extract were added.

The same final numbers were obtained in transformations conducted at 60° C. (140° F.) It will be remembered that, according to O'Sullivan, the conversion at these temperatures should take place according to equation A—i. e., the mixed products should have values of $[a]_j = 170°.6$ and $\kappa = 44·1$. Brown and Heron were, however, unable to obtain any indications of a stoppage of the reactions at this point, and they conclude that the normal transformation of starch-paste at 60° (140° F.) and at all lower temperatures is that given above, and is a close approximation to the reaction represented by the equation—

$$10C_{12}H_{20}O_{10} + 8H_2O = 8C_{12}H_{22}O_{11} + 4C_6H_{10}O_6.$$

in other words, the transformation products at 60° C. (140° F.) and below consist of four-fifths maltose, and one fifth dextrin. When the transformation has reached this point, it is not appreciably altered either by the addition of further quantities of malt-extract or by continued digestion for some hours, the former being a proof that the cessation of the action is not due to a weakening of the diastatic power of the malt-extract. That the cessation was not due to the converting agent being paralysed by the excess of maltose produced, was shown by the fact that at the end of such a conversion the liquid was able to reduce to its lowest terms a further quantity of starch. The facts were established by a large number of experiments, and the fixed and constant character of the reaction is of the utmost importance in the later developments of the theory of the transformation of starch.

Under favourable conditions the conversion rapidly runs down to this point, and when once reached the further formation of maltose is exceedingly slow, even after the addition of successive quantities of malt-extract. The following experiment illustrates the progress of the conversion considered as a function of the time. 5 grams of starch were gelatinised in 100 cc. of water and 10 cc. of malt-extract, previously heated to 60° C. (140° F.), were added; the temperature of the conversion was 60° C. (140° F.)

Time.	$[a]_{j \cdot 88}$	$\kappa_{j \cdot 88}$	Iodine reaction.
1 minute	191·7°	—	Pure blue
2½ minutes	175·6	—	Brown
5 ,, ...	166·8	—	Very light brown
15 ,, ...	165·7	—	None
30 ,, ...	163·7	—	,,
60 ,, ...	162·9	49·6	,,

More malt-extract added at rate of 4·5 cc. per 100 cc.

90 minutes	161·3°	—	

More malt-extract at rate of 4·2 cc. per 100 cc.

120 minutes	160·1°	52·0	

At temperatures above 60° C. (140° F.), Brown and Heron found that the transformation was more or less restricted, but generally speaking they were unable to confirm O'Sullivan's equations for the transformation of starch at these temperatures. They considered that their experiments substantiated the existence of at least four well-defined molecular transformations of starch, of which the one described above is the best defined. The next fixed point is the disappearance of the iodine reaction for erythrodextrin when the conversion is made with malt-extract heated to 66° (151° F.) This takes place when the value for $[a]_{j \cdot 88}$ is 188·5, and for $\kappa_{j \cdot 88}$ 25·0 or thereabouts, and is very constant with varying quantities of malt-extract. The next strongly-marked reaction is that obtained with malt-extract heated to 75°, and also with malt-extract heated to 66°, and rendered alkaline with sodium carbonate; the numbers obtained being

$[a]_{j \, \text{s·ss}}$ 195-196,° and $\kappa_{\text{s·ss}}$ 18·9.   The highest stable transformation is obtained by the action of malt-extract heated to 66° (151° F.), and made very slightly alkaline with soda, the same reaction being also marked in all transformations above 66° (151° F.) by the appearance of the maximum iodine colouration for erythrodextrin.   The numbers for this transformation, $[a]_{j \, \text{s·ss}}$ 202-203°, and $\kappa_{\text{s·ss}}$ 12·7, are very nearly those of O'Sullivan's equation C, the dextrin being, however, an erythrodextrin, and not an achroodextrin, as described by O'Sullivan.

It was also found that when higher conversions are acted upon by unheated malt-extract below 60° (140° F.), they are at once converted into the lower form, and are not further altered on continued digestion ; thus a conversion which had—

$$[a]_{j \, \text{s·ss}} \quad .. \quad .. \quad .. \quad .. \quad .. \quad .. \quad .. \quad .. \quad 187·8$$
$$\kappa_{\text{s·ss}} \quad .. \quad .. \quad .. \quad .. \quad .. \quad .. \quad .. \quad .. \quad 28·9$$

was converted in two minutes by 5 c.c. of malt-extract into—

$$[a]_{j \, \text{s·ss}} \quad .. \quad .. \quad .. \quad .. \quad .. \quad .. \quad .. \quad .. \quad 162·6$$
$$\kappa_{\text{s·ss}} \quad .. \quad .. \quad .. \quad .. \quad .. \quad .. \quad .. \quad .. \quad 49·3$$

which corresponds to a percentage composition of—

Maltose	80·8
Dextrin	19·2
	100·0

From the above, and other considerations, Brown and Heron concluded "that the dextrins are not metameric but polymeric bodies ; those corresponding to transformations of high optical activity being of greater molecular complexity than those yielded by transformations of lower optical activity, the latter being produced from the former by a partial act of hydration with consequent elimination of maltose."

Their experiments led Brown and Heron to adopt the formula $10C_{12}H_{20}O_{10}$ for soluble-starch.   The first action of the transforming agent upon this complex and unstable molecule was assumed to result in the removal by hydration of one of the groups $C_{12}H_{20}O_{10}$, thus producing maltose, whilst the remaining nine groups of $C_{12}H_{20}O_{10}$ constituted the first dextrin of the series, erythrodextrin a. This action was supposed to continue; at each step a further $C_{12}H_{20}O_{10}$ group being split off and hydrated to maltose, whilst the remaining $C_{12}H_{20}O_{10}$ groups formed a lower dextrin. On these lines Brown and Heron drew out a table for the nine theoretical transformations, with the values of $[a]_j$ and κ for each, as follows :—

No. of transformation.	$[a]_j$ 3·86	κ 3·86	Resulting dextrin.
Soluble starch ..	Deg. 216·0	0	—
(1)	209·0	6·4	Erythrodextrin a
(2)	202·0	12·7	,, β
(3)	195·4	18·9	Achroodextrin a
(4)	188·7	25·2	,, β
(5)	182·1	31·3	,, γ
(6)	175·6	37·3	,, δ
(7)	169·0	43·3	,, ϵ
(8)	162·6	49·3	,, ζ
(9)	156·3	55·1	,, η
Maltose	150·0	61·0	—

Of these the most stable is No. 8, alluded to above, and which is often spoken of in the literature of the subject as the " No. 8 equation." This transformation assumes a most important place in later work, which entered upon a new phase with the recognition of maltodextrin and the part it plays in starch transformations.

The following table indicates in a graphic form the conversion of starch-paste by malt-extract under various conditions. The malt-extract in the first case is unheated ; in the others, it has been heated to different temperatures previous to the conversions being made. It will be noted

I

firstly, that the maximum conversion takes place almost entirely within the first five minutes, after which, it becomes

Fig. 2.

TABLE SHOWING AVERAGE CURVES FOR STARCH TRANSFORMATIONS
UNDER DIFFERENT CONDITIONS.

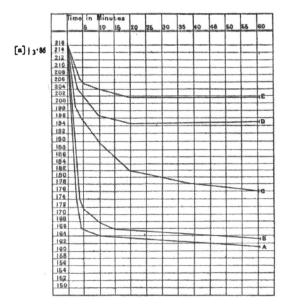

A, transformations with unheated malt-extract at 40–50° C. (104–122° F.)

B, transformations with malt-extract heated to 60° C. (140° F.)

C, transformations with malt-extract heated to 66° C. (151° F.)

D, transformations with malt-extract heated to 75° C. (167° F.), and also with malt-extract heated to 66° C. (151° F.), and rendered slightly alkaline with sodium carbonate.

E, transformations with malt-extract heated to 66° C. (151° F.), and rendered slightly alkaline with caustic soda.

gradual; and secondly, that the higher the temperature to which the malt-extract has been heated prior to conversion, the less sudden is the fall of angle at first, and the less complete the total conversion ; in other words, the conversion stops at a much higher point.

The curve C is the most interesting from our standpoint, as it relates to a limit of temperature near which actual brewing operations are performed. It is true that at this temperature the conversion is not so sudden as it is in the case of conversions at other temperatures, but even here, it is nearly complete in 30 minutes, the hydration after that time being only gradual.

We may here digress for a moment, to draw attention to another set of curves, deduced, not from the conversion of starch-paste by malt-extract, but from the conversion with malt itself. In one case, the malt was mashed in the laboratory at 66° C. (151° F.), in the other, it was mashed in the brewery at the same temperature. The same quantity of water was used in each case, and polariscopic readings were taken every fifteen minutes up to two hours.[*]

It will be noticed that the curves are very similar to each other, and also to the curve expressing the conversion of a starch-paste at the same temperature. The drop of angle during the first 30 minutes, is also very marked.

In 1879, Herzfeld presented, at the University of Halle, his inaugural dissertation on " Maltodextrin."[†] In this he adopts the view of Musculus and Gruber, that the transformation of starch by diastase is not a splitting up of the molecule of the former, but is the result of the gradual degradation of the starch through the stages of soluble-starch, erythrodextrin achroodextrin, maltodextrin, and maltose. To the first three and the last of these Herzfeld ascribed the properties with which we are now familiar, whilst maltodextrin, which he regards as being intermediate between achroodextrin and maltose, is described as being a substance soluble in dilute alcohol, but insoluble in alcohol of 90 per cent., and *completely*

[*] Heron, Journ. Soc. Chem. Industry, 1888, p. 268.

[†] " Ueber Maltodextrin " (Halle, 1879) ; see also Berlin Berichte, 1879, 2120.

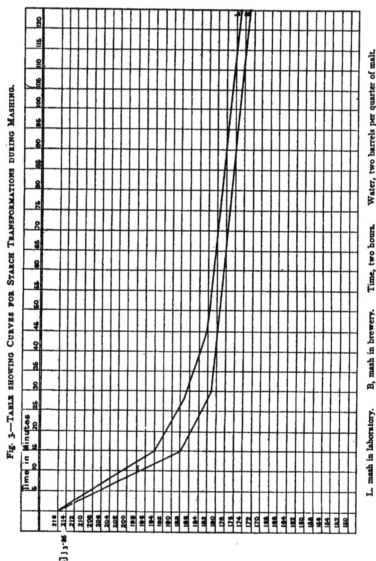

Fig. 3—Table showing Curves for Starch Transformations during Mashing.

L, mash in laboratory.   B, mash in brewery.   Time, two hours.   Water, two barrels per quarter of malt.

fermentable by yeast; it is said to have an optical activity of $[a]_1 171 \cdot 6°$, and a cupric-reducing power of $23 \cdot 5$.

After the year 1879, nothing of any moment was published until 1885, when Brown and Morris read a paper before the Chemical Society,[*] which was virtually a continuation of the paper by Brown and Heron referred to above.

In this communication Brown and Morris first show that the law that the products of the transformation of starch-paste by malt-extract at temperatures above 40° C. (104° F.) can be expressed in the terms of a mixture of maltose and non-reducing dextrin applies not only to the conversion as a whole, but also to all the different portions of the conversion when fractionally precipitated by alcohol. The authors adopt this law as a criterion of purity for the products of the action of malt-extract or diastase upon starch, and in any cases in which the work of other observers does not conform to this rule, they ascribe the discrepancy either to the presence of impurities or to errors in analysis.

We have already fully shown that the tendency of all transformations carried on with malt-extract between 40° (104° F.) and 60° C. (140° F.) is speedily to attain a point of equilibrium beyond which further progress is relatively very slow.

All transformations of starch which show a higher percentage of dextrin and a lower percentage of maltose than this, are at once brought down to this point by treatment with a little fresh malt-extract at 50° C. (122° F.) Now we know that maltose is not acted on by this treatment, therefore Brown and Morris say the cause of the change must be looked for in the degradation of the dextrin, and it is most important to ascertain if the dextrin would behave in a similar manner and be affected to the same extent when isolated, for if this should be the case it ought then to be possible to determine the position which any dextrin may occupy in the

[*] Journ. Chem. Soc., 1885, pp. 527-570.

polymeric series by determining the amount of degradation it undergoes with malt-extract at 50–60° C. (122–140° F.)

In the first place the authors proved by careful quantitative experiments that the dextrins during the ordinary processes of separation and purification by means of precipitation with alcohol, evaporation, &c., underwent no hydrolysis. They then showed that when the products of a starch-transformation are fractionated with alcohol, the mean maximum hydrolysing effect of malt-extract, acting separately on the portions soluble and insoluble in alcohol, is the same as the maximum hydrolytic action exercised under similar conditions on the original transformation-products.

Having proved that the action of malt-extract at 50–60° C. (122–140° F.) on the mixed products of transformation always brings the dextrin down to a given point, and that the total action is not altered by previously separating the starch-products by alcohol, Brown and Morris consider that there exists here a valuable means of differentiating the dextrins, of testing their homogeneity, and of ascertaining the position they occupy in the polymeric series. Subsequent research on these lines, with different conversions, satisfied the authors that, with the exception of the No. 8 equation, no conversion yielded a homogeneous dextrin—that is to say, that although a conversion made under suitable conditions of temperature might give a product the analytical numbers of which would closely correspond to one of the equations described by O'Sullivan or by Brown and Heron, yet the dextrin of this conversion would never be homogeneous but would consist of those corresponding to both higher and lower equations.

In the above and many other experiments, Brown and Morris were confronted with certain curious and anomalous facts which were without any adequate explanation. In the first place, it was invariably noticed in high-angled conversions, that although the erythrodextrins are undoubtedly highest in the series and the least soluble in alcohol, yet it is the dextrins which are soluble in the strongest alcohol which yield the largest amount of maltose on degradation with malt-extract; then again, it is extremely difficult to crystallise maltose from the products of a conversion higher than the No. 8, and it is equally difficult to ferment the whole of the maltose from a similar conversion; thus it is easy to prepare solutions from the higher conversions which, judging from their cupric-reducing power, appear to contain from 30-40 per cent. of maltose among the solid products, and yet are perfectly unfermentable with ordinary yeast. But, on the other hand, if the conversion is run down to the No. 8 limit, the whole of the maltose within 1 per cent., or even less, can be easily fermented. The accumulation of evidence on these and other points led Brown and Morris to the conclusion that there existed in high-angled conversions, in addition to maltose and dextrin, a third substance, which possessed an optical activity and cupric-reducing power corresponding to a mixture of maltose and dextrin, but which was not decomposed by treatment with ordinary solvents, was not fermentable under ordinary conditions, and was readily and entirely converted into crystallisable and fermentable maltose by the action of malt-extract. For this substance they adopted Herzfeld's name of *maltodextrin* (see p. 130), although it will be seen that the properties of the substance, as described by Brown and Morris, are in some points directly opposite to those described by Herzfeld; and the former workers assign to it an entirely different position among the starch transformation products to that given by Herzfeld.

The authors rely upon the following facts to prove that maltodextrin is not a mixture of maltose and dextrin :—

(a) A *mixture* of maltose and dextrin prepared so as to have the same optical activity and reducing power as maltodextrin, is separable into its constituents by a single judicious treatment with alcohol.

Maltodextrin is not separable into maltose and dextrin by any possible treatment with alcohol, but is dissolved and precipitated as a homogeneous substance.

(*b*) From a *mixture* of maltose and dextrin it is possible, by means of the *S. cerevisiæ* of the "high fermentation," to ferment the maltose, leaving the dextrin untouched.

Maltodextrin treated in a similar manner is entirely unfermentable.

(*c*) When a *mixture* of maltose and dextrin is submitted to the action of malt-extract at 50–60° C. (122–140° F.) a residue of No. 8 dextrin is always left, the amount being greater the lower in the series the dextrin of the mixture is.

Maltodextrin when acted on by malt-extract under similar conditions, is entirely convered into maltose, its amylin or dextrin constituent leaving no residual dextrin.

Whilst maltodextrin is unfermentable by yeast (*S. cerevisiæ* of the high fermentation), it is converted into fermentable, crystallisable maltose by malt-extract, or by the slow action of certain forms of Saccharomyces (notably *Sacch. ellipsoïdeus* and *Pastorianus*), which accompany the secondary fermentation.[*]

* This statement has been the subject of much adverse criticism and misunderstanding; the principal objection being that the yeasts employed were not pure cultures, and therefore that the experiments are valueless. It must be remembered that at the time the experiments were made, the methods of pure cultivation were in their infancy, and the investigations of Hansen on the species and races of yeast were but little known outside Copenhagen, and therefore when Brown and Morris speak of *Sacch. ellipsoïdeus* and *Pastorianus* they use these terms in the sense in which they are employed by Reess and Pasteur.

The following experiment, which was one of many, may be quoted as illustrative of the facts upon which the statement was based. A conversion was made in the usual way, and the free maltose was fermented out. The fermented liquid was freed from alcohol, and a solution of the residue made of sp. gr. 1093·0,

Brown and Morris considered that maltodextrin is not, as supposed by Herzfeld, a mere hydration-product of achroodextrin, but that it is produced from starch and the polymeric dextrins by the fixation of a molecule of water upon the

---

corresponding to 24·093 grams of solids per 100 c.c. Analysis showed that these solids were made up of—

Maltose	6·986 grams.
Dextrin..	16·792 ,,
Inactive matter	0·315 ,,
	24·093

and the dextrin on degradation yielded 66·8 per cent. of maltose.

The authors then say : " Some yeast was again added to this solution, the solid matter of which, it will be remembered, had already been fermented once before, and the whole was submitted to a temperature of 30° C. (86° F.) The solution was narrowly watched, and a little of the sediment examined daily under the microscope."

" The original yeast-cells, consisting of the *Sacch. cerevisiæ* of the 'high fermentation,' not finding any free maltose in the solution (although, judging from the reducing power, there should have been nearly 7 grams per 100 c.c.) shrivelled up, and to all appearances died. At the end of six or seven days, not a trace of fermentation having previously been observed, a few cells of *Sacch. ellipticus* and *Pastorianus* began to appear. Concurrently with this alteration in microscopical appearance, fermentation commenced, at first with extreme slowness, but gradually becoming much more rapid as the new forms of *Saccharomyces* increased in abundance. At the end of forty days, when the experiment was stopped, fermentation was still going on slowly."

" The amount of maltose which had fermented was determined in two ways : (1) By the decreased specific gravity of the solution after distilling off the alcohol, and (2) by estimating the alcohol formed, and calculating it as maltose, due allowance being made in each case for the non-volatile products of fermentation.

Method 1 gave a disappearance of 14·398 grams maltose per 100 c.c.
,, 2 ,, ,, 14·410 ,, ,, ,,

Now, when we consider that the original solution only contained maltose equal to 6·986 grams per 100 c.c., it is evident that the remaining 7·414 grams must have been produced at the expense of the dextrin."

Brown and Morris consider the new forms of yeast to have been derived from the brewing yeast employed, which, as is well known, always contains more or less of the so-called " wild " forms. Nothing could be plainer than the description here given, which points to the death of the original *Sacch. cerevisiæ* and the gradual growth of the admixed wild forms ; but the authors guard against the sweeping assertion, for which they have sometimes been made responsible, that *Sacch. cerevisiæ* cannot ferment maltodextrin under any circumstances, for they ternary group $(C_{12}H_{20}O_{10})_3$, of which they assumed that there could not be less than *five* in the starch molecule. This action would result in the separation from the dextrin residue of maltodextrin $\begin{cases} C_{12}H_{22}O_{11} \\ (C_{12}H_{20}O_{10})_2. \end{cases}$ This by the fixation of two more molecules of water gives rise to freely fermentable and crystallisable maltose.

The recognition of maltodextrin among the products of the conversion of starch rendered a modification of the theory of the transformation of starch necessary.

It will be remembered that Brown and Heron adopted the formula $10C_{12}H_{20}O_{10}$ for soluble-starch, and supposed the hydrolysis of starch to take place in ten stages, at each of which a $C_{12}H_{20}O_{10}$ group was broken off and hydrolysed to maltose. This theory required the existence of nine polymeric dextrins between soluble starch and maltose.

It is obvious that this theory will not allow of the formation of a substance with the properties and constitution of maltodextrin ; Brown and Morris therefore assumed that the molecule of starch could not be less than five times $(C_{12}H_{20}O_{10})_3$. They supposed the action of malt-extract in hydrolysing this molecule to consist of a successive hydration and removal of the $(C_{12}H_{20}O_{10})_3$ groups, leaving a dextrin residue of decreasing complexity, until the last remaining group of $(C_{12}H_{20}O_{10})_3$ is reached—this is one-fifth of the original starch-molecule, and constitutes the dextrin of the No. 8 equation to which we have referred above. The removal of each $(C_{12}H_{20}O_{10})_3$ group is, however, effected prior to its complete hydration, the ternary group being split off in the form of maltodextrin, when one

_____

say, " It would not be safe to assume that typical *Sacch. cerevisiæ* is incapable, *under all conditions*, of hydrolysing the dextrins and maltodextrins, and thus supplying itself with food."

We have authority for saying that prolonged experience has in no way caused Brown and Morris to alter the view expressed in the above extract, and experiments carried out with pure cultures of yeast fully confirm the general conclusions arrived at in 1885.

only of its amylin or dextrin groups has been converted into
an amylon or maltose group, thus—

$$(C_{12}H_{20}O_{10})_2 + H_2O = \begin{cases} C_{12}H_{22}O_{11} \\ (C_{12}H_{20}O_{10})_2 \end{cases}$$
Starch group.   Maltodextrin.

And maltodextrin, under the further action of diastase is
completely converted into maltose, thus—

$$\begin{cases} C_{12}H_{22}O_{11} \\ (C_{12}H_{20}O_{10})_2 \end{cases} + 2H_2O = 3C_{12}H_{22}O_{11}$$
Maltodextrin.   Maltose.

More recently Brown and Morris have had occasion to
very considerably modify this theory.[*]

In the course of a series of determinations of the molecular
weights of various carbohydrates by Raoult's method, Brown
and Morris submitted soluble-starch and the dextrins to examination. The general results of this examination were that
soluble-starch had an extremely large molecule, so large in fact
as to be incapable of determination by the method in question,
whilst several preparations of dextrins, obtained by stopping
starch transformations at different points, and corresponding
to various so-called high and low dextrins, all gave practically
the same result. A carefully purified dextrin from a conversion
carried down to the No. 8 equation, gave results which pointed
unmistakably to the formula $(C_{12}H_{20}O_{10})_{20}$, and as there can
be no doubt that the molecule of this stable dextrin is one-
fifth of the size of the soluble-starch molecule from which it
has been derived, the authors were forced to the conclusion
that the formula of the latter is $5(C_{12}H_{20}O_{10})_{20}$.

The fact of the so-called high dextrins giving numbers
pointing to molecular weights almost identical with that
obtained for the dextrin of the No. 8 equation, appears to
leave little doubt that the so-called high and low dextrins do
not form a polymeric series, but are, at the most, metameric.
In this connection it is important to remember that evidence

[*] Journ. Chem. Soc., 1889, pp. 449 and 462.

was obtained of a most decided difference in the size of the soluble-starch and dextrin molecules.

A consideration of these facts, and of others elicited in the same inquiry, together with the recognition of several substances with the properties of maltodextrin, but of varying composition, in restricted starch conversions, caused Brown and Morris to frame the following theory for the transformation of starch by diastase. This may be taken to represent their most recent views on the subject.

The starch-molecule may be regarded as consisting of four complex amylin groups arranged round a fifth similar group, constituting a molecular nucleus. The first action of diastase is to break up this complex group, and liberate all the five amylin groups. The central amylin nucleus, consequent on a closing-up of the molecule, withstands the further influence of hydrolysing agents, and constitutes the stable dextrin of the No. 8 equation. The four outer amylin groups are capable, when liberated, of being rapidly and completely converted into maltose by successive hydrolysations through a series of amyloïns, whose number is only limited by the size of the original amylin group.

It appears most probable that these outer amylin groups cannot exist as such; but immediately on separation from the central nucleus are partially hydrolysed, yielding amyloïns of possibly the very highest type. A supposition of this kind will enable us to understand the slight amount of cupric-reducing power which even the very highest conversions exhibit.

In conversions that are allowed to take their normal course, we have a gradual hydrolysis of these high amyloïns, which continues until the final stage is reached, when the whole of the four outer amylin groups have been converted into maltose, and the fifth group, the nucleus of the original soluble-starch molecule, forms the residual dextrin. The further hydrolysis of the amyloïns is also undoubtedly accompanied by the splitting up of the original groups into smaller aggregations, as is evidenced by the formation of malto-dextrin.

The action may be expressed by the following equations, in which m represents the number of amylin groups converted into amylon groups, and n the number of unchanged amylin groups in the amyloïns: in the first equation the sum of n and m equals 20. We may then express the very earliest stage of the hydrolysis thus—

$$\left\{ \begin{array}{l} (C_{12}H_{20}O_{10})_{20} \\ (C_{12}H_{20}O_{10})_{20} \\ (C_{12}H_{20}O_{10})_{20} \\ (C_{12}H_{20}O_{10})_{20} \\ (C_{12}H_{20}O_{10})_{20} \end{array} \right. + nH_2O = \underset{\text{Stable dextrin.}}{(C_{12}H_{20}O_{10})_{20}} + 4 \left\{ \begin{array}{l} (C_{12}H_{22}O_{11})_m \\ (C_{12}H_{20}O_{10})_n. \end{array} \right.$$

$$\text{Starch molecule.} \qquad \qquad \underset{\text{varying type.}}{\text{Amyloïns of}}$$

At an intermediate point in the hydrolysis, the following probably represents the reaction :—

$$\left\{ \begin{array}{l} (C_{12}H_{20}O_{10})_{20} \\ (C_{12}H_{20}O_{10})_{20} \\ (C_{12}H_{20}O_{10})_{20} \\ (C_{12}H_{20}O_{10})_{20} \\ (C_{12}H_{20}O_{10})_{20} \end{array} \right. + \left(x + \frac{m(80-x)}{n+m} \right) H_2O = \underset{\text{Stable dextrin.}}{(C_{12}H_{20}O_{10})_{20}}$$

$$\text{Starch molecule.} \qquad + x\, \underset{\text{Maltose.}}{C_{12}H_{22}O_{11}} + \frac{80-x}{n+m} \left\{ \begin{array}{l} (C_{12}H_{22}O_{11})_m \\ (C_{12}H_{20}O_{10})_n. \end{array} \right.$$

$$\underset{\substack{\text{composition.}}}{\text{Amyloïns of varying}}$$

This theory appears to embrace and explain all the known facts in connection with starch transformations; it enables us to understand why it is impossible to separate the whole of the maltose of a restricted conversion, either by solution in alcohol or by fermentation; it also offers a complete explanation of the observed facts in connection with fractional degradations, and other questions of a like nature.

In another chapter (Chap. V., p. 207) we have fully dealt with this theory from a practical standpoint, and shown the important bearing it has upon many points which were formerly obscure.

It is now necessary that we shall study the properties of the various products more in detail.

Soluble-starch.

We have already referred to soluble-starch as being the first stage in the degradation of starch, and as being obtained by the limited action of heated malt-extract or of acid upon starch-paste heated to a suitable temperature.

Obtained by either of these methods, soluble-starch separates out on long standing of its concentrated solutions, or immediately on the addition of alcohol to dilute solutions, as a white substance, which, when washed with alcohol and dried, forms a friable, amorphous body. Examined under the microscope it is seen to be entirely without structure, and in its solid state without action on polarised light. The substance itself and its solutions are coloured an intense blue by iodine. It is almost insoluble in cold water, but readily dissolves in boiling water, and is thrown down again on cooling as a white, flocculent, amorphous precipitate.

The specific rotatory power of both forms of soluble-starch in solution is $[a]\,_{j\,._{..}} = 216\cdot0°$ ($[a]_D = 195\cdot0°$), and it has no cupric-reducing power. (Brown and Heron).

The molecular weight of soluble-starch, determined by Raoult's freezing method, is 32,400, and its formula is consequently $5(C_{12}H_{20}O_{10})_{20}$. (Brown and Morris).

When acted upon by malt-extract, it is converted into dextrin and maltose, according to the equation:—

$$5(C_{12}H_{20}O_{10})_{20} + 8oH_2O = 8oC_{12}H_{22}O_{11} + (C_{12}H_{20}O_{10})_{20}$$

Soluble-starch.   Maltose.   Dextrin.

Soluble-starch forms soluble compounds with potassium and sodium hydroxides, and insoluble compounds with the hydrates of barium and lead.

Soluble-starch may also be prepared by digesting intact starch-granules in the cold with dilute hydrochloric or sulphuric acid for a few days. The granules retain their shape, and cannot be distinguished, when examined microscopically, from untreated starch-granules, and the altered substance has exactly the same action on polarised light as the unaltered starch. They dissolve, however, readily in hot water without gelatinisation. This form of starch is usually known as Lintner's soluble-starch, and is that used for the determination of diastatic power (see p. 452). The precise action of the acid in producing this change is very obscure, and the time required for producing the alteration is very small; Brown and Morris stating that when potato-starch is digested with 12 per cent. hydrochloric acid for twenty-four hours the power of gelatinisation is completely lost, whilst the amount of matter which goes into solution is excessively little.

Dextrin.

In treating of the transformation-products of starch, we have mentioned at some length the views of past workers on the nature and properties of dextrin, or, more correctly speaking, the dextrins: it having been the universal opinion that there was a series of dextrins, the members differing, according to some authors, in their optical activities and cupric-reducing powers; according to others, simply in their molecular size, and colouration with iodine solution. In addition to the dextrins proper, various other substances were described : thus, W. Nägeli[*] mentions two amylodextrins (see p. 129), Herzfeld[†] describes maltodextrin (see p. 130), and other writers mention other substances intermediate between starch and maltose.

Recent work by Brown and Morris leads us, however, to the conclusion that there exists but one true dextrin, which has a specific rotatory power of $[a]_{j, 8a} = 216 \cdot 0°$ ($[a]_D = 195 \cdot 0°$)[‡], is without action upon Fehling's solution, and does not give any colouration with iodine solution. In all starch-conver-

---

[*] 'Beiträge zur naheren Kenntniss der Stärkegruppe,' Leipsig, 1874.
[†] *Loc. cit.*   [‡] See footnote, p. 102.

sions, made under normal conditions, it forms one-fifth of the original starch, and is produced in the very earliest stages of the hydrolysis (see p. 124), and is not further acted upon by diastase within the ordinary time-limits of a starch-conversion. It is unfermentable by the normal *Sacch. cerevisiæ* of the primary fermentation, but in the combined presence of diastase and yeast it is quickly and entirely fermented. Dextrin is soluble in water and dilute alcohol, but in strong alcohol it is almost entirely insoluble; advantage is taken of the latter fact to separate it from maltose in a starch conversion carried to its lowest point.

We attribute all the discrepancies which exist in the literature of the dextrins regarding their nature and properties, to the unrecognised presence in the conversions of amyloïns of varying type. The presence of these substances will account for the colouration with iodine which the *erythro-dextrins* are said to give, for the varying optical and reducing powers which different observers have assigned to the dextrins, and for the varying behaviour with diastase noted by different authors.

Soxhlet[*] has stated that the dextrins produced by the action of acid, and by the action of diastase upon starch, differ very markedly in their properties, the former not being further changed by diastase, whilst the latter are slowly converted into maltose by the same agent. This statement is undoubtedly based upon incomplete experimental evidence, for experience has shown us that conversions with acids closely resemble those with diastase, and that it is perfectly possible to obtain amyloïns of varying type, and, of course degradable by diastase, from conversions with acid. These substances would formerly have been classed as higher dextrins, that is to say, as dextrins capable of undergoing further hydrolysis in the presence of diastase.

[*] Zeitschrift f. Spiritus-industrie, 1884, No. 11.

Amylodextrin.

We have already referred in several places to the substance described by W. Nägeli, under the name of *amylodextrin*. It has been confused by some authors with starch-cellulose, but by the majority of recent workers, including Musculus and Gruber, and A. Meyer, with soluble-starch. It is obtained by the long-continued action of dilute acids upon starch-granules in the cold ; after some weeks the latter lose their original structure and become completely disintegrated. This residue consists of crude amylodextrin, which can be purified by solution in hot water, and subsequent precipitation, either by the action of cold or by the addition of alcohol. Thus prepared, amylodextrin consists of crystalline spherules, made up of minute needles arranged radially ; these crystalline aggregates closely resemble the well-known spherules of inulin.

Brown and Morris have recently[*] more closely examined this substance. They show that it is not identical with starch-cellulose or soluble-starch, but that it is a definite chemical compound derived from the latter by hydrolysis. In solution amylodextrin gives an intense reddish-brown colouration with iodine ; it is absolutely unfermentable by primary *Sacch. cerevisiæ*, and is slowly diffusible in an unaltered form. The specific rotatory power of amylodextrin is $[a]_{13 \cdot 86} = 206 \cdot 11°$ ($[a]_D = 186 \cdot 8°$). (Nägeli gave $[a] = 175°$ to $177°$), and its cupric-reducing power is represented by $\kappa_{3 \cdot 86} = 9 \cdot 08$.

In composition, amylodextrin is analogous to maltodextrin (see p. 130) ; it may be represented by the formula :—

$$\begin{cases} C_{15}H_{72}O_{11} \\ (C_{15}H_{26}O_{10})_6 \end{cases}$$

that is, as constituted of one *amylon* or maltose-group in combination with six *amylin* or dextrin-groups. The molecular weight, determined by Raoult's method, is 2220, which closely agrees with the above formula.

[*] Journ. Chem. Soc., 1889, pp. 449–461.

K

When acted upon by diastase, amylodextrin is completely converted into maltose.

Amylodextrin forms one of the class of substances now known as amyloïns, and it is on this account that we have given its properties somewhat in detail.

Maltodextrin.

We have stated (p. 115) that Herzfeld described in 1879 a substance among the transformation products of starch which he named maltodextrin. The principal properties he assigned to it were $[a]_j = 171 \cdot 6°$, $\kappa = 23 \cdot 5$, and complete fermentability with yeast.

Brown and Morris in 1885 found[*] a substance among the products of the action of diastase upon starch, which agreed in some of its properties with the body described by Herzfeld, and for which they adopted the same name. It differed, however, in several most essential particulars from Herzfeld's maltodextrin.

Maltodextrin exists to a greater or lesser extent in all high-angled conversions, and may be obtained from these by the fermentation of the free maltose with *Sacch. cerevisiæ*, and the extraction and solution of the residue with alcohol of 85 per cent., in which maltodextrin is freely soluble. When purified the substance has an optical activity of $[a]_{j\,s\cdot\,ss} = 193 \cdot 6°$ ($[a]_D = 174 \cdot 7°$), and a cupric-reducing power of $\kappa_{3\cdot\,ss} = 20 \cdot 7$; these numbers exactly correspond to those required for the formula :—

$$\begin{cases} C_{1s}H_{ss}O_{11} \\ (C_{1s}H_{ss}O_{1s})_s \end{cases}$$

that is to say, to those required for a compound consisting of one amylon or maltose-group combined with two amylin or dextrin-groups. The numbers obtained by the determination of the molecular weight of maltodextrin by Raoult's method

* Journ. Chem. Soc., 1885, pp. 527-570.

fully confirm this formula, Brown and Morris[*] obtaining a value of 965, as against 990 required by theory.

Maltodextrin is absolutely unfermentable by ordinary *Sacch. cerevisiæ* of the high fermentation, but is slowly fermented by certain forms of secondary yeast. It is slowly diffusible without alteration, and is completely and rapidly converted into maltose by treatment with diastase.

We have already explained (p. 122) the position which maltodextrin occupies among the transformation products of starch.

Amyloïns.

Brown and Morris have recently[†] adopted the term *amyloïns* for the class of bodies of which the two preceding substances—amylodextrin and maltodextrin—are the type. The word was suggested by Prof. Armstrong, and denotes the constitution of this class of substances, which these authors regard as composed of varying proportions of amylon and amylin groups.

The distinguishing characteristics of these substances are :—

1. That they give numbers on analysis which allow their composition to be expressed in terms of a mixture of maltose and dextrin.

2. That they cannot be separated by any known means into maltose and dextrin, and are therefore compound bodies.

3. That they are completely converted into maltose by the action of malt-extract or diastase.

4. That they are unfermentable during the primary fermentation.

The last of these statements probably requires to be modified ; for while there can be no doubt that it is perfectly true of maltodextrin and the higher amyloïns generally, yet recent

[*] Journ. Chem. Soc., 1889, p. 465.
[†] Transactions of the Laboratory Club, 1890, p. 83.

experiments point to the probability of the lower amyloïns, that is, those which are only very little removed from maltose, being to some extent fermentable by the primary *Sacch. cere-visiæ*. We have referred to this in another place (p. 386). The position which the amyloïns occupy in the starch trans-formation products has been already explained (p. 124).

Maltose.

Maltose was discovered by Dubrunfaut in 1847, who first showed that the sugar formed by the saccharification of starch was not, as generally supposed, dextrose, but a new sugar, with distinctly different properties, which he named maltose. The discovery was overlooked, and it was not until O'Sullivan rediscovered the sugar in 1876 that the existence of maltose attracted any attention.

Maltose crystallises from absolute or very concentrated alcohol in the anhydrous state in the form of fine pointed needles; from aqueous solutions it crystallises with one molecule of water, $C_{12}H_{22}O_{11} + H_2O$. It is freely soluble in water, but only slightly so in alcohol. Its solutions have a slightly sweet taste. The optical activity of maltose is $[a]_{j,\cdot\,ss} = 150\cdot4°$ ($[a]_D = 135\cdot4°$), and its cupric-reducing power $\kappa_{s\cdot\,ss} = 61\cdot0$ (Brown and Heron).

The opticity of maltose is somewhat less directly after solution, than it is 24 hours later. When a freshly prepared solution is boiled, the opticity becomes constant at once. This is an instance of the phenomenon of half-rotation.

Boiled with dilute acids, maltose is completely converted into dextrose, thus :—

$$\underset{\text{Maltose.}}{C_{12}H_{22}O_{11}} + H_2O = \underset{\text{Dextrose.}}{2C_6H_{12}O_6}$$

Diastase is without action upon maltose, and, unlike cane-sugar, maltose apparently undergoes no inversion before fermentation, all the evidence pointing to the direct and complete fermentability of this sugar.

Maltose forms salts with sodium, strontium, barium and calcium.

Maltose, as we have already stated, is the end-product of the action of diastase upon malt, and methods for its preparation in the pure state from this source have been described by Soxhlet [*] and Herzfeld. [†]

THE READY-FORMED SUGARS OF BARLEY AND MALT.

We have already referred to the presence of sugars in barley and malt, and to their increase during the malting process. Although Kuhnemann, Kjeldahl, and notably O'Sullivan, have shown the presence of sugars in considerable quantity in malt, we find even now that their presence is neglected, and much work rendered useless by the omission to recognise the influence which they may play in the analysis of malts and worts. O'Sullivan [‡] gives the following numbers as the results obtained from upwards of 20 analyses of malts :—

From 2·8 to 6·0 per cent. of cane-sugar,
  „   1·3 „ 5·0   „   maltose,
  „   1·5 „ 3·0   „   dextrose, and
  „   0·7 „ 1·5   „   levulose.

And, in addition to these sugars, he has identified raffinose among the sugars of barley.

(a.) *Cane-sugar.*

Cane-sugar (see also p. 172) forms large monoclinic prisms, without water of crystallisation. It is easily soluble in water, and has a specific rotatory power of $[a]_{j2 \cdot 88} = 73 \cdot 8°$ ($[a]_D = 66 \cdot 5°$); it is entirely without action upon Fehling's solution. Treated with dilute acids, or invertase, cane-sugar

[*] Journ. f. prakt. Chemie, cxxix. p. 277 ; Zeitschrift f. d. gesammte Brauwesen, 1880, iii. p. 249.
[†] Zeitschrift f. d. gesammte Brauwesen, 1882, v. p. 338.
[‡] Journ. Chem. Soc., 1886, p. 69.

is converted into invert-sugar. It is not directly fermentable by yeast, but is first converted into invert-sugar, which is then fermented.

Cane-sugar forms well-defined compounds with the alkaline earths ($2C_{12}H_{22}O_{11}$, $3CaO$) and ($C_{12}H_{22}O_{11}$, BaO); these compounds are decomposed by carbonic acid yielding unchanged cane-sugar.

When heated to 160° C. (320° F.) cane-sugar melts without decomposition, but when the temperature is raised to 210–230° C. (410–446° F.) it becomes dark in colour and gives off water, so forming sugar-colouring or caramel (see p. 415). The molecular weight of cane-sugar has been shown by Brown and Morris to be 342, the number required for the formula $C_{12}H_{22}O_{11}$.

(b.) *Dextrose.*

In the following chapter we shall refer to dextrose as a brewing sugar (p. 179). Dextrose crystallises from 95 per cent. alcohol as small anhydrous needles; whilst from an aqueous solution it separates in warty masses containing one molecule of water. It is soluble in water and alcohol, and its aqueous solutions are sweeter than those of cane-sugar. The optical activity of dextrose is $[a]_{13.86} = 58\cdot6°$ ($[a]_D = 52\cdot8°$), and its cupric-reducing power, $\kappa = 100$. Heated, it first melts and afterwards darkens and becomes decomposed, forming caramel. The molecular weight of dextrose has been shown by Brown and Morris to be 180, corresponding to the formula $C_6H_{12}O_6$. Dextrose is completely and directly fermentable by yeast.

Dextrose exhibits in a marked degree the phenomenon of bi-rotation, that is to say, its optical activity immediately after solution is very nearly twice what it is after standing. The opticity falls very rapidly at first, but it does not become constant until 24 hours after solution in the cold. The same change is, however, brought about by heating to 100° C. for 15 minutes.

(c.) *Levulose.*

Levulose is a constituent of honey and many fruits, and forms one-half of the product of the inversion of cane-sugar with acid or invertase. Levulose has been recently prepared in the crystalline state, and, after much discussion, its optical activity has been fixed as $[a]_{13.88} = -105.98°$ ($[a]_D = -95.65°$), and its cupric-reducing power as $\kappa_{3.88} = 92.4$. The specific rotatory power of levulose varies very greatly with the temperature, the rotation decreasing with an increase of temperature.

Levulose is completely fermentable by yeast, but more slowly than dextrose. Its molecular weight corresponds to the formula $C_6H_{12}O_6$.

A mixture in equal proportions of levulose and dextrose constitutes invert-sugar; this is obtained by inverting cane-sugar either with invertase or dilute acid. The action consists in the splitting-up of the cane-sugar molecule into two equal, smaller molecules, accompanied by the fixation of a molecule of water, thus :—

$$\underset{\text{Cane sugar.}}{C_{12}H_{22}O_{11}} + H_2O = \underset{\text{Dextrose.}}{C_6H_{12}O_6} + \underset{\text{Levulose.}}{C_6H_{12}O_6}$$

Invert-sugar has an optical activity of $[a]_{13.88} = -23.6°$ ($[a]_D = -28.3°$), and a cupric-reducing power of $\kappa_{3.88} = 96.6$. The dextrose in invert-sugar, freshly prepared by the action of invertase, exhibits the phenomenon of bi-rotation.

(d.) *Raffinose.*

We have stated that O'Sullivan found raffinose in barley. It has a specific rotatory power of $[a]_D = 104.5°$, and is without action on Fehling's solution. Brown and Morris found its molecular weight to be 528.0, corresponding to the formula

$C_{18}H_{32}O_{16}$, $5H_2O$; they also found that raffinose is capable of affording nourishment to the young plant,[*] and therefore, when present in barley, it serves as nutriment for the embryo in the early stages of its growth.

In addition to the foregoing there are certain substances belonging to the carbohydrates, which occur in barley and malt. Chief among these are :—

a- AND β-AMYLAN.

O'Sullivan [†] found these two substances in barley, wheat, and rye. They are present in the former to the extent of about 2 per cent. of a-amylan, and 0·3 per cent. of β-amylan. They may be obtained by extracting barley, first with alcohol, and then with water, evaporating the aqueous solution and precipitating with alcohol ; cold water then dissolves β-amylan from this precipitate, and a-amylan may be obtained from the residue by extracting with dilute hydrochloric acid, and again precipitating with alcohol.

Both have a left-handed rotation ; the following being the values given by O'Sullivan :—

$$a\text{-Amylan} \quad .. \quad .. \quad [a]_j = -24^0.$$
$$\beta\text{-Amylan} \quad .. \quad .. \quad [a]_j = -73^\circ.$$

Both the amylans are converted into dextrose by dilute acids.

NON-NITROGENOUS EXTRACTIVE MATTERS.

Barley and malt have long been known to contain a certain small percentage (2 to 2·5 per cent) of substances which were classed under the general name of non-nitrogenous extractive matters. Quite recently C. J. Lintner[‡] has succeeded in separating from barley, malt, and beers an amorphous white substance, which has all the characteristics of a gum, and which he pro-

---

[*] Journ. Chem. Soc. lvii. (1890), p. 486.
[†] Journ. Chem. Soc., xli. (1882), pp. 24, 32.
[‡] Wochenschrift für Brauerei, 1890, pp. 961-963.

visionally names *barley-gum.* It is apparently a polysaccharide of the group of carbohydrates containing five carbon atoms, and Lintner assigns to it the formula $n(C_5H_{10}O_5)$. It has a left-handed rotation of about $[a]_D = -26 \cdot 8°$, and Lintner suggests that it may be identical with the amylans of O'Sullivan, described above.

FAT.

Fat exists to a greater or less extent in all seeds; in some, for instance rape-seed and linseed, it forms the greater part of the reserve material stored for the nourishment of the young plant; in others, the cereals for instance, it forms but a very small part of the total weight of the seed. In barley, the fat is situated almost entirely in the aleurone-cells, and in the cells of the scutellum, the remaining parts of the embryo containing but little fat. Barley contains on an average about 2–3 per cent. of fat.

Lintner found cholesterin in barley-fat, and Ritthausen [*] identified this substance, together with tripalmitin and palmitic acid, in the fat of wheat and rye.

Stein found that the fat obtained from barley had the same smell which is noticed on the grain when lying in bulk, whilst that from green- and kilned-malt had the same odour as the malt from which it was derived. From this, Stein concludes that the fat of barley undergoes changes during germination and during kilning, which are marked by the alteration of smell.

Barley-fat is obtained from ground-barley by extraction with ether, and subsequent evaporation of the ether. It is a thin, yellow liquid of pleasant odour; on standing exposed to the air, it assumes an offensive smell.

More recently Stellwaag [†] has investigated the fat of barley. He found that the product extracted by ether,

* 'Die Eiweisskörper der Getreidearten,' &c.
† Zeitschrift f. d. gesammte Brauwesen, 1886, p. 175.

separates on long standing into about equal quantities of a solid crystalline fat and a fluid oil. The composition of barley-fat he gives as follows :—

Free fatty acids	=	13·62	per cent.
Neutral fats	=	77·78	,,
Lecithin	=	4·24	,,
Cholesterin	=	6·08	,,

Beckmann has obtained a fat, crystallising from ether in plates, by distilling grains with dilute sulphuric acid. He terms it hordeinic acid.

(II.) NITROGENOUS ORGANIC SUBSTANCES.

The nitrogenous organic substances in barley and malt have been the subject of much speculation in the past, and much has been written of their beneficial or injurious influence upon worts and beers. There is, however, no one subject, perhaps, on which we have so little accurate knowledge, as on the desirability or non-desirability of the various nitrogenous constituents of wort. Many methods have been proposed for the separation and estimation of the different classes of compounds in which nitrogen occurs, but all of them are more or less defective, and give unreliable results.

For the present we may divide the nitrogenous organic substances of barley, malt, wort, beer, and yeast into the following classes :—

> (*a.*) Albuminoids.
> (*b.*) Peptones.
> (*c.*) Amide and amido-acids.
> (*d.*) Soluble-ferments, or enzymes.

(a.) *Albuminoids.*

The term albuminoids is a very elastic one, covering a variety of substances, which are non-crystalline, without taste and smell, and generally inactive towards chemical agents. There are a number of bodies of this nature, derived from animal and vegetable sources, but we have only space to discuss those substances belonging to this group, which occur in barley and malt.*

Our knowledge of the albuminoids, which are also known as proteids or protein-substances, is very vague and limited, and we are almost entirely in the dark as to their true nature. This is to a large extent accounted for by the great difficulty attending their investigation. Certain substances have been isolated from barley, malt, and beer, and it appears certain, as we have mentioned in another place (p. 71), that changes take place during malting and mashing, which are analogous to the changes taking place in the degradation of starch. What these changes are precisely, we do not know. The total amount of nitrogen in a malt, or in a wort, gives us but little information. At one time it was customary to condemn a barley merely on the ground of an excess over an arbitrary standard, and to give the preference to that malt which contained least nitrogen. This, however, is an idea which is becoming exploded, and it must be admitted that the percentage of nitrogen is no guide to the qualities of a barley, malt, or wort. The nitrogenous substances in a malt, calculated as albuminoids, range from 7 to 14 per cent., and the soluble nitrogenous substances, calculated on the same basis, from 3 to 8 per cent.†

The true albuminoids of barley are partly soluble and partly insoluble in water, partly coagulable and partly non-coagulable. According to Ritthausen, barley contains,—

Glutencasein ⎫
Glutenfibrin ⎭ insoluble in water, but soluble in alcohol.

---

* For a more complete description of these substances, we would refer our readers to Ritthausen, 'Die Eiweisskörper der Getreidearten,' Bonn, 1872 ; and Sachsse, 'Die Chemie und Physiologie der Farbstoffe, Kohlehydrate und Proteinsubstanzen,' Leipzig, 1889.

† The albuminoids in grain or wort are determined by multiplying the amount of nitrogen by the factor 6·25. This factor is founded on the assumption that all albuminoids contain 16 per cent. of nitrogen.

Mucedin, slightly soluble in cold water, more soluble on boiling ; consequently solutions prepared with hot water become turbid on cooling.

Albumin, readily soluble in cold water, and thrown out on heating the bright solution as a flocculent precipitate, which is insoluble in dilute acids and alkalis.

The proportions in which these albuminoids occur in barley are naturally very varied, but at present we know of no reliable method for separating and determining these substances.

Brown and Heron have shown that the diastatic power is largely a function of the coagulable albuminoids ; they state that every stage in the coagulation of malt- (or barley-) extract by heat is attended by a distinct modification of its diastatic power, and conversely every modification of starch-transforming power is attended by distinct coagulation.

Recently Reychler has shown that certain of the albuminoids of grain, notably gluten or mucedin, can be converted by the action of dilute acid into a substance possessing many of the properties of diastase. This so-called "artificial diastase" has been since examined by Lintner and Eckhardt, who confirm, to a large extent, Reychler's experiments. We shall have occasion to refer to this more in detail in speaking of diastase (p. 144.)

(b.) *Peptones.*

When the substances belonging to the preceding class are acted upon by a special ferment, peptase or pepsin (see p. 160) they undergo a marked alteration, which is apparently closely analogous to the degradation of starch under the influence of diastase. The exact nature of the change which takes place is unknown, but we find that the non-diffusible, insoluble albuminoids become converted into soluble and easily diffusible substances, which are known as *peptones* and *para-peptones.*

The peptones are readily precipitated by alcohol, tannin,

&c., but not by metallic salts or acids. They are also not precipitated by boiling.

Griessmeyer isolated peptone and parapeptone from wort and beer,[*] and he considered these bodies to be distinct from the similar substances isolated from animal sources; more recently, however, Szymanski has investigated the peptones of malt and beer, and he proves their identity with the animal peptones.[†]

We have already referred to the alterations which take place in the nitrogenous constituents of barley during germination (p. 71); a somewhat similar process is supposed to take place during mashing, the soluble-ferment of the malt degrading and peptonising the albuminoids of the grain. We have, however, no means at present of determining the conditions under which this change proceeds; or the extent to which it is desirable, or the reverse, that the degradation of the albuminoids should take place.

There is reason to believe that the peptones are a very favourable medium for the growth of bacteria and yeast.

(c.) *Amides and Amido-acids.*

The ultimate products of the action of peptase upon the albuminoids are substances belonging to the classes of chemical compounds, known as amides and amido-acids.

The former may be regarded as derivatives of ammonia (NH_3), in which one or more of the hydrogen atoms are replaced by organic radicles. They are crystalline substances, soluble in water, alcohol, and ether, and highly diffusible. The best known members of this class are asparagin and glutamin, which, together with others of the class, are found in malt. Recent experiments by A. J. Brown point to these substances possessing a low nutritive value for yeast, vastly

* Der Bayerische Bierbrauer, 1877, p. 121 ; Zeitschrift f. d. gesammte Brauwesen, 1878, p. 137.

† Zeitschrift f. d. gesammte Brauwesen, 1886, p. 105.

inferior to that of yeast-water. The amido-acids are formed by the action of boiling hydrochloric acid upon the amides, ammonia being separated and combining with the hydrochloric acid ; thus aspartic acid is formed from asparagin according to the following equation :—

$$C_4H_5(NH_2) O_2 (OHNH_2) + H_2O = C_4H_5 (NH_2) O_2 (OH)_2 + NH_3.$$
Asparagin.   Aspartic acid.

The amido-acids are crystalline and resemble the amides in their general properties. The best known of this class of substances are *leucin* and *tyrosin*, both of which are found among the products of the decomposition of the albuminoids.

We have already mentioned (p. 72) that several methods have been proposed for the separation and estimation of these different classes of nitrogenous substances in barley, malt, wort, and beer. The method in most general use is that of Ullik, but we are bound to state that it is unreliable, and uncertain in its results. We are at present quite in the dark regarding the influence which the different classes of nitrogenous substances exert in brewing operations, and we shall remain so until the different bodies have been isolated in a state of purity, and their properties examined, and methods for their determination devised.

(*d.*) *The Soluble-Ferments or Enzymes.*

Very great confusion has arisen in the past from the indiscriminate way in which the word " ferment " has been applied both to organisms which produce changes as the result of their vital activity, and thus give rise to true fermentation, and also to chemical agents which, although the product of living cells, yet act outside and independent of these cells, and simply induce the hydrolysis of the substances upon which they act. It is with the latter class that we have to do in this chapter. They are often referred to as unorganised or soluble-ferments, both of which names serve to differentiate them from the members of the former class. Recently, however, it has been suggested that it is unadvisable to employ the word "ferment" in any way in connection with agents which are incapable of producing true fermentation, and, therefore, the term "enzyme" has been suggested in place of soluble-ferment. This term is a very appropriate one, since it serves to indicate the vital origin of the agent, and thus differentiate it from agents like the acids, which bring about similar changes.

Very little is known of the mode of action of the soluble-ferments or enzymes; we know that in all cases in which a chemical change is brought about by an enzyme, hydrolysis has taken place, that is to say, the molecule of the original substance has been broken down into other substances accompanied by the fixation of water, and the latter substances always bear a simple relation to the original substance. We know also that a very minute quantity of the enzyme is capable of effecting this change in a relatively large amount of the substance, in fact it appears that the enzyme is capable, under favourable conditions, of promoting an indefinite amount of change without undergoing any deterioration itself. Many theories have been enunciated to explain the action of enzymes. Brown and Heron regarded it as a function of the coagulable albuminoids capable of being exerted under favourable conditions; Liebig, and later Nägeli, considered it to be due to molecular vibrations set up under certain conditions (see p. 297).

All enzymes have certain limits of temperature outside which they are incapable of exerting their influence; at higher temperatures they are destroyed, and at lower temperatures their action is, in many cases, very different. Heated in the dry state, however, enzymes can withstand a temperature of 100° C. (212° F.) and upwards, although their activity is slowly destroyed if they are kept for any length of time at these temperatures. Alcohol precipitates all enzymes from their solutions, and some are decomposed by standing in contact with alcohol, whilst certain substances influence their activity in a marked manner. A very large number of enzymes exist in the vegetable and animal kingdoms, and are concerned in the various processes of constructive and destructive metabolism, but we shall only˙ refer here to those which are of importance to our industry.

1. DIASTASE.—We have already mentioned the fact that early in the present century, Payen and Persoz recognised the presence of a substance in malt-extract, which had the power of converting starch into dextrin and maltose. They gave the name *diastase* to this substance, and attempted to prepare it in a state of purity, but without any great success. Later, Dubrunfaut described the preparation of an active substance from malt, and gave to it the name " maltin." We now know that a starch-transforming agent is very widely distributed in the vegetable kingdom, and occurs in all cereals, whether malted or unmalted. We shall, however, first describe diastase as it exists in barley-malt, reserving the consideration of its wider distribution until later.

The earlier methods for the preparation of diastase in a pure state were extremely unsatisfactory, the product being, as a rule, of inferior activity, owing to the method of preparation having involved the use of elevated temperatures or of unsuitable reagents. O'Sullivan[*] described the first satisfactory method, which is as follows : Finely ground pale barley-malt is taken, and sufficient water added to completely saturate and slightly cover it. After standing three or four hours, as much as possible of the extract is separated by a filter-press, and the liquid filtered until bright. To the bright solution alcohol of sp. gr. 0·83 is added, so long as a flocculent precipitate forms, the addition of the alcohol being discontinued as soon as the supernatant liquid becomes opalescent or milky. The precipitate is washed with alcohol of 0·86–0·88 sp. gr., then dehydrated with absolute alcohol, pressed between cloth to free it as much as possible from that liquid, and finally dried *in vacuo* over sulphuric acid. Prepared in this way, diastase

* Journal of the Chemical Society, xlv. (1884), p. 2.

is a white friable powder, easily soluble in water, and retaining
its activity for a considerable time. Diastase is precipitated
by tannic acid, with which it, like the albuminoids, forms an
insoluble compound.

C. J. Lintner[*] and Szilágyi[†] have also published methods
for the preparation of diastase, which do not differ from the
above except in the fact that the malt is extracted with
alcohol of 20 to 30 per cent. respectively, instead of with
water. This procedure yields the diastase in a purer state,
since the alcohol does not dissolve so much of the matter,
which is soluble in water, and afterwards precipitated by
alcohol with the diastase. Prepared by either method, the
diastase is somewhat impure, albuminous and dextrinous
substances and salts being precipitated by the alcohol. In
order to remove these, it is necessary to re-dissolve the crude
product in water, and again precipitate with alcohol. This
process should be repeated several times, and the aqueous
solution of the purified product finally submitted to dialysis
in order to further remove the inorganic salts. During purification the percentage of ash decreases, while the percentage
of nitrogen and the diastatic power both increase. This is
shown by the following numbers given by Lintner :—

	Percentage of Nitrogen.	Diastatic Power.[‡]
Crude diastase	8·3	96
After two precipitations	9·06	100
After dialysis	9·9	100

	Percentage of Nitrogen.	Percentage of Ash.
After one precipitation	6·84	13·25
After two precipitations	8·12	9·94
After three precipitations	9·49	4·9

[*] For C. J. Lintner's valuable contributions on the nature of diastase, see
Journal für praktische Chemie, xxxiv. p. 378; xxxvi. p. 481; xli. p. 91; and
Brewing Trade Review, 1887, p. 204; 1888, p. 209; 1889, p. 376.
[†] *Wochenschrift für Brauerei*, 1891, p. 366.
[‡] For an explanation of the mode of determining diastatic power, see p. 452.

The composition of diastase has been very variously represented by different observers, the results given depending greatly upon the degree of purity of the preparation. We give below the numbers obtained by four different workers, but only the last two can be regarded as analyses of fairly pure preparations, and it will be noticed that these two independent analyses agree very closely.

	Krauch.	Zulkowsky.	Lintner.	Szilágyi.
Carbon	45·68	47·57	44·33	44·50
Hydrogen	6·90	6·49	6·98	7·08
Nitrogen	4·57	5·14	9·92	9·49
Sulphur	36·77	37·64	1·07	1·08
Oxygen			32·91	32·95
Ash	6·08	3·16	4·79	4·90

Calculated on the ash-free substance, the results of Lintner and Szilágyi are :—

	Lintner.	Szilágyi.
Carbon	46·66	46·80
Hydrogen	7·35	7·44
Nitrogen	10·42	9·98
Sulphur	1·12	1·14

Compared with the numbers obtained for the albuminoids of barley, diastase is considerably richer in nitrogen and poorer in oxygen than these substances. The ash of diastase consists, according to Zulkowsky, of the phosphates of potassium, calcium, and magnesium ; according to Lintner, entirely of neutral calcium phosphate. Carefully purified diastase gives all the reactions for albuminoids, but does not give the characteristic biuret reaction for peptones. It gives, however, a blue coloration with tincture of guaiacum and hydrogen peroxide, which, according to Lintner, is given by no other proteid substance, and appears to be characteristic of diastase. When pure, diastase does not reduce Fehling's solution.

The distinguishing feature of diastase is, of course, its power of converting starch, under suitable conditions, into a mixture of dextrin and maltose. We have already referred at length to the products of this action and the conditions under which they are produced. We must now examine the nature of the agent which produces the change.

The action of diastase, or, what is practically the same thing, cold-water malt-extract, is greatly modified by the conditions under which it acts, namely, the quantity employed, the temperature, the concentration, and the presence of foreign substances.

When different quantities of diastase solution are allowed to act upon the same amount of starch for the same length of time under certain specific conditions, it is found that the production of sugar is proportional to the diastase employed. Brown and Heron first pointed this out, and proposed to ascertain the diastastic powers of two solutions by plotting a time-curve for each reaction and measuring the amount of degradation occurring in each case in a given time. About the same time Kjeldahl[*] showed that, in order to make this comparison strictly accurate, it was necessary that the amount of diastase employed should not be more than would suffice to produce a certain quantity of sugar. On this he founded what he terms the "law of proportionality," which may be briefly expressed as follows :—When different volumes of a solution of diastase are allowed to act upon solutions of soluble-starch under identical conditions of time and temperature, then the amount of sugar formed, as shown by the cupric-reducing power of the solutions, is strictly proportional to the amount of diastase employed, *provided the cupric-reducing power is not allowed to exceed* κ 25–30.

[*] Résumé du compte-rendu des travaux du Laboratoire de Carlsberg, 1879, p. 109.

Kjeldahl employed for his determination, a starch solution prepared by acting upon starch-paste with a little malt-extract, and arresting the action by boiling as soon as limpidity was produced.   C. J. Lintner improved upon this, by employing a solution of soluble-starch prepared by the action of dilute acid upon starch-granules (see p. 126), and by using a volumetric method for determining the reducing power. With these modifications, he founded the method for the determination of diastatic power which bears his name, and which is fully described in the analytical section (Chapter X., p. 452).

As we have before stated, the action of diastase upon starch-paste is greatly influenced by the temperature at which the reaction proceeds. According to Kjeldahl the action rapidly increases with the temperature until it reaches 54° C. (129° F.), from 54° C. to 63° C. (145° F.) it remains fairly constant, and above 63° C. it rapidly decreases as the temperature rises. Lintner places the maximum temperature at from 50–55° C. (122–131° F.); his experiments were made with soluble-starch, which probably accounts to some extent for the difference. We shall return to this in considering the homogeneity of diastase.

The action of heat on diastase itself in solution is well-shown in the following results obtained by Lintner, who heated similar quantities of diastase dissolved in water at 55° C. (131° F.) for varying times, and then determined the quantity of each solution required to effect the same amount of change in solutions of starch :—

Of the unheated solution	..	..	0·55 c.c. of solution was required.
After 20 minutes at 55° C..	..	..	1·10 ,,	,,	,,
,, 40 ,, ,,	..	..	1·75 ,,	,,	,,
,, 60 ,, ,,	..	..	2·22 ,,	,,	,,

Thus, after 60 minutes at 55° C. (131° F.), four times as much diastase solution was required to bring about a result equal to that effected by the unheated solution.

Similar experiments performed in starch solutions instead of water, showed that much less influence was exerted by heat in the presence of starch and the products of its transformation than in water alone. In the former case the activity was reduced by only about one-half the extent to which it was reduced in the latter case.

The influence of time upon the reaction depends entirely upon the quantity of diastase employed and the temperature. With a sufficiency of diastase and at temperatures below 55° C. (131° F.), the maximum action takes place very rapidly, the higher the temperature the quicker the maximum is attained. With a small quantity of diastase and a temperature above 55° C. (131° F.) the action proceeds more slowly, as will be seen from the previous section (p. 93). Concentration, within very wide limits, appears to exercise no influence on diastatic action, equal quantities of diastase, according to Kjeldahl, acting for the same time and at the same temperature, producing equal quantities of sugar in solutions of very different concentrations. This applies only to starch-paste conversions in the laboratory, where dilution is always greater than it is in practice. In practical operations, concentration does play a part, if only an indirect one (see p. 224). Foreign substances have an important influence on the action of diastase. Payen and others considered that the maltose produced in the reaction, itself exercised a retarding action on diastatic activity. This, however, was shown by Kjeldahl to be incorrect, and we now know that, under favourable conditions of temperature, diastase rapidly degrades a starch solution to its lowest limits, in spite of the presence of excess of sugar.[*]

Kjeldahl states that very small quantities of sulphuric,

* Lindet (Compt. rend, 108, p. 453), has recently revived this old idea of Payen, and endeavoured to prove it by showing that when the maltose present in a mash is removed by phenylhydrazin, a further quantity is produced by the diastase present. The author's conclusions are, however, entirely incorrect, for he appears to ignore the fact that in the strongly acid liquid which results from the action of the phenylhydrazin hydrochloride and sodium acetate, degradation would take place under the action of the acid at the temperature he employs. Further, since he experimented with a wort which in all probability contained amyloins, the presence of them would serve to explain his anomalous results.

hydrochloric, and organic acids accelerate the action of diastase; larger quantities, however, retard or destroy it. Lintner
disputes this, and although allowing that a very minute quantity of acid very slightly favours the action, yet, having regard
to the very slightly increased activity induced by small amounts
of acid (0·002 per cent. of sulphuric acid), and to the retarding action of 0·01 per cent. of acid, and to the totally destructive action of 0·10 per cent., he considers that acids must be
regarded as retarding or destroying diastatic action. The
same must be said of alkalies, 0·001 per cent. of ammonia
distinctly retards diastatic action, 0·005 per cent. almost stops
the action, and 0·2 per cent. entirely stops the action. Copper
sulphate and the salts of the heavy metals also limit or altogether stop the action of diastase. Lintner also finds that
sodium and potassium chlorides are, in dilute solutions, without
influence on the diastatic power: concentrations of from
4 to 8 per cent. of these salts increase the action. Calcium
chloride is also without influence in dilute solution. The last
result, however, applies only to pure chloride of calcium,
which is neutral; the calcium chloride used by brewers is
always slightly alkaline, and the alkalinity would probably
reduce the action of the diastase. Lintner's results on these
points do not agree with the experience of one of us, and the
explanation of the differences is, perhaps, to be found in the
fact that Lintner worked with prepared diastase and pure
soluble-starch, whereas our experiments were made with malt
itself. These experiments were referred to in a former
chapter. Diastatic action is greatly increased, according to
Müller, by pressure and by carbonic anhydride; these facts are
of practical importance, for the great diastatic activity of a
comparatively small amount of malt-flour added to fermentations, and the degrading influence of secondary yeasts upon
the malto-dextrins, would both possibly depend upon the
carbonic anhydride and the pressure which play a prominent
part in both stages, but especially in cask.

In an earlier section of this chapter, we have said that diastatic activity consists of two kinds, namely, liquefying and saccharifying, and that there is every reason to believe two distinct enzymes give rise to these respective actions. Diastase from barley-malt exhibits both forms of activity, but the two actions are influenced by temperature to a very different extent—thus, the saccharifying action reaches, as we have seen, its maximum at 50–55° C. (122–131° F.), above which it rapidly decreases, and is totally destroyed at 80° C. (176° F.); whilst the liquefying action is at its maximum at 70° C. (158° F.) and is not entirely destroyed below 93° C. (200° F.).

It is only within comparatively recent times that ungerminated barley has been found to contain a starch-transforming agent. This is undoubtedly due to the fact that in all of the earlier work the experiments were made with starch-paste instead of with soluble-starch, as is now usually the case, for the diastase of barley is entirely without action upon the former, whilst it acts vigorously upon the latter. The explanation of this is that the diastase of barley is entirely without liquefying action, although it possesses saccharifying power in a marked degree. This was first noticed by Kjeldahl,[*] and more recently Bungener and Fries[†] pointed out that in some cases the diastatic power of barley was greater before germination than that of the malt prepared from it.

It was not at first supposed that the enzymes from the two sources differed, except in regard to liquefying action on starch-paste, but recent work of Lintner and Eckhardt[‡] has shown a radical difference in the nature of the two diastases. They caused equal quantities of extracts of barley and of malt, having equal hydrolytic powers, to act upon soluble-starch at varying temperatures for the same length of time, and then determined the reducing power in each case. The results of the actions are expressed in the following curves, of

[*] Loc. cit.   [†] Zeitschrift f. d. gesammte Brauwesen, 1886, p. 261.
[‡] Journal f. praktische Chemie, xli. (1889), p. 91.

which the abscissæ represent the temperature, and the ordinates the reducing power.

Fig. 4.

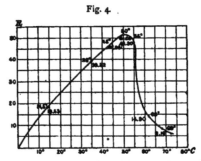

Curve representing the action of Malt-Diastase on Soluble-Starch.
R = reducing power.

Fig. 5.

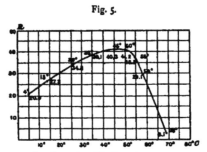

Curve representing the action of Barley-Diastase on Soluble-Starch.
R = reducing power.

As will be seen, the forms of the two curves are very different, for, whilst the optimum temperature for a conversion with malt-extract lies between 50–55° C. (122–131° F.), that for barley-diastase is between 45–50° C. (113–122° F.); and, moreover, the hydrolytic power of barley-diastase at 4° C. (39° F.) is as high as that of malt-diastase at 14·5° C. (58° F.).[*]

[*] Lintner and Eckhardt considered that the so-called "artificial diastase" of Reychler (p. 155), produced by the action of dilute acid upon the gluten of wheat, might be simply the natural diastase of the ungerminated grain; they therefore

These results unmistakably point to the non-identity of the enzymes from the two sources, and this has been further confirmed by the work of Brown and Morris, before referred to.

These workers found that the excised embryos of immature barley were quite incapable of secreting any enzyme which would attack starch-granules, and that this was the case even with embryos from morphologically mature barley, provided that the ripening had not advanced to the point of partial desiccation.* It is clear, then, that the presence of diastase in raw grain cannot be due to a secretion by the epithelium of the same kind as that observed in germination (p. 69). It is, however, closely connected with the growth of the embryo, and is also largely confined to the germ end of the grain, as the following determinations show :—

(1) Endosperm about one-half developed.
(2)   „   two-thirds developed.
(3)   „   fully developed, but grain not ripened.

Diastase estimated in 50 Corns.

(1) Equivalent to .. .. 4·390 grams CuO.†
(2)   „   „   .. .. 7·833 „   „
(3)   „   „   .. .. 9·675 „   „

In 50 corns of ripe barley, after steeping in water :—

Diastase in half-endosperms (germ-end).. .. .. 1·715 grams CuO.
    „   „   (remote from germ) .. 0.610 „   „   .
            Total .. .. 2·325 „   „

examined it closely and found that although the substance acted upon did possess diastatic power, yet this was greatly increased by the action of dilute acids, yielding a substance which, when examined under the above conditions, gave a curve closely agreeing with that of barley-extract. They conclude that the gluten or mucedin of wheat contains a *fermentogen* or *zymogen* (see p. 63), which is acted upon by the acid, but they consider that Reychler's contention, that the action of acid may be compared with the physiological action which takes place during germination, is quite incorrect.

* We have previously referred to the practical importance of this observation (see p. 79).

† See footnote, p. 70.

We have already referred to the secretion of diastase by the epithelium during germination, and to its accumulation in the endosperm. The following experiment shows the distribution of diastase in the germinated grain, and the increase brought about by germination :—

Distribution of Diastase in Barley germinated for Seven Days under the ordinary conditions of the Malting Process.

			Grams CuO.
Diastase in 50 half-endosperms (germ-end)			9·7970
,, ,, ,, (remote from germ)..			3·5310
,, radicles of 50 corns			0·0681
,, plumules ,, ...			0·0456
,, scutella ,, ...			0·5469
Sum of diastase in separate estimations			13·9886
Diastase determined in 50 *whole* corns of the same barley ..			13·7260

The accumulation of diastase in the endosperm is well shown in this experiment, as is also the distribution of diastase in the organs of the embryo. The scutellum contains by far the largest proportion of the total diastase of the embryo, and it was also found that the enzyme of the scutellum is capable of liquefying starch-paste with great rapidity, whilst that of the plumules and radicles is incapable of acting upon starch-paste, thus resembling the diastase of raw grain. Another point in common is the inability of the enzymes from the latter sources to erode and disintegrate starch-granules. Brown and Morris found that the diastase of raw grain is entirely without this power, and they state that in all cases it was found that the inability to erode starch-granules and to liquefy starch-paste went hand in hand.

The granules of transitory starch, to which we referred in an earlier section, always disappeared by solution, as distinct from erosion, and Brown and Morris attribute this solution to the action of an enzyme which they term "diastase of translocation," or "translocation diastase"; this translocation diastase they suppose to be produced within the cell for the purpose of facilitating the translocation of the starch. This is the diastase of raw grain and of certain tissues of plants, and is distinguished by its inability to erode starch-granules or to liquefy starch-paste, and by the difference in its action on soluble-starch, as shown above. It is also probably identical with Reychler's "artificial diastase." The diastase secreted by the epithelium is termed by Brown and Morris "diastase of secretion," and is regarded by them as the direct glandular secretion of the columnar epithelium of the embryo. The "diastase of secretion" possesses all the properties we usually attribute to malt-diastase; it erodes, disintegrates, and dissolves starch-granules, producing maltose, and it liquefies and saccharifies starch-paste.[*]

The other cereals and grasses behave in much the same way as barley, and what we have said regarding barley and malt-diastase may be taken as applicable to the diastases contained in the ungerminated and germinated cereals generally.

II. INVERTASE.—The name invertase was given by Donath to the enzyme prepared from beer-yeast, and which had the power of converting cane-sugar into the mixture of dextrose and levulose, known as invert-sugar. The action, like that brought about by diastase, is one of hydrolysis, and is represented by the following equation :—

$$C_{12}H_{22}O_{11} + H_2O = C_6H_{12}O_6 + C_6H_{12}O_6.$$
$$\text{Cane-sugar.} \qquad \text{Dextrose.} \quad \text{Levulose.}$$

Invertase is present in malt to a considerable extent; this was first shown by Brown and Heron[†]; later Kjeldahl pointed out that it was chiefly located in the rootlets of germinated barley, and more recently J. O'Sullivan[‡] has shown it to be

---

[*] Quite recently Wijsmann (Rec. Trav. Chim., ix., p. 1) has endeavoured to prove the compound nature of malt-diastase. He starts with the assumption that diastase is composed of a mixture of two enzymes—*maltase* and *dextrinase;* and that the former converts starch into a mixture of maltose and eythrodextrin, whilst the latter converts starch into maltodextrin. The last named is converted by *maltase* into maltose, and the eythrodextrin is further degraded by *dextrinase* into leucodextrin. The experiments he quotes in support of this theory are, however, inconclusive, and allow of an explanation in accordance with the generally received theories.

[†] Journal of the Chemical Society, 1879, p. 610.

[‡] Transactions Laboratory Club, iii. (1890), p. 104.

confined to the growing embryo. This is where we should naturally expect to find it, for we have seen that the assimilable carbohydrate enters the embryo as maltose, and is there transformed into cane-sugar and the products of its inversion.

From the brewing standpoint the chief source of invertase is yeast. In this it is present, according to C. O'Sullivan and Tompson,[*] to the extent of from 2 to 6 per cent., calculated on the dry solid matter.

C. O'Sullivan and Tompson have recently made a very exhaustive examination of invertase and its properties. The chief results may be summed up as follows :—

The method adopted in determining these properties of invertase consisted in mixing a solution of cane-sugar with a measured amount of invertase, and allowing the action to take place at a known temperature during a definite time ; action was then stopped by the addition of alkali, and its extent determined.

The rate of hydrolysis of cane-sugar by means of invertase may always be expressed by a definite time curve ; and whatever the conditions may be under which hydrolysis is taking place, as long as these conditions remain unchanged, this curve is adhered to.

When the degree of acidity is that most favourable for the action of invertase, the rapidity of the action is in proportion to the amount of invertase present. The most favourable concentration of the sugar solution at a temperature of 54° C. (129° F.) is about 20 per cent. Below that there is a rapid decline in the speed of hydrolysis. Greater concentrations are only slightly less favourable until about 40 grams of sugar per 100 cc. is reached. In saturated solutions hydrolysis only proceeds with extreme slowness.

The speed of hydrolysis increases rapidly with the temperature until 55–60° C. (131–140° F.) is reached. At 65° C. (149° F.) the invertase is slowly destroyed, and at 75° C.

[*] Journal of the Chemical Society, lvii. (1890), pp. 834–931.

(167° F.) it is immediately destroyed. At the lower temperatures, the speed of the action increases with the rise of temperature. Elevated temperatures have no permanent effect on the activity of the invertase so long as they are not sufficiently high to destroy it.

The caustic alkalis, even in very small proportions, are instantly destructive to invertase.

Minute quantities of sulphuric acid are exceedingly favourable to the action, but a slight increase of acidity beyond the most favourable point is very detrimental. The most favourable amount of acid increases to some extent with the proportion of invertase, and increases with rise of temperature. The influence of alcohol varies in direct proportion with the amount present. Five per cent. of alcohol decreases the speed of the action about one-half.

The dextrose formed by the action of invertase is initially in a bi-rotary state, and consequently the optical activity of a solution undergoing hydrolysis is no guide to the amount of hydrolysis that has taken place.

If a caustic alkali be added to a solution undergoing hydrolysis, and the optical activity be allowed sufficient time to become constant, it is a true indicator of the amount of inversion that had taken place at the moment of adding the alkali.

A sample of invertase which had induced hydrolysis of 100,000 times its own weight of cane-sugar was still active.

Invertase itself is not injured or destroyed during its action on cane-sugar; and there is no limit to the amount of sugar which can be hydrolysed with the aid of invertase.

The hydrolysis of cane-sugar by means of invertase is a simple chemical change differing in no important particular from those which inorganic substances undergo. The products of hydrolysis have no influence on the rate of the action.

A solution of invertase will withstand a temperature of

25° C. higher in the presence of cane-sugar than in its absence. O'Sullivan and Tompson are of opinion that, when invertase inverts cane-sugar, combination takes place between the two substances, and that the invertase remains in combination with the invert-sugar.   This combination breaks up in the presence of excess of cane-sugar.

Invertase, when it approaches a state of purity, is a very unstable substance.*   The products of its decomposition have been carefully examined by O'Sullivan and Tompson, and are found to constitute a new series of substances constituting the *invertan* series.   This is a series of substances which on analysis yield numbers that may be expressed in terms of an albuminoid and a carbohydrate.   Seven members of the series are described in which the nitrogen gradually decreases from 8·3 per cent. in a-invertan to 0·76 per cent. in η-invertan, the lowest member of the series, whilst the optical activity increases.   Invertase itself is a member of the series, occupying the position of β-invertase.

The following table shows their relation : A represents 1 part by weight of albuminoid ; and S, 1 part by weight of carbohydrate ; a represents a-*invertan*, that is, A_4S_3 ; η repre-

---

* The following is the method of preparation of invertase given by O'Sullivan and Tompson :—If sound brewers' yeast be pressed and then kept at the ordinary temperature for a month or two, it does not undergo putrefaction, but changes into a heavy, yellow liquid ; the product possesses no power of fermentation, but an apparent increase takes place in the invertive power.   From such liquefied yeast it is easy to filter off a bright solution of high hydrolytic power.   It is shown that all the invertase of yeast is in this solution, which O'Sullivan and Tompson term "yeast liquor" ; it has a density of about 1080, and will remain for a long time unaltered, excepting that the colour darkens.   If exposed to the air it may slowly become covered with mould.   If spirit be added to yeast-liquor until the mixture contains 47 per cent. of alcohol, the whole of the invertase separates with only a slight loss of power.   The precipitated invertase may be washed with spirit of the same strength, and the residue either dehydrated with strong alcohol, and dried in vacuo, or it may be extracted by means of 10 to 20 per cent. alcohol, and then filtered ; the filtrate then contains the invertase. O'Sullivan and Tompson have not succeeded in further purifying invertase preparations carefully made in this manner.   The slightest attempt at purification destroys the invertase.

sents three times η-*invertan*, that is, $3AS_{18}$; η thus represents 57 parts by weight, and a 7 parts by weight :—

Name of Substance.	Inverting Power. $\pm o =$	Optical Activity. $[a]j =$	Constitution.	Composition. Parts of Weight.
a invertan	∞	$-15°$ (approx.)	a	A_4S_3
β invertan..	$22\cdot5'$ (approx.)	$+80°$ (approx.)	ηa_3	$A_{23}S_{60}$
γ invertan..	∞	$+45°$	ηa_4	$A_{18}S_{64}$
δ invertan..	∞	$+54°$	ηa_5	$A_{16}S_{63}$
ϵ invertan..	∞	$+65°$	ηa_3	$A_{11}S_{60}$
ζ invertan	∞	$+75°$	ηa	A_7S_{37}
η invertan..	∞	$+75°$ (?)	η	A_3S_{44}

Name of Substance.	Composition. According to Theory.			Composition Found.		
	C.	H.	N.	C.	H.	N.
a invertan..	49·41	6·90	8·30	48·06	6·65	8·35
β invertan..	45·93	6·57	3·63	46·41	6·63	3·69
γ invertan..	45·64	6·54	3·25	45·62	6·55	3·15
δ invertan..	45·30	6·51	2·79	46·50	6·82	2·43
ϵ invertan..	44·90	6·47	2·25	44·45	6·36	2·07
ζ invertan..	44·40	6·42	1·59	44·73	6·40	1·61
η invertan..	43·78	6·36	0·76	—	—	1·85

Invertan is present in the majority of the species of *Saccharomyces*, but not in all; thus *Sacch. membranæfaciens* and *apiculatus* are unable to invert cane-sugar. Among the Torula, too, there are species which are unable to invert cane-sugar. Certain of the moulds, notably *Aspergillus niger, Penicillium glaucum*, and *Mucor racemosus*, secrete invertase, and therefore invert cane-sugar when grown in solutions of this substance.

Fernbach [*] has also investigated the nature and properties of invertan, but his results do not markedly differ from those of O'Sullivan and Tompson.

[*] Annales de l'Institut Pasteur, iv. (1890), pp. 1 and 641.

III. PEPTASE.—A peptonising ferment is very widely distributed in the animal kingdom, and occurs in the form of " pepsin " in the pancreatic juice, &c., of all animals.   It has the power of liquefying and degrading albuminoids, but only in an acid medium.   Gorup-Besanez, Green, and others have shown that a similar enzyme exists in various vegetable tissues, and has the power of converting the albuminoids into peptones, leucin, and tyrosin.   We have already mentioned that malt, and malt-extract also, contains an enzyme of this nature, which is capable of converting the separated coagulable albuminoids of malt into crystallisable compounds.   The action takes place more readily in an acid fluid, and at a temperature of about 45–50° C. (113–122° F.).   At present we have no definite knowledge of the peptase or proteo-hydrolytic enzyme of malt, or of the nature of its action, beyond the fact that the albuminoids are degraded under its influence in much the same way as starch is under the influence of diastase.

IV. CYTASE.—This is the name which we have authority for giving to the enzyme that brings about the dissolution of the cellulosic walls of the endosperm of barley.   We have, in a previous section (p. 64), fully referred to the importance of the part it plays in germination and malting.   It, like diastase, is secreted by the epithelium during germination and accumulates in the growing grain; it may be found in considerable quantity in green or air-dried malt, but as it is destroyed at a comparatively low temperature, it is only found to a very slight extent, if at all, in kiln-dried malt.

Brown and Morris clearly proved by the growth of excised embryos upon sections of barley, potato, &c., the secretion of this enzyme by the epithelium of the scutellum, and the dissolution of the walls of the starch-containing cells by its action ; this dissolution always taking place prior to the attack of the diastase upon the starch-granules.

A cold-water extract of air-dried malt, or the diastase prepared from it, possesses the power of dissolving the cellulosic-walls of the starch-containing cells of all cereals and grasses. That this action is not due to the diastase is proved by the fact that previous heating of the solution to a point (60° C.) which does not affect the amylo-hydrolytic action, entirely destroys the cyto-hydrolytic or cellulose-dissolving power of the solution.

The most favourable temperature for the action of cytase is 40-45° C. (104-111° F.); at 50° C. (122° F.) it is markedly retarded, and at 60° C. (140° F.), it is completely destroyed.

The enzyme has not been isolated in a state of purity, or the product of its action determined, but it undoubtedly yields a soluble carbohydrate, which is readily assimilable by the growing embryo.

III. MINERAL SALTS (AND ASH).

We have already referred to the mineral salts of barley and the changes they undergo during germination (p. 75). The mineral salts of barley naturally vary much according to soil, manuring, climate, and weather, but in all cases it is found that potassium and phosphoric acid are the largest constituents. In 57 analyses[*] it was found that barley contained 1·9 to 3·1 per cent. of ash, the mean being 2·61 per cent.; 100 parts of this ash contain the following :—

	Minimum.	Maximum.	Mean.
Potash	11·40	32·20	20·92
Soda	0·00	6·00	2·39
Lime	1·20	5·60	2·64
Magnesia	5·00	12·70	8·33
Oxide of iron	0·00	4·70	1·19
Phosphoric acid	26·00	46·00	35·10
Sulphuric acid	0·00	3·90	1·80
Silica	3·70	36·70	25·91
Chlorine	0·00	5·20	1·02

The mineral salts of the malt partly (1·5 to 1·7 per cent.) pass into the wort and thence into the beer, where their presence

[*] E. Wolff, ' Ash Analysis,' Part II., Berlin, 1880.

is necessary for the nourishment of the yeast. For this purpose the potassium and phosphoric acids are the most important, and here again it is found that these substances are the chief constituents of the ash of beer, Thausing giving the composition of the soluble salts as :—

Potash	41·9 per cent.
Soda	0·03 ,,
Lime	4·5 ,,
Magnesia	2·2 ,,
Phosphoric acid	31·5 ,,
Silica	20·4 ,,

It is clear from the above analyses that the main mineral constituent is some kind of potassium phosphate. Phosphates are necessary for the development of yeast, the yeast itself containing and therefore demanding them. In wort the phosphates, calculated as P_2O_5, range from 0·074 to 0·197 per 100 c.c. of wort. The phosphates naturally extracted from a malt are as a rule sufficient (if not sometimes excessive) for the development of the yeast. The exact manner in which the phosphoric acid and potassium exist is unknown : the bulk would apparently be combined as a potassium phosphate ; whilst it has been stated that part of the phosphorus is loosely combined with some of the nitrogenous constituents of the wort. During boiling in the copper the loose combinations are split up, the liberated phosphoric acid going to form acid phosphates. Thausing ascribes to this the increase of acidity during boiling.

W. Schultze[*] gives the following analyses of 29 barleys examined by him.

	Maximum.	Minimum.	Mean.
Water	17·03	13·43	15·11
Dry substance .. ,. ..	86·57	82·97	84·89

In 100 parts of dry substance :—

Starch value (starch, dextrin, &c.)	67·72	61·97	64·14
Nitrogen	2·094	1·388	1·794

[*] Zeitschrift f. d. gesammte Brauwesen, 1881, p. 90.

	Maximum.	Minimum.	Mean.
Albuminoids (N × 6·25) ..	13·09	8·98	11·21
Potash	0·697	0·435	0·610
Soda	0·247	0·035	0·107
Lime	0·052	0·029	0·041
Magnesia	0·363	0·209	0·229
Oxide of iron	0·072	0·012	0·019.
Phosphoric acid	1·202	0·872	0·995
Silica	0·894	0·540	0·712

And 79 barleys analysed in 1885 in the Munich Laboratory gave the following results :—

Starch	71·00	54·21	63·00
Albuminoids	13·75	8·87	10·79
Phosphoric acid	1·170	0·777	1·003

THE CHEMICAL VALUATION OF MALT.

In valuing malts upon the results of chemical analysis we are strongly of opinion that the main points to which attention should be drawn are, the amount of residual diastase in the malt, and the manner in which the starch is broken down under standard conditions of conversion. So far as the former point is concerned, the amount of residual diastase (or the diastatic capacity of the sample) will depend primarily upon the duration of kilning and the temperature at which it was conducted, the diastase, of course, varying inversely with duration and temperature. The approximate diastatic capacity required will vary according to the use to which the malt is intended to be put. Thus, in the brewing of pale ales, where higher mash-tun heats are customary, we must have a greater amount of diastase than in malts intended for mild running ales. Colour of wort would in itself determine such a choice, since the paleness required for pale ales is, as a rule, only compatible with relatively light curing, and light curing must necessarily leave in the malt a relatively high amount of diastase.

In the case of pale ale malts, we have found a diastatic capacity of 35 to 48 to give the best results. With lower capacities we are unable to use the relatively high mashing heats necessary for the production of the required type of maltodextrin (see p. 232); with higher diastatic capacities there must necessarily be a tendency to poverty of malto-dextrins, and consequent thinness and instability of the resulting beer.

For mild ale malt, fair limits are 18 to 28; lower capacities rendering difficult the attainment of sufficient maltose in the wort; higher capacities necessitating higher mashing heats than are desirable for the production of the lower types of maltodextrin necessary in mild ale worts.

The above capacities are those obtained by Lintner's method, which is fully described in the analytical section.

The diastatic capacities as given above are taken on pure soluble-starch of constant quality, and are therefore entirely independent of the behaviour, in respect to gelatinisation, of the starch contained in the same malt. But, as we know, the starch of a well-grown and that of an ill-grown malt will behave very differently with the same amount of diastase: the former being converted in unison with the amount of diastase, whilst the latter is stubborn, and is liable to yield, in spite of abundant diastase, an excess of maltodextrins, frequently of unduly high type.

Now, if after having determined the diastase, irrespective of the accompanying starch, we cause the diastase to act upon that starch, and then determine the conversion products, we shall be put into possession of very useful knowledge concerning the state of that starch. If, for instance, we find that under standard conditions a given amount of diastase should yield, in the case of a well-grown malt, a given amount of maltodextrin ; then, if we find in another sample the same or a higher amount of diastase, and yet a higher proportion of maltodextrin in the resulting wort made under the same mashing conditions, we are able to say that the starch in the second sample is stubborn, owing to imperfect changes on the maltster's floors; and since such stubborn starch is productive of serious troubles, we cannot lay too great a stress upon the importance of this point.

In the case of well-grown pale ale malts, the starch of which is sensitive to diastase, we find that we get from 9 to 12 per cent. of maltodextrin calculated on wort-solids under the mashing conditions described in the analytical section. If, therefore, we find, with a diastatic capacity of 35 to 48, the maltodextrins appreciably exceeding 12 per cent., we at once get evidence of stubborn starch, and the evidence will be more complete when the diastase exceeds the limits given. In pale ales the normal type of maltodextrin is generally $\begin{cases} 1 & M \\ 1\cdot5 & D \end{cases}$.

In the case of well-grown mild ale malts, we find the maltodextrin yielded in accordance with the diastatic limits given above, amounts normally to from 15 to 21 per cent. on the wort-solids; and again, any excess on the above percentages would indicate the presence of refractory starch. The normal type in this class of malts is generally $\begin{cases} 1 & M \\ 2 & D \end{cases}$.

Sometimes the maltodextrins will come out lower than would be anticipated from the diastatic capacity. This is relatively seldom the case, but when it happens it will be generally found to be due to an excessive amount of "ready-formed sugars" in the malt: such an excess naturally tending to decrease the starch, and therefore the starch-products, maltodextrin among them. The normal percentage of these ready-formed sugars is about 12 to 14 per cent. (calculated on the malt), but we have known them amount to 20 and even 24 per cent. In the latter case the starch would be reduced by about 12 per cent., and the maltodextrins by 2 or 3 per cent. We have found that such abnormally high rates of ready-formed sugars point to a forced growth on floors, that is to say, to a vegetation conducted at excessive temperatures and with excessive sprinkling liquor, especially towards the close of the process.   When malt is loaded too wet, this action seems to be continued on the kilns for some little time. Malts containing this excess of ready-formed sugars give bad results in brewing ; whether their defects are entirely due to the substitution of a portion of the dextrin and maltodextrin by fermentable sugars, or whether they are in part due to an excessive change in the nitrogenous constituents, accompanying an excessive change in the starch, we are not in a position to say.   But experience has shown us that when the ready-formed sugars of a malt exceed 16 per cent., the results in practice are unsatisfactory.

An exception must be made, however, in the case of very highly dried malts which have suffered caramelisation.   Such malts yield large amounts of ready-formed sugars, without a forced growth being necessarily indicated.   For in this case the heat will have affected a portion of the starch, the resulting caramelisation products being soluble in cold water, and hence they would be included in the ready-formed sugars.

When these constituents are abnormally low (say under 10 per cent., calculated on the malt), we get evidence of an insufficient vegetation.   But where this is the case better evidence of it is afforded by the determination of the malto-dextrin in the wort made from the malt under standard conditions.

So far as the albuminoids are concerned, we are strongly of opinion that in the present state of our knowledge no importance can be attached to the percentage of either the soluble or the insoluble portions of them.   Nor do we regard the methods at present at our disposal for separating the nitrogenous constituents into various groups as sufficiently reliable to serve for purposes of valuation or interpretation. We generally find, however, that the amount of diastase, and the manner in which the starch breaks down under it, affords a sufficient indication of the suitability or unsuitability of a malt for brewing purposes, irrespective of reference to the albuminoid question; whether it is that the albuminoids are really of minor importance, or whether it is that their harmlessness or harmfulness is really connoted by satisfactory or unsatisfactory starch conversion, we regard as a matter of only academic interest.

Doubtless due regard must be had to mould, and obvious defects of that sort; for a mouldy malt must be bad, affording nutriment to bacteria introduced at the later stages, whether the diastase and the starch conversions are satisfactory or not.

The percentage of water in a malt will vary with its age. A new malt just off the kiln should contain not more than $1\frac{1}{2}$ per cent. of water, and it will generally contain less; and well stored old malt should not yield more than 3 to 4 per cent. The higher limit here shows incipient slackness, and such malt should be redried. When the water exceeds the above limit it will certainly be slack, and redrying will be imperative.

The acidity of malt does not vary as a rule beyond very narrow limits; about 0·18 per cent., calculated on the malt being an average for the total acid calculated as lactic acid. When the acidity greatly exceeds this, it raises suspicion as to the storage-conditions of the malt, and its original quality, but we have no reason to think that the acid *qua* acid is operative for harm. The acid is usually greater in a slack malt than in a dry one. While any excess of acidity will invariably be accompanied by other results indicating defects, the majority of really defective malts contain no excessive acidity.

To the other constituents, as determined by analysis, we attach no importance; the ash, for instance, is merely useful in calculating the ready-formed sugar, and much or little of it is of no apparent moment.

We attach the greatest weight to the determinations of the diastase, and the analysis of the conversion products, and we believe that no accurate opinion of a malt can be formed without some such analytical test, however great the experi-

ence of the purely practical man who endeavours to value it.
At the same time we are free to confess that we do require
practical experience in addition to analytical data, if we are
to judge malt in the most perfect way. The friability, the
appearance, the freedom from mould, uniformity of vegetation
and colour, &c., are all obviously important points which do
not perhaps demand very much experience; but the flavour
of a malt is, perhaps, of all practical points the most important,
and it does certainly require experience to decide definitely
as to the satisfactory or unsatisfactory nature of a malt in this
respect. To attempt to explain in words what constitutes a
good and what a bad flavour is a quite impossible task, and
one we shall not attempt; but we cannot too strongly urge
brewers and analysts to carefully taste every sample with
which they may have to do, and then collate for themselves
their own record of flavours, and what these flavours mean
in practice.

CHAPTER III.

MALT SUBSTITUTES.

INTRODUCTORY.

THE majority of English brewers displace a certain proportion of their malt by substances which are known as malt substitutes or malt adjuncts. These consist principally of sugars, and prepared and unprepared cereals. Taking the sugars first, those in most general use are, *invert-sugar*, *glucose*, and *cane-sugar*. The object of using these sugars is not, as is usually supposed by the general public, an economical one, since, taken as a class, they are more expensive than their equivalent of malt. Their popularity must rather be ascribed to the improvement in the finished beer with which their employment is credited. That they do, as a rule, effect an improvement is an undoubted fact; but its exact reason is by no means thoroughly understood. The most obvious difference between them and malt lies in the fact that they contain either no nitrogen at all, or only traces of it; and it has been long supposed that the reduction of the percentage of nitrogenous bodies in a wort results in its increased soundness. We are not prepared, however, to consider that this in itself affords a sufficient explanation, if any, of the benefits to which the use of sugar leads. In the first place, the nitrogen percentage in different worts varies within exceedingly wide limits; and, secondly, many all-malt worts, which have given inferior beers, have shown on analysis far less nitrogen than worts partially brewed with sugar, and which gave beers of good character. The theory that the superiority of beer varies indirectly with the nitrogen percentage of the wort is in fact entirely objectionable.

It is usually asserted that a proportion of sugar gives sounder beers than malt alone. We are, however, not prepared to say that such is the case. It is at any rate significant that many brewers who habitually use sugar for their general run of ales, still brew their stock ales from malt only ; and in so doing we consider that they act wisely. The use of ordinary sugar reduces the proportion of maltodextrin, and for stock beers any such reduction is to be deprecated for reasons that will be fully set out in Chapter V. It seems to us, however, that the use of sugar is capable of conferring special benefits in the brewing of mild ales, semi-stock ales, and stouts. In the first place, sugar-brewed beer will clarify more rapidly than all-malt beers. That such is the case is a matter of everyday practical experience, and the reason for it probably lies in the reduction in the viscosity of the beer and of the wort, owing to the substitution of some of the dextrin and maltodextrin by fermentable sugars. This reduction in viscosity means a more ready elimination of yeast in the fermenting vessel, and afterwards in the cask, as well as greater freedom for the operation of the artificial finings when these come to be added.

But the main point about sugars is that they give beers of a distinctive flavour—a fuller, sweeter flavour, and one which happens to be liked at present by the beer-drinking public.

In talking of sugar in this way we refer to them as brewing materials, and not as priming. As priming they have a special and distinctive use, so far as rapid conditioning is concerned, which has nothing directly to do with brilliancy or stability.

It has appeared to us that, to be able to use sugars without reducing stability, they should contain at least such a proportion of maltodextrin as is usually contained in beer wort. In that case they could be used in stock ales, and

probably to the benefit of such ales; since the partial dis-
placement of what is after all a more or less uncontrollable
malt-extract, by a controllable carbohydrate mixture of
fermentable sugars and maltodextrins, could not fail to give
the brewer greater knowledge as to what he was producing.
It is with that object that researches were instigated, which
have resulted in the production (under the patent of Messrs.
H. T. Brown and the authors) of brewing sugars containing
varying proportions of maltodextrins or amyloïns.

In the ordinary way the malt substitutes are added either
to the under-back, the copper, or the hop-back, and they
thus become an integral part of the wort submitted to fermentation. But it is also a very common practice to add sugar
solutions to the finished beer in cask. This addition constitutes priming, the object of which is to artificially promote
a rapid conditioning of the beer. During the main fermentation
practically the whole of the maltose and other fermentable
sugars have been fermented away; the carbohydrate residue
consisting of maltodextrins and dextrin. The former are
slowly hydrolysed under the influence of certain types of
secondary bottom yeasts, which operate during storage.
Maltose is produced, and, as it is produced, it ferments, thus
giving rise to the necessary condition. This is "condition"
naturally produced. But to obtain it we require the secondary types of yeast to be present in sufficient quantity; this is
nearly invariably the case, but we have to allow time for
them to come into operation. When, however, we add
fermentable matter to the beer, we obtain a fermentation
of the added sugar through the agency of the residuary
primary yeast. In that case we make no call upon the
secondary yeasts, either in respect of their degrading influence
upon the maltodextrins, or upon their subsequent fermenting
activity; and, as might be anticipated, we get a very rapid, if
somewhat transient, cask fermentation from the addition
described.

Priming is adopted in some parts of the country to impart greater fulness and sweetness to beers. This, however, is merely demanded by considerations of local palate and local economy, and need not concern us further. For purposes of promoting condition, priming is deservedly popular. Indeed some brewers who use no sugar before fermentation, freely avail themselves of sugar added as priming syrup.

CANE-SUGAR.

The cane-sugar in general use by brewers is derived from the sugar-cane, and from the white beet. Pure cane-sugar, derived from either source, is necessarily of the same constitution, and possesses the same properties. In brewing, cane-sugars are not, as a rule, used in a state of chemical purity, but only in a more or less refined condition. In this state the sugars are associated with substances derived from the raw material, and from which the refining process has not entirely freed them. The foreign substances which accompany partially refined sugars derived from the cane do not seem to possess such a deleterious influence as the substances which similarly accompany beet-sugars in the same stage of partial refinement.

Cane-sugar, as such, is not fermentable, and when cane-sugar is used as a malt substitute, it is converted by the yeast which it meets in the fermenting vessel into another sugar called invert-sugar, which consists of equal molecules of levulose and dextrose : both of these sugars are readily fermentable. The change of cane-sugar into invert-sugar by means of yeast is due to the action of a soluble-ferment secreted by the yeast, and termed *invertase* (p. 155), which acts upon cane-sugar in a manner analogous to the action of diastase upon starch-paste. The inversion of cane-sugar precedes the fermentation of the invert-sugar. It will thus be seen that when cane-sugar is used as a malt substitute, the brewer is really using invert-sugar. It cannot be denied that, except in very special cases, it is better to have the cane-sugar inverted prior to fermentation, either by the manufacturer, or on the brewery premises.

Cane-sugar is, however, popular in certain classes of beers, especially in the production of full and heavy mild ales and stouts. In such cases, only low-grade sugars are employed, and their popularity in this respect is ascribable to the full lusciousness which sugar of this kind gives. The lusciousness is, however, a property of the impurities which accompany the sugars, and not of the sugars themselves.

Unrefined and partially refined cane- and beet-sugars are distinguishable in several ways, flavour being one of them; beet-sugars imparting a salty and somewhat unpleasant taste. In an analytical examination the following points indicate beet-sugar :—

(1) Little or no glucose as compared to cane-sugar.

(2) Low percentage of phosphoric acid in the ash.

(3) High percentage of soda in the ash.

The following are some analyses of cane- and beet-sugars and molasses :—*

	Cane-sugar.	Glucose.	Extractive matters.	Insoluble matters.	Ash.	Water.
West India	94·4	2·2	0·3	0·1	0·2	2·8
,, pieces	87·7	6·0	0·5	—	0·8	5·0
,, bastards	68·3	15·0	1·2	—	1·5	14·0
Beet-sugar	89·15	—	3·96	—	2·63	4·26
,, ...	95·7	0·3	0·4	—	1·6	2·0
Date sugar	95·4	1·8	0·4	1·7	0·2	0·8
,, lump	97·3	0·5	—	—	0·2	2·0
Golden syrup	39·6	33·0	2·8	—	2·5	22·7
Green syrup	62·7	8·0	0·6	—	1·0	27·7
Molasses	48·0	18·0	1·5	—	1·4	31·1
Treacle	32·5	37·2	3·5	—	3·5	23·4
Beetroot molasses	50·9	1·1	16·1	—	12·9	19·0

* Allen, 'Commercial Organic Analysis.'

The following are analyses of the *sulphated* ashes of beet- and cane-sugars :—

	Cane.[*]	Beet.[†]
Potash	28·79	34·19
Soda	0·87	11·12
Lime..	8·83	3·60
Magnesia	2·73	0·16
Ferric oxide and alumina	6·90	0·28
Sulphuric anhydride	43·65	48·85
Sand and silica..	8·29	1·78
	100·06	99·98

The difference in phosphoric acid in the ash of cane-sugar and beet-sugar is shown in the following analyses :—

	Cane.[‡]	Beet.[§]
Phosphoric acid (anhydrous)..	5·59	—
Potash	29·10	25·65
Soda	1·94	21·62
Lime	15·10	6·53
Magnesia	3·76	—
Sulphuric acid (anhydrous)	23·75	17·63
Carbonic acid	4·06	22·87
Chlorine	4·15	4·48
Silica	12·38	0·72
Iron and alumina	1·20	—
	101·03	99·50

INVERT-SUGAR.

Invert-sugar is the most largely used of all malt substitutes. On the commercial scale it is prepared by dissolving raw cane-sugar in water, adding a small quantity of sulphuric acid, and boiling the mixture, either with or without pressure, until the cane-sugar is almost completely converted into invert-sugar. The sulphuric acid is then neutralised with whiting, and the main portion of the gypsum

---

[*] Scheibler, Stammer's Jahresbericht, iv. p. 225.
[†] Wallace, Chemical News, xxxvii. p. 76.
[‡] Macdonald, Chemical News, xxxvii. p. 127.
[§] Ibid.

so formed allowed to subside. The remainder of the gypsum and other impurities are then removed by evaporation and passage through animal charcoal. The final product is a bright, light-coloured syrup, which, after a time, sets into a more or less colourless, opaque mass. The change that takes place when cane-sugar is boiled with acid, and when it is subjected to the action of invertase, is one and the same, and is an example of hydrolysis; i.e., an assimilation of water attended by a splitting-up of the molecule. The change may be expressed by the following equation :—

$$C_{12}H_{22}O_{11} + OH_2 = C_6H_{12}O_6 + C_6H_{12}O_6$$
$$\text{Cane-sugar.} \qquad\qquad \text{Dextrose.} \quad \text{Levulose.}$$

The " setting" of the liquid invert-sugar to a honey-like mass is due to the separation of dextrose in a crystalline form. It is, however, only relatively high-class sugar that will readily set in this way. When the sugar contains an excess of potash salts, the dextrose refuses to crystallise, and the invert-sugar retains its syrupy consistence, at any rate for some time. Abundance of potash salts indicates the abstraction of sugar crystals prior to inversion; in other words, the use of inferior raw material.

Invert-sugar reduces Fehling's solution. Its cupric-oxide reducing power is $\kappa = 96\cdot6$; i.e., it reduces a slightly smaller amount of Fehling's solution than does dextrose. It is lævo-rotary, since levulose has a stronger lævo-rotary action than dextrose, which is dextro-rotary, and the sugars are formed in equal proportions. Both dextrose and levulose are fermentable, but they are not equally so. As will be seen from the table given below, dextrose is distinctly more fermentable than levulose. Bourquelot has proved that dextrose is more diffusible through a moist membrane than levulose, and it will therefore be seen how this explains its more ready fermentability by yeast. The sugar to be fermented must diffuse through the cell-wall of the yeast, and consequently *that* sugar which the more readily diffuses, will be the more readily fermented.

Bourquelot has also proved that, although inferior to dextrose in diffusibility and fermentability, levulose is more diffusible, and therefore more fermentable, than maltose. This disposes of the assertion that the levulose of invert-sugar survives the fermentation and passes into the beer, imparting fulness to it; for since during the fermentation the maltose has disappeared, there can certainly be no levulose left in the fermented beer.

Time in hours.	Rotation.	Dextrose fermented per 100 c.c. (mgms.)	Levulose fermented per 100 c.c. (mgms.)	Difference (mgms.)
	Degrees.			
0	−112	0	0	0
20	−114	557	279	278
40	−110	968	532	436
64	−102	1196	721	475
86	−88	1392	942	450
108	−72	1616	1194	422
130	−56	1796	1423	373
158	−26	1970	1767	203

Used as priming, we can understand how the levulose of invert-sugar would only slowly undergo fermentation, and thus give rise to a more persistent and gradual fermentation, accompanied in the early stages by considerably more fulness and sweetness, than when dextrose only is used as the priming agent. It has been said that invert-sugar is almost completely fermentable. In its pure state it would be completely so, but during the process of manufacture certain unfermentable decomposition products are formed, which in low-class samples constitute an appreciable percentage of the total.

These decomposition products vary according to the material used and the skill displayed in manufacture. Even in the best samples they occur in small quantities, due to a slight decomposition of the levulose into levulinic acid. This in itself, is perhaps unimportant, but Conrad and Gutzeit have pointed out that prolonged boiling with acid further reduces the levulose to humin bodies, which if not harmful, are certainly absolutely valueless. These bodies would be specially formed when the material used was stubborn in inversion, and when, therefore, the boiling had to be protracted, or more acid used. This is the case when low-class syrups and beet-sugars are taken as the raw material; the formation of these humin bodies is therefore a reflection on the quality of the raw material used for the manufacture of the invert-sugar.

In choosing samples of invert-sugar, the brewer generally has a selection of different shades, varying in price, the palest being the most expensive. This paleness is obtained by more efficient charcoal filtration, which, besides eliminating the colour, also eliminates impurities, more especially the nitrogenous impurities of the original sugar. As has been before stated, samples of invert-sugar which will set are preferable to those which will not. Care must, however, be exercised to discriminate between the true setting of a pure invert-sugar, and an artificial solidification produced by mixing this substance with glucose. Analysis is required to detect this intermixture.

Apart from colour and consistence, commercial invert-sugars vary in their constituents to an appreciable degree, and analysis can, within certain limits, determine the proportions which indicate good or bad quality. In good samples there should not be more than 2–3 per cent. of uninverted cane-sugar —larger proportions, though not detrimental in themselves, denoting the employment of slowly-inverting inferior beet—or low-class cane-syrups, from which the more readily invertible crystals have been abstracted. In samples made from syrups we should also expect to find an excess of mineral substances, because these are contained in far larger proportion in the syrups than in good raw sugars, and it is safe to say that the ash of

N

a good invert-sugar should not exceed 2·5 per cent. It is not possible in commercial analysis to determine the humin substances to which we have referred, but some idea as to their presence may be arrived at by adding up the percentages of cane-sugar, invert-sugar, ash, albuminoids, and water, and subtracting the total from 100. This difference in good samples should not exceed 3 per cent., and in the best samples there is hardly any difference at all. At the same time no hard and fast rule can be laid down as regards these figures. The quantity which would not denote inferiority in one sugar might do so in another. These considerations must be weighed conjointly with the percentage of ash, albuminoids, cane-sugar, &c., and must not be taken by themselves. The following are typical analyses of two classes of invert-sugar :—

——	Good Invert-sugar.	Inferior Invert-sugar.
Invert-sugar	75·23	60·53
Cane-sugar	0·95	8·56
Ash	1·16	5·53
Albuminoids	0·78	1·89
Water	19·23	13·77
Other matter	2·65	9·72
	100·00	100·00

It has lately become somewhat common amongst brewers to invert their cane-sugar themselves. In so doing there are some advantages, especially of an economical order. There are two processes in vogue, one of which is a copy of the manufacturing process on a small scale, but in which, of course, filtration through charcoal, and evaporation, are omitted. The second, introduced by Tompson, consists in utilising the invertive capacity of yeast for the purpose of inversion, suitable temperatures being chosen in order to gain a maximum inversion in a minimum of time, and to prevent fermentation. The invertase is not separated from the yeast, but ordinary brewers' yeast is simply added to the sugar solution at a suitable temperature (133° F.), and after inversion the fluid is run into the copper to be boiled with the wort. The yeast so added is precipitated during the boiling of the wort in the copper, and is therefore left behind there, or in the hop-back. As no boiling is necessary during inversion, there is no deepening of the colour of the solution. The process is certainly far simpler and more effective than the inversion of cane-sugar on the brewery premises by acid, and one of its advantages is that there can be no formation of the humin products previously referred to. The humin bodies are abundantly produced, however, when the brewer inverts the sugar by the acid process, and in the absence of skilled superintendence. Not only is much of the sugar lost in this way, but there is a considerable increase of colour, due to the transformation of a portion of the sugar into caramel. Any increase of colour is to be avoided, for it demands the use of paler and therefore in many cases less sound malts.

GLUCOSE.

The word glucose is really a generic term given to members of a class of sugars which comprise both dextrose and levulose, but in commerce it is applied to the various commercial forms of dextrose, and which, as we have seen, constitutes one-half of invert-sugar. Commercial glucose is prepared by a process somewhat similar to that for the preparation of invert-sugar ; but instead of cane-sugar being employed as a raw material some form of commercial starch is used.

When starch is boiled with dilute sulphuric acid, either with or without pressure, it is rapidly hydrolysed into dextrin and maltose, which in turn, are rapidly further hydrated into glucose. The following equations represent these changes :—

$$nC_{12}H_{20}O_{10} + H_2O = C_{12}H_{22}O_{11} + nC_{12}H_{20}O_{10}$$

Starch.   Maltose.   Dextrin.

$$nC_{12}H_{20}O_{10} + 2nH_2O = 2nC_6H_{12}O_6$$

Dextrin.   Glucose (dextrose).

$$C_{12}H_{22}O_{11} + H_2O = 2C_6H_{12}O_6$$

Maltose.   Glucose (dextrose).

The proportion of these carbohydrates in commercial samples depends upon the concentration, the quantity of acid, the temperature, and the time employed in the conversion. It is possible, by stopping the reaction at an early stage, to obtain a product containing but little else than dextrin, whilst a product consisting almost entirely of dextrose is obtained by allowing the conversion to proceed to its ultimate limits. As in the case of invert-sugar, the acid is neutralised by whiting, the gypsum separated by filtration, the liquid passed through charcoal and then evaporated ; the concentration in the case of glucose being as a rule greater than with invert-sugar. The materials used for the preparation of glucose in this country are maize, sago, and rice, and these cereals are, as a rule, subjected to some purifying treatment before being converted, in order to extract as far as possible nitrogenous, oily, and other foreign matters. This is particularly necessary in the case of maize, which in its natural state contains about 7 per cent. of oil. The oil and other impurities, if allowed to enter the converting vessel, are acted upon by the sulphuric acid, with the formation of substances of an unpleasant flavour, which would be imparted to the beer when the glucose was used in brewing. The preparatory purification of the cereal is all-important, and it is due to a want of sufficient care in this direction that many commercial glucoses are far from satisfactory.

Glucose, as it is sold to the brewer, is in small solid blocks, which contain, as a rule, besides dextrose, a certain proportion of maltose, from 2–8 per cent. of dextrin, about 1 per cent. of nitrogenous matter, and 1·5 per cent. of ash. Dextrose is not decomposed, under ordinary circumstances, by the action of acids as much as levulose. By the protracted action of boiling acid, however, humin substances are formed, and under certain ill-defined conditions, a particular carbohydrate, called gallisin, is produced, to which we shall subsequently refer.

Potatoes are also used as the raw material for glucose; either mixed with the cereals, or alone. Continental glucoses are especially likely to be prepared from this source. Potato glucoses are to be condemned, for no treatment seems to entirely free them from injurious alkaloidal bodies derived from the potatoes.

From the researches of Bourquelot, we know that dextrose is exceedingly fermentable, being more so than levulose, and as levulose is more so than maltose, it would therefore be assumed, at first sight, that the use of glucose as a malt substitute would lead to exceedingly low attenuations. This is, however, not borne out by practical experience, and it seems without doubt that the high attenuations occasionally obtained, when using large proportions of some forms of glucose, are due to the large quantities of the unfermentable gallisin the samples contain. It might, at first sight, be thought that the dextrin in glucoses would account for this observation, but dextrin is, as a rule, only present in small amounts, far less than ordinary analysis would lead us to suppose. In the ordinary commercial analysis of glucose, the maltose is not determined, and it becomes erroneously included, partly in the percentage of dextrin, partly in that of dextrose. When the maltose is separately determined, the dextrin percentage is shown to be small, as the following analysis of glucoses by Valentin will show.

—	No. 1.	No. 2.	No. 3.	No. 4.	No. 5.
Dextrose	80·00	58·85	67·44	63·42	61·46
Maltose	None.	14·11	10·96	13·50	13·20
Dextrin	None.	1·70	None.	None.	None.
Unfermentable carbohydrates with a little albuminoids ..	8·20	9·38	4·30	8·40	8·60
Mineral matter..	1·30	1·40	1·60	1·50	1·60
Water	10·50	14·56	15·70	13·18	15·20
	100·00	100·00	100·00	100·00	100·00
Total solids	89·50	85·44	84·30	86·82	84·80
Matter of use to the brewer ..	80·00	74·66	78·40	76·92	74·60

The following analyses (quoted by Allen in his 'Commercial Organic Analysis,' page 346) of the same sample of glucose, the first analysed in the usual inaccurate way, and the second including the determination of maltose, will clearly show to what serious errors in the dextrin proportions, omission of the maltose determination may give rise to :—

	A.	B.
Water	17·77	17·77
Mineral matter	0·63	0·63
Glucose	72·60	64·94
Maltose	—	12·35
Dextrin	9·00	4·31
	100·00	100·00
Total solid matter	82·33	82·33

From these figures it is very evident that the small amount of dextrin is not responsible for the unfermentability of some glucoses ; it is rather to the gallisin that we must look for an explanation of this fact.

The following analyses by one of us show the composition of maize glucoses, when due allowance is made for the presence of gallisin :—

	A.	B.	C.
Dextrose	50·58	48·84	47·71
Maltose	14·19	14·88	12·29
Dextrin	1·76	1·80	2·98
Gallisin .. .. ·· ..	15·59	14·71	15·90
Ash	1·44	1·36	1·39
Albuminoids	1·18	0·86	0·81
Water	16·49	18·84	20·77
	101·23	101·29	101·85

The quality of a commercial glucose can be judged by the following analytical standards : the dextrose and maltose should together exceed 80 per cent., the dextrin should not exceed 3 per cent., the albuminoids 1·5 per cent., the correct proportion of gallisin can only follow a much more extensive knowledge of the properties of this substance than now exists, and there should not be more than a trace of fatty matter ; the solution should be bright, and throw down a flocculent rather than a powdery deposit, and the moisture percentage should not exceed 11 per cent. ; the colour should be pale, and the mineral matter, albuminoids, maltose, dextrose, dextrin, gallisin, and water should add up to within 1–2 per cent. of 100.

When glucose is used for priming purposes, and is free from gallisin, it ferments very readily indeed, giving, therefore, condition in a very short time. The fermentation is, however, soon over, for there is no carbohydrate reserve, gallisin, so far as we know, not being attacked by yeast, either primary or secondary ; or if it be so, it is affected only after very protracted storage.

RAW AND PREPARED GRAINS.

The repeal of the malt-tax and the consequent inauguration of the free mash-tun, gave to brewers the right of employing a number of malt substitutes which before this time were inadmissible ; and some brewers have taken advan-

tage of this permission to brew with raw, unmalted grain, and
with grain which has been subjected to some form of preparatory treatment. At the present time, barley, rice, and
maize, are the only forms of grain at all largely employed by
brewers. Rice, on account of its cheapness, its high proportion
of starch, and its freedom from soluble nitrogenous matter and
oil, being by far the most popular of the three. Untreated rice
is, of course, the husked rice as it enters this country; the
average percentage composition of which is as follows:—

Water..	14·41 per cent.
Nitrogenous matter	6·94 ,,
Fat	0·51 ,,
Ash	0·45 ,,
Crude fibre	0·08 ,,
Starch	77·61 ,,

Rice may, therefore, be regarded as containing but little
matter other than starch, and it becomes necessary to see
how this starch is utilised for the brewery as an extract-
yielding material. In a preceding chapter it was stated
that the starch-granules are contained in cells, the cellulosic
walls of which it is necessary to rupture before the granules
can be acted upon by diastase, and it is also necessary to
rupture the granules themselves by gelatinisation with hot
water, or other means. It will also be remembered that the
process of malting liberates the starch-granules from the cells
by the dissolution of their cell-walls, and renders them more
sensitive to the influence of heated water, and that the comparatively low mashing heats adopted by brewers are
sufficiently high to effect the gelatinisation of malt-starch. In
treating raw rice, however, higher temperatures are necessary.
Pure rice-starch, as before shown, gelatinises at 142° F.; but
in order to rupture the walls of the starch-containing cells
and gelatinise the rice-starch as it exists in the grain, boiling
water is necessary.

Raw rice is generally used as follows :—The rice is mashed

in a vessel provided with rake machinery, and in such a position as to command the mash-tun. The rice is mashed with hot water, and the whole heated to the boiling temperature by free steam. After the rice is gelatinised, the mass is allowed to cool to 190° F., a little malt-flour is added, and the whole then run into the mash-tun as soon as possible after the malt-mash is made. The conjoint malt- and rice-mashes are now allowed to stand together.

The malt-flour is added to prevent the gelatinised starch from "setting." At the temperature at which it is added the saccharifying action of the contained diastase must be destroyed; but we know that at 190–200° its liquefying action is still appreciable. The quantity of malt-flour added is small, and the liquefaction of the rice only incomplete; so that in practice it is necessary to run the rice-mash into the mash-tun at about 200° F., to prevent it setting, which would render its removal to the mash-tun impossible. The hot rice-mash is now mixed with the malt-mash, the temperature after mixing being about 158–160°; the gelatinised rice-starch is then acted upon by the superfluous diastase of the malt, and converted into the usual products.

It cannot be denied that this system is one which has given very excellent results, and which in all cases must have certain specific advantages. But on the whole we believe that it possesses serious disadvantages, which in the case of many breweries are so marked as to counterbalance the benefits which may accrue from its adoption. The advantages are these :—The rice-mash is exceedingly pale, and we may, therefore, either produce with normal malt exceedingly pale beer, or we may produce a normally pale beer from higher dried, and therefore presumably sounder, malt. Beers brewed on this system, too, fine very rapidly, while the flavour is exceedingly clean; so clean, indeed, that it almost amounts to a thinness which in many parts of the country would constitute a serious disadvantage. This, however, is a disadvantage of

a purely local character, and must not be confused with the
disadvantages which must always, more or less, apply.   It
will be remembered that it is an absolute necessity to run
the rice-mash into the mash-tun at a temperature not appreciably below 190–200° F., it therefore follows that unless
our malt-mash is at an abnormally low temperature we shall
obtain a mixed mash at so high a temperature that the
diastase cannot get through the extra work imposed upon
it by the rice-starch.   It is therefore clear that the malt must
be mashed at a low temperature, and in fact a temperature
verging on 140–143° F. is that actually chosen as the goods
heat of the malt-mash.

Now, in the last chapter special stress was laid upon the
fact that the conditions of the first twenty minutes really
decide the type of wort we are to get ; and as the rice-mash
cannot practically be got into the mash-tun under twenty
minutes from the time of starting the malt-mash, it follows
that the type of wort we get is really one decided by the
condition of the malt-mash, that is, a wort mashed at a low
temperature.

The result of these conditions is, then, that the malt
diastase being relatively lightly restricted, owing to the lowness
of temperature, will convert the malt-starch so completely as
to leave but very little maltodextrin in the wort.   The residual
diastase will then attack the rice-starch now run in, which,
owing to its previous treatment in the upper vessel, will offer
no resistance to its converting power.   As we shall presently
see (Chapter V. p. 240), this starch will be converted by the
residual diastase to the same low point as was the original
malt-starch ; and since the malt-mash was so completely
converted as to leave but an insufficient amount of maltodextrin, it will follow that the maltodextrin in the rice-
mash will be similarly deficient.   The result then is a final
wort defective in maltodextrins.   This defectiveness is fatal
to the stability of the beer, as will be fully explained in

Chapter V.; and it is to this ground that we attribute the inferior stability of rice-brewed beers, made on the above lines.

When maize is used as a malt substitute, it is necessary to remove from it the oil, which, from the following analysis, it will be seen to contain in a prohibitive proportion :—

Water	17·10 per cent.	
Starch	59·00	„
Albumin	12·80	„
Oil	7·00	„
Dextrin and sugar	1·50	„
Cellulose	1·50	„
Ash	1·10	„

The oil is mostly contained in the germ, and if this is removed, we consequently almost entirely remove the oil. The maize used in brewing, therefore, is mostly de-germed by an ingenious milling process. While giving a fuller-tasting beer than rice, its use is always attended by the same disadvantages to which we referred at length in treating of that cereal.

Sago and millet, and unmalted barley, are also used in a similar way, but they are not liked so well as rice, the advantages and disadvantages of which apply equally to these other cereals.

TREATED GRAIN.

Within the last year or two the use of unmalted grain in brewing has been facilitated by some preparatory form of treatment, which consists in the previous steeping and torrefaction of the grain ; by this means the starch-granules are modified, and rendered more sensitive to the action of diastase.

Rice and maize are the two cereals which have been prepared in this way, both by torrefying and by flaking. By torrefaction the walls of the starch-containing cells are ruptured sufficiently to allow the starch to undergo gelatinisation at the mashing temperatures ; consequently there is no need to gelatinise it in a separate vessel, no need to run it in at high temperatures, and no need, therefore, to risk the instability which follows low initial mashing heats. This form of grain is used in the mash-tun, and so long as the malt is sufficiently diastatic to convert it at the ordinary mashing heat without loss through unconverted starch, and without loss owing to bad drainage, then this form of substitute is good, as it gives a good extract, and its paleness permits the use of higher kilned malts.

In comparing these forms of treated and untreated rice, the preference must be unhesitatingly given to the treated. At the same time, it is probable that the defects incident to the employment of untreated rice could be overcome if the malt-mash and the rice-mash were entirely separated, and only allowed to mix in the copper.

Untreated grain, is, we believe, converted by a few brewers by the acid process, and does not mix with the malt-mash until diastatic conversion is over. In this case, the starch of the cereal is converted into glucose—maltose and dextrin being only present in small proportions; but the objections to the process are the flavours and impurities which are extracted from the grain and pass into the beer, and which in the manufacture of glucose are in great part removed by efficient charcoal filtration. On the brewing premises charcoal filtration is not feasible, or only feasible in an incomplete manner. Both on the ground of its cumbrousness, and of the impurities which are extracted and unremoved, this process does not seem, in its present state, likely to command much attention.

Note.—It seems probable that the recent researches of Brown and Morris on Germination will put into the brewer's hands a better method of manipulating raw cereals than that which has been referred to, and it would appear that the method in question will enable him to dispense with the previous preparation of the grain by torrefaction or otherwise. It will be remembered that the investigations in question have taught us that during germination an enzyme is developed which has the property of dissolving the cellulosic tissue surrounding the starch granules. It is due to this solution that malt-starch is gelatinisable at the heat of the mash-tun, while raw grain starch requires a far higher temperature. This cellulose-dissolving ferment is absent in ordinary malt, for it is quite killed in the early stages of kiln-drying. But green-malt (that is unkilned malt) contains it in abundance, and if this be extracted with cold water, the infusion will possess the property of dissolving the cellulose.

By the employment of such an infusion it is clear that to a very great extent we can artificially *malt* raw cereals ; at any rate we can so alter them in the direction of malting that the contained starch shall be gelatinisable at, ordinary mashing heats. The action of the infusion is comparatively rapid, and in a few hours a mash of raw grain, digested at about 120° with the infusion of green malt, will be ready for intermixture with the malt-mash. By this means, boiling of the grain-mash and the consequent lowering in heat of the malt-mash are avoided, while the treatment in question does (and at far less expense) all that can be claimed for the torrefaction of grain as it is carried out by the manufacturers of prepared brewers' cereals. At the moment of going to press, the process is only in the experimental stage, and it is therefore not possible to say more than we have already done. The process has been protected by patents in the name of Mr. Horace T. Brown and the authors.

CHAPTER IV.

HOPS.

INTRODUCTORY.

ALTHOUGH it must be granted that in recent years we have got to know something precise as to many of the constituents of the hop, yet its chemistry, like all botanical chemistry, is surrounded by difficulties in regard to the isolation and investigation of the various constituents, difficulties far exceeding those of the study of other materials used in brewing. It will not be possible to touch on the cultivation of the hop, on the soils best suited to its development, on the many diseases to which it is liable, or on the manner in which these diseases may be to some extent checked. These questions, though important, do not form a part of the brewing operation ; and it will be our duty, therefore, to take the hops as they come to the brewer, to describe their structure, and the properties of those constituents which play a part in brewing operations, and to shortly describe the means by which we can judge of their relative value for brewing operations.

Hops are added to beer for the following reasons :—

1. To give the beer the distinctive bitter flavour and aroma.

2. To precipitate certain nitrogenous constituents of the wort.

3. To clarify the wort, not only by the separation of the above constituents, but by the mechanical clarifying property of the hop leaves when agitated in the copper, and by the formation of a filter-bed for the filtration of the wort in the hop-back.

4. To preserve the beer by the antiseptic influence of some of their constituents.

5. To assist in the sterilisation of the wort.

The bitter flavour is imparted by some of the resins and the so-called hop acid (" hopfenbittersäure ") ; the aroma by the volatile oils ; the precipitation of the nitrogenous matters by the tannic acid ; and the antiseptic properties by certain of the resins. It is therefore essential that hops, to be of value, should contain these substances in due proportions. The mechanical fining and filtering of the beers is a function of the insoluble leaves of the hop—and as in all hops, bad and good, there is an abundance of them, we need not further consider this point ; in some very old hops, however, the leaves become so disintegrated during boiling as to form a very unsatisfactory filter-bed.

It must be at once admitted that the commercial valuation of a sample of hops cannot, at present, be entirely based upon a complete determination of all these substances. It is relatively easy to determine the volatile oils, and the tannic acid, but to examine each sample of hops for the different resins would be a work of several weeks, if not months, and for practical purposes is out of the question. Fortunately, however, the physical properties of the flowers afford some means of judging the richness and effectiveness of these resins, for by practical experience we know that certain outward characteristics imply certain effects in brewing which are due to the activity of these resins ; and although these modes of judgment are necessarily rough and ready, they are at present the only ones practically possible. Nor do these outward characteristics apply only to the quantity and nature of the resins, for we obtain some idea from them as to the oil, though so far as we know there is no clue to the percentage of tannic acid. As, however, fair proportions of tannic acid appear to accompany the presence in due quantity of resins and oils, as indicated by favourable outward indications, it is but seldom that hops are subjected to any chemical determination of tannic acid; in fact, as a general rule, hops are bought and sold on the strength of their physical characters, unchecked by chemical analysis.

THE STRUCTURE OF HOPS.

The structural formation of hops has a special importance for brewers, since upon their physical attributes (formation among the number) hops are generally valued.

Hops, as they are known to the brewer, are the female flowers of the hop plant; hop-cones, or strobiles, being synonymous terms. In each cone there are from forty to one hundred leaves (bracts), and these lie close above each other, grouped round what is termed the stalk or rachis. This stem is knotted (geniculated) in eight or ten different places, and from these knots spring the small branches to which the hop leaves are attached. The leaves (bracts), are turned over on the lower side of the inner edge, and enclose the fruit developed from the flower. The fruits of the hop are the round, hard granules which the brewer terms seeds, and which we see floating in the hopped wort. These granules are enclosed in a thin, transparent scale (the sepal or perigon). Freed from the scale, the granule has an ash-grey colour, changing on both sides to a brownish-violet, bordered by a light yellowish-red. Enclosed in this granule is the true seed of the hop, which is devoid of endosperm. Wild and inferior hops bear large fruit, the finer hops have shrivelled and generally sterile fruit. Attached to the hop fruit, the stem, and the inner side of the leaves, is the fine yellowish resinous dust, which is known as hop-flour, lupulin, or "condition." This powder, when magnified, is found to consist of granules, consisting of glands made up of complete cells. Their form may be described as that of a pear, where the top is broader and flatter, and the bottom more rounded than the ordinary form of pear. The cells forming the upper part and corresponding to the broad top of the pear (operculum) are thicker than those forming the sides and bottom, and are also differently formed. The glands, which are hollow, contain the resins and oils, and, when bruised, they shed their substance in the form of small oily drops. In fresh hops these expressed drops are thin and of a yellowish colour; in old hops they are of a thick consistency, and of a darker colour.

The hop-meal or flour, therefore, contains all the valuable· principles of the hop except the tannic acid, which is contained in the leaves. It is, therefore, clearly essential that hops should be thickly dusted over with the flour; and in valuing hops the quantity of yellow dust and its consistency are regarded as prime factors. Haberlandt recommends the mechanical separation of the flour, and its estimation by weight as a means of judging hops; experiments by him show that the flour ranges from 5·33 to 17·5 per cent. It is not likely, however, that the process will ever play a part in the valuation of hops, for the complete separation of the flour is difficult and tedious, and we have not sufficient data before us to enable reliable conclusions being arrived at from the results. Southby recommends that hops should contain from 15 to 16 per cent. of flour.

TANNIC ACID.

In a communication quoted by Thausing, from the Agricultural Laboratory of Vienna, we find the tannic acid, estimated in 100 parts of the natural hops, to range from 1·38 to 5·13 per cent., the average being between 3 and 3·5 per cent., and this, so far as we know, is the normal amount found in good hops. In the hops analysed at Vienna the highest amount of tannic acid appears to have been present in the samples which had been sulphured, and Thausing quotes from the report on these analyses to the effect that sulphuring does seem to raise the tannic acid. This, however, requires confirmation. Other things being equal, the more tannic acid we have the better, in order to bring about the separation of certain nitrogenous constituents of the wort, with which it combines, and which boiling alone will not serve to remove by coagulation. As we shall show in a later chapter, this separation is never complete, in fact there is always an excess in the wort of the bodies precipitable by tannin. The addition of tannic acid, in the form of catechu or other bitter drugs containing it, has never been found to be of service, probably because the form of tannic acid in them is quite different to hop-tannin.

VOLATILE OILS.

The volatile oils of the hop may be obtained by passing steam into an infusion of the hops, which is kept warm the while. The steam carries over with it the oil, which will float on the condensed water, from which it can be separated by ether.

The percentage of hop-oil present in the samples the analysis of which was given in the Austrian report above referred to, ranged from 0·15 to 0·48; the average appears to be about 0·25 to 0·35. To these oils we owe the aroma and delicate flavour of the beer. A considerable portion of them is dispelled by boiling hops in the copper, for they are freely volatile at this temperature; while a certain amount is also lost during cooling, and carried off with the carbonic acid gas during fermentation. The more we have of the oil the more aroma and delicate flavour shall we get in our beer, and in the ordinary way the importance laid by brewers on the odour of the hops depends upon their richness or poverty in these oils. Hop-oil is only very slightly soluble in water (1 in 600), but it is readily soluble in dilute alcohol. In spite of the very slight solubility of the oils in wort, the quantity added, compared to the quantity of wort, is so small that there is no reason to suppose they are not freely dissolved in the copper. If not dissolved in the copper, they would probably dissolve as alcohol is formed during the fermentation. These oils exist in the hop-flowers combined with a portion of the resin in the form of a balsam. It has been asserted that the valerianic acid which we know to be produced in old hops is due to the oxidation of these oils. Bungener, however, asserts that this acid is due to the oxidation of the bitter principles. Which ever view is correct, it is certain that old hops contain less volatile oils than new, but whether they are oxidised to valerianic acid or pass off as other compounds, is not known.

Thausing regards the presence of hop-oil as of minor importance in view of its expulsion during boiling, cooling, and fermentation, and on this ground thinks that the weight laid upon the odour of hops as indicating the presence of these oils is a mistake. This may apply to the Continental system of brewing, but for us it is of the highest importance that hops, especially those used for addition to finished beer and for adding to copper shortly before turning out, should be rich in this constituent; as a consequence, the odour of a sample remains a factor of much importance.

BITTER PRINCIPLES.

The bittering principles of hops are still the subject of considerable divergence of opinion. According to Hayduck the resins are the essential bittering principle, and as Hayduck's researches are the most recent and are characterised by completeness and definiteness, it is probable that his views are more worthy of credence than those of the older investigations. Among these is Lermer, who claims to have separated a crystalline bitter acid from hops, to which he attributes its bittering properties. The acid is insoluble in water, but soluble in dilute alcohol, imparting to the solution an intensely bitter taste.

Julich separated an intensely bitter substance from hops, which was easily soluble in water. Bungener attributes the bitter to a substance partially of an acid, partly of an aldehydic character. The substance is insoluble in water but easily soluble in alcohol, ether, &c. It is easily oxidised to valerianic acid, and Bungener attributes the presence of this acid in old hops to this cause.

HOP RESINS.

Hayduck's researches on the resins show that there are at least three of these bodies in hops ; the first, a soft resin precipitated by lead acetate ; the second, also a soft resin, but not precipitated by lead acetate ; the third, a harder resin, also unprecipitable by lead acetate, but insoluble in petroleum ether, while the first two resins are freely soluble in that medium. Hayduck thinks that the second resin is probably identical with the soft resinous substance obtained when Bungener's and Lermer's bitter acid crystals are oxidised. The solubility of the resins in water is as follows :—

Soft resin precipitated by lead	0·042 per cent.
,, ,, not ,, ,, ...	0·048 ,, ,,
Hard resin	0·058 ,, ,,

The first two resins give an intensely bitter taste, the third, an unpleasant feeble bitter. Experiments made with these resins, as to their antiseptic influence, showed that the first resin materially restricted the lactic fermentation, the second resin restricts it to only a limited extent, while the third resin had no appreciable restricting influence ; the acid fermentation was retarded, but the ultimate amount of acid produced was the same as in the case where no resin at all was used. Hayduck also found that as the soft resins were repeatedly extracted with water, the solution decreased in strength and antiseptic influence; this is to be ascribed to a change in the resins, the active being partially converted into the inactive variety as extraction proceeds. It would thus appear that, as regards these resins, boiling hops for the second time is useless; for the first boiling will extract the greater part of the resin, the remainder becoming inactive during the protracted extraction. Hayduck had shown previously that the resins do not subsequently affect the acetic ferment, and he has shown that Sarcina (*Pediococcus*) is equally unaffected by them.

ALKALOIDS OF HOPS.

In this country Graham first directed attention to the narcotic influence of hops, and its bearing upon the consumer of beer. He stated that the stupefying nature of English beers, as compared with Lager beer, was rather due to the greater amount of hops used by us than to the only slightly larger amount of alcohol contained in our beers. This view is reasonable, although it is not at all improbable that the higher alcohols, developed at the higher temperatures prevalent in our top fermentations, are also factors in this difference. The narcotic influence of the hops has of course been acknowledged for a long while, and recent researches have fully demonstrated this to be due to the presence of an alkaloid. Researches on this matter have been carried out by Griessmeyer, Williamson, Griess and Harrow, and others. Griessmeyer styles the alkaloid lupulin. Griess and Harrow separated the base cholin (or neurin) from beer, a base which also occurs in the brain. Griessmeyer denies the presence of cholin, as such, in hops, stating that it exists combined with other bodies as lecithin, a body of very complicated constitution. The alkaloids are bitter substances, and it has been suggested that hop alkaloid, combined with resin, forms the bitter principle soluble in water.

Besides the substances above referred to, hops contain fibrous matter, mineral substances, nitrogenous bodies, sugars and gums. The fibrous matter has a distinct function in brewing, for by being agitated in copper and afterwards forming a filter bed in the hop-back it helps to clarify the wort; we also think that the ready clarification of dry-hopped beers, is in part due to the assistance given to the finings by the surface of the hop leaves. With regard to the nitrogenous matters, we know but little of them. Bungener has found the amide, asparagin; what effect this would have in the small quantities in which it exists is not clear. The gums and sugars, so far as we know, have no direct bearing in brewing. To the mineral substances some importance attaches, inasmuch as their nature and character show us what minerals are demanded for the growth of the hop, and they therefore indicate the nature of the manure to be used. From the following analyses by Wolff it will be seen that they are specially rich in potash and phosphoric acid. As the result of twenty-six analyses Wolff found that hop leaves contained on an average 7·54 per cent. of mineral matter—the maximum was 15·3, the minimum 5·3; 100 parts of ash contained :—

	Minimum.	Maximum.	Average.
Potash	16·30	51·60	34·61
Soda	0·00	8·80	2·20
Lime	9·80	24·60	16·85
Magnesia	1·50	13·40	5·47
Oxide of iron	—	3·20	1·40
Phosphoric acid	9·20	22·60	16·80
Sulphuric acid	0·00	12·20	3·59
Silica	10·30	26·10	16·36
Chlorine..	1·00	7·00	3·19

SULPHURING HOPS.

The object of sulphuring hops, i. e., using sulphur with the fuel when drying the hops, is, as a rule, to cover defects in the hops, or, at any rate, by bleaching and rendering them brighter, to make a sample appear of higher market value than it really is. In some respects sulphuring is of use—it would in a way retard the deterioration of hops in regard to putrefactive change and mildew-development, and it is claimed to yield hops richer in tannic acid. This latter point requires confirmation, but even admitting its effect in this respect and in respect to its preservative influence, it must be said that sulphuring hops is a practice which brewers rightly condemn, and for two reasons. First (and this is by far the more important reason), sulphuring is only applied to those hops the defective appearance of which would render them less valuable if not submitted to this process. As the sulphur in part covers the defects, the brewer is paying a higher price for his article than he should do, while again the defects which it covers may be real defects, defects which would give rise to defective beer, and which, therefore, the brewer prefers he should have the opportunity of observing before finally purchasing his hop stock. The sulphur itself, in the hops, and therefore in the beer, exists as sulphurous acid, and is probably of minor importance. In cask it might become reduced to sulphuretted hydrogen by the hydrogen evolved by micro-organisms, and so give rise to what brewers know as "stench." This change, however, is one which is not likely to occur except under very protracted storage, and when there is matter present which encourages the development of these organisms. The main objection is certainly not to the sulphur, but to the possible defects hidden by the sulphur, and it is on this ground that the brewer is anxious to know whether or not sulphur has been used. Sulphur is also used as a dressing upon the plants during their growth in the fields, in order to protect them from insects. This form of sulphuring is in many respects justifiable, and it would be far less likely to give beer stench than that used during kilning.

The detection of sulphur is very easy, and can be carried out in two ways, which will be treated of in the analytical section (p. 495).

MICRO-ORGANISMS ON HOPS.

We think that the micro-organisms adherent to the surface of hops play a more important part than they are generally credited with doing.   The main portion of hops being boiled in the copper, the various bacteria, wild-yeasts, mould-spores, &c., which so thickly coat their surface, would in all probability have only an indirect bearing, namely, that as they do not survive the boiling process, they would simply form putrescent food for the development of organisms introduced during the subsequent stages.   Apart from this source of adding to the sum total of putrescible matters, we must not forget that deteriorated hops—hops which have suffered changes analogous to malt that has moulded, must likewise increase the amount of these bodies.   On this ground hops should be carefully chosen, and they should be free from mildew and other signs of similar deterioration.   The hops introduced shortly before turning out—and this, as we shall see, is a commendable practice—would have the adherent ferments killed in the same way, but it is always desirable that hops added at this stage should be good and new.

The question of these adherent organisms is, however, of far greater importance in the case of the hops added to cask—a practice which is general with all ales stored for considerable periods.   Here, naturally, they are submitted to no conditions tending to their destruction, and they are necessarily free to develop in the beer.   When the beer is well brewed, i. e., when it is free from readily putrescible bodies, and when the hops used at this stage are healthy, we are inclined to regard the organisms adherent to dry hops as on the whole favourable.   From Matthew's researches we know that secondary types of yeast are included among their number, and they would therefore promote the desired after-fermentation by degrading the maltodextrins, and then fermenting them.   Apart from this, we think the peculiar, characteristic, but undescribable flavours of mature old ales are partly due to the products of the development of these various ferments. It is, however, of the highest importance that the hops added at this stage should not introduce readily putrescible matter into the beer, nor should they introduce ferments tending to produce acidity and putrid flavours. The chief object of dry hopping is to induce the gradual solution of the hop flavours and to preserve the beer, and these objects are clearly negatived when the hops so used are likely to lead to acid and putrid beer. What we have said in regard to the degradation of maltodextrins by the secondary yeasts adherent to hops is, we think, borne out not only by the condition which we know to result from the judicious employment of hops in cask, but also by the decided attenuation which brewers know to result from this practice. The attenuation and condition have been attributed to the aëration of the beer when introducing dry hops after it is racked. But we are convinced that aëration alone could not induce the decided effect which we know to result from the practice, though it might assist the vigour of the secondary yeasts in their degrading and fermentative functions.

VALUATION OF HOPS.

In the valuation of hops, the following points are to be observed :—The hop must certainly contain a large amount of flour, and it is of importance to closely observe the comparative richness or poverty in this material. The hops must also be free from mildew or mould. This is generally discernible by mere inspection, but in all doubtful cases it should be supplemented by microscopic examination. The odour of the hop is also of the utmost importance, an aromatic delicate smell being requisite ; whilst hops defective in odour, or still worse with an odour tainted with the characteristic faint smell of valerianic acid, are completely unsuitable for good ales. Both the amount of "condition" and odour are best observed by rubbing the hops between the hands; the flour (condition) and all its aromatic constituents will adhere to the skin, where they can be better judged than in the cones themselves.

The colour of the hops is another important point. They should be ripe (in the brewer's sense of that word; not botanically ripe), but not over-ripe; the correct degree of ripeness is indicated by a greenish yellow tint. Hops which are a light green, are unripened, and contain less flour, and are therefore the less valuable for brewers' purposes. Over-ripeness is indicated by a slight red or red-brown colouring, and means loss of flour, through the agency of insects and wind. A darker dull brown indicates over-drying, or heating in the hop-bales; and from such hops we expect very bad brewing properties. What has been said about colour must not be taken too strictly. For very pale ales, very pale hops are required. The paleness is sometimes obtained at the expense of sufficient maturity, and therefore at the expense of the valuable constituents of the hop; while again it may be artificially produced by sulphuring. Many hops are well suited for brewing excellent beers of any but the lightest shades, simply on the score of possessing a little extra colour. Colour is one guide to quality, but not the only one, and the relative importance attached to it by the brewer is necessarily as much a question of the paleness of the beer it is desired to produce, as it is of the inherent soundness of the hop.

The taste of the hops should be an agreeable bitter: not harsh and grating, but delicate. The taste should be free from acid and other foreign flavours.

Apart from requisite colour, richness in flour, absence of mould, light yellowish tint of flour, and delicacy of odour and taste, it is desirable to see that the leaves should lie tightly packed above each other; open cones are indicative of unripeness. A good sign is also the balling together of the cones when pressed; this affords some indication of the richness of the hops in resin.

Hitherto we have mentioned a few guides to the valuation of hops irrespective of age. But brewers have long known that the storage of hops for a period of more than 7-8 months means distinct deterioration in regard to brewing value. The volatile oils disappear, and according to Hayduck the preservative qualities of the antiseptic resins are gradually lost, the bitterness becomes less delicate—in fact, under our present method of storage, the hops entirely lose all their valuable chemical properties (except those due to the tannin), if the storage is long enough; while mould and mildew develop, and a gradual transformation of some of the constituents (either the oil or resins) into the disagreeably smelling valerianic acid takes place. Age of hops shows itself as follows:—The colour deepens, hops stored for many years becoming brown. The aromatic oil is decreased, and consequently there is less odour; in fact, very old hops have either no odour or a cheesy disagreeable smell. The leaves easily separate from the stem, the flour becomes of a darker colour—reddish to reddish-brown, and the hops do not stick to each other as before, nor do they yield, when rubbed between the hands, the same viscous resin. The hop-flour should be examined under the microscope; when fresh the glands are glossy, smooth, and, when bruised, discharge a light greenish-yellow matter. The glands of old hops are shrivelled and corrugated; the matter discharged on pressing is thicker and darker.

Added to the observation of these physical attributes should be the detection of sulphur, and the estimation of hop-tannin; for no inspection, however skilled, will afford information as to these points. The hop-oil is also easy of direct determination, but we gather some information as to this from the odour.

There is no branch of brewing science so unsatisfactory to

treat of as the hop. The analyses are contradictory, and our knowledge as to the necessary constituents vague and indefinite, in fact, the whole subject of hops is still shrouded in ignorance. Hops are bought and sold almost entirely by their appearance, flavour, and odour. As a rule, perhaps, these indications suffice, but it would be desirable to check these indications by analysis or microscopic examination. But before analysis or microscopical examination can be of any real value it will certainly be necessary to have more definite information before us. Quite apart, however, from checking physical indications, exact scientific valuations would, were they possible, put an end to an immense amount of humbug.

As a rule brewers buy readily of the well-known brands, and are ready to give exorbitant prices for them, while unknown brands, however good intrinsically, are hardly looked at. In mentioning these facts no charge is implied against the hop factors. It does not follow that because a favourite brand is bad one year that the hop factor gave less for it. Probably he gave a high price for it, and requires a proportionately high price from the brewer ; but that enormous prices are given for favourite brands of hops, quite irrespective of the fluctuations in quality year by year, is a matter of fact ; and it would be only by having correct methods of valuation, such as we have for malts and sugars, that hops, irrespective of repute and locality, could be priced according to their real deserts. Certain districts are admittedly particularly favourable for hop cultivation, and the hops from these districts have been known to give good results for many years. This, however, is also the case with malting barleys. But with correct methods of valuation at our disposal we do not hesitate to freely condemn defective malts made from barleys grown in the most favoured districts ; while we also are able to judge of the merit of malts from absolutely new and untried districts. The case is very different with hops.

HOP SUBSTITUTES.

Hop substitutes and malt substitutes are not at all on the same footing. Malt substitutes are used to improve certain characteristics of beer, hop substitutes to cheapen its production. The advocates for so-called "pure beer" seem, however, to make no distinction between them, except perhaps that those who represent hop counties see special wickedness in using hop substitutes, while to those from a barley district the employment of malt substitutes is particularly distasteful.

With an ordinary crop of hops, no brewer ever dreams of using a hop substitute; among the very many breweries the ins and outs of which are well known to us, we do not know of one where any hop substitute is being employed now, or has been employed since the year of the hop famine (1882). In these remarks we except catechu or cutch. This material is used in some breweries, but not as a hop substitute. It is used not because of its bitterness, but because of the supposed preservative influence of the large amount of tannic acid it contains. In fact its bitterness, being a disagreeable bitterness, is a disadvantage, and limits its employment to very small rates. We may regard it, therefore, roughly as tannic acid, and as its purpose is to supplement the deficient tannic acid of the hops, it stands on much the same footing as the gypsum added to a water deficient in this constituent. The so-called hop substitutes—gentian, camomile, quassia, &c., are bitter substances, innocuous to the system, but yielding harsh, grating, clinging bitters very different to that of the hop. They are devoid of the delicate oils, the antiseptic resins, and the other characteristic constituents of the hop. They can in no way entirely displace the hop, and, as Thausing points out, they are not really substitutes in the true sense of the word.

There is no objection to their employment when hops are so scarce and dear as to prohibit the brewer from using them in the right quality and quantity; and their use under these circumstances is so far of benefit that it does away with the necessity of using damaged old and rank hops, only too likely to be rich in narcotic, but poor in preservative qualities and delicate flavour.   This is their only justification, and as a matter of fact this is the light in which brewers regard them.   With hops at his command no brewer would think of using these vegetable bitters ; competition would soon show him how dangerous a policy this would be.   In charging brewers with their continual employment, the "beer-purists" show themselves ignorant of the facts of the case.   In spite of specific and emphatic denial they however cling to the charge ; it is repeated merely to attract the interest of that section of the agricultural class which seeks to relieve the depression due to faulty technical attainments, by means of ill-considered legislative protection.

—·—

CHAPTER V.

MASHING AND SPARGING.

INTRODUCTORY.

THE first real process in the brewery is the mashing process. Preparatory to this, however, the water has to be heated to the required temperature. This is done in the liquor back, a vessel provided with suitable heating apparatus, and which preferably should be large enough to take the whole of the mashing and sparging liquor, which together may be roughly reckoned at six barrels to the quarter of malt mashed. In this vessel the water is either merely heated up to the required mashing temperature; or, when required, it is previously boiled, either with or without the addition of saline or antiseptic substances. The manner of treating the water has already been explained (Chapter I. pp. 11–26), we will therefore assume that our water stands at the required heat, and that it is contained in a vessel which commands the mash-tun.

The mash-tun is a vessel provided with a false bottom perforated with holes and slots. The actual bottom of the vessel is fitted with several spend-pipes, either all converging into one large pipe fitted with a cock, or each fitted with its own cock. The tun should also be fitted with rake machinery; for although not always necessary, it may at any moment be wanted for special requirements.

There are two systems used in this country for making the mash. On the one, the tun is filled with the requisite

volume of water, and the grist is then run into it, the rakes
being worked the while to ensure intermixture.   On the other
system the grist and water are mixed in a special appliance
fixed outside the tun.   Into this vessel the grist and water
simultaneously enter, and are there mixed in their approximately correct proportions.   The mash therefore falls in the
tun ready mixed, and rake machinery is, under these conditions, not called into requisition.   These vessels are called
"outside mashers" or "mixers."   They are of two kinds, those
in which the intermixture of grist and water is effected by
machinery, and those in which the grist and water mix
automatically.   The former, though more costly, are preferable ;
but even the less satisfactory automatic mashers are more to
be relied upon for a proper intermixture of grist and water
than the method first alluded to, in which the grist is run into
the water lying in the tun, and the intermixture is then
effected by rake machinery.

Most mash-tuns are fitted with an underlet, i.e., a pipe
admitting water into the bottom of the tun.   Hot water is
frequently introduced by this pipe at some period of the
mash, and the temperature thereby raised.   Such an
appliance is most desirable, and in the production of mild
ales may be regarded as necessary.   During the admission
of the underlet, the rakes should be gently revolved to secure
uniformity of mash-heat.   This is one of the reasons why
rake machinery should always be provided.

All mash-tuns are fitted with sparging apparatus.   This
consists of a perforated pipe revolving horizontally just above
the level of the mash, and distributing water upon the goods
in a fine shower.   Spargers are either automatic or driven.
The former are the more common, and revolve by the im-
petus of the escaping water ; the driven spargers, however, are
perhaps preferable, since the water can be delivered in a very
fine spray, without impeding the rate of revolution.

The mash being made, it is usually allowed to stand for

about two hours ; at the end of this time the taps are opened and the sparge liquor showered on. Sometimes an underlet will be applied shortly after mashing ; sometimes a second mash becomes desirable. A second mash is made in this way: when the first mash has stood its appointed time, the taps are opened, and a certain quantity of wort withdrawn, generally about one barrel to the quarter. The taps are then closed, and from a barrel to a barrel and a half run in by the underlet at a slightly higher heat than that of the original mash liquor ; the rakes being revolved the while. The mash is then generally allowed to stand about forty to fifty minutes ; the taps are then opened, and sparging liquor applied either at once or shortly afterwards.

If by chance the brewer desires to make a second mash, and possesses no rake machinery, he should run under two-thirds of his second mash-liquor, and sparge over the remaining third. By this means a fair degree of uniformity may be ensured. Second mashes are not necessary or desirable with good malts ; when, however, malts contain refractory starch, they are of decided service.

Some brewers drain their worts closely after mashing, before applying the sparge liquor. This, however, is probably a mistake ; for although the grains will always hold about a barrel of liquor to the quarter, yet it is well not to lose too much of the diastase, since this is wanted to some extent during sparging operations ; again, by allowing the grains to fall down too closely on the plates, they are apt to clog the perforations, and so interfere with subsequent drainage.

In a well-conducted mash the grains are buoyant, lying on the surface of the wort, and they are dry and firm to the touch. When they fall to the bottom, and the liquor makes its appearance above them, they clog up the plates, interfere with drainage, and reduce the ultimate yield or extract. This state of affairs is what brewers term a

P

"sloppy" or "dead" mash. It may arise from several causes. The principal one is unsatisfactory malt—malt, the starch-containing cells of which have been incompletely modified during vegetation; an appreciable proportion of the starch in such malts will refuse to gelatinise at the usual mashing temperature, and will only come into solution at the higher temperatures prevalent during the later stages of mashing and during sparging, when the residual diastase is enfeebled by the high temperatures to which it is then exposed. Grinding the malt too fine, may also induce this condition of affairs, especially when the husk has been separated from the fine flour by elevating the grist in Jacob's ladders. Another fruitful source of this defect is an over-deep tun; in this case the sparge liquor, falling from a distance, forces down the goods. Excessively low mashing heats and excessively high mashing heats will also lead to the same thing, the former because much of the starch is left un-gelatinised and unconverted; the latter because the diastase is so enfeebled as to be unable to convert the starch with sufficient rapidity.

The grinding of the malt is no unimportant matter. Each corn should be cracked, but not ground fine. To leave corns unbroken is obviously to lose extract: to grind too finely is to clog the perforations of the false bottom with the particles of unconverted flour, and so impede drainage and obtain lower extracts. It is always well to have two sets of malt-rolls; the one adjusted to the large, the other to the small corns. Fluted rolls, we understand, give good results. When possible it is always preferable to have the grinding mills immediately above the grist cases. Where, as so often happens, the malt is ground below and then elevated to the grist cases, the grist is agitated during elevation. This induces a separation of husk and flour, which distinctly impedes drainage and extract.

The shape of the mash-tun is of some importance; an over-deep tun leads to a sinking of the goods as has already been

explained; an over-shallow tun means an excessive loss of heat
by radiation; but it is wiser to err on the side of shallowness
than of depth.* The tun should be provided with well-fitting
covers; and it should not be exposed to draughts, nor should
any portion of it be in contact with the outside air. When,
however, the tun is in an exposed position, it is well to lag it
with felt, or some similar material, for loss of heat during
mashing means loss of extract.

THE TEMPERATURE OF THE MASHING LIQUOR.

From a consideration of the influence of temperature upon
the diastatic conversion of starch, which was discussed in
its scientific aspects in Chapter II., we know that by altering
this temperature we can produce different proportions of the
conversion products. What applies to the diastatic conversion
of starch-paste is equally applicable to the diastatic conversion
of the malt-starch; the temperature at which the conversion
is carried out is, in fact, the most important factor in settling
the type of wort produced.

In dealing with malt on the practical scale, the temperature of conversion will depend upon several factors which have
not previously required consideration. Primarily it will depend

* Southby gives the following dimensions ('Practical Brewing,' p. 71) for
mash-tuns :—

Quarters.	Without Outside Masher.		With Outside Masher.	
	Diameter.	Depth.	Diameter.	Depth.
	ft. in.	ft. in.	ft. in.	ft. in.
5	5 10	4 0	5 2	5 1
10	7 8	4 6	7 0	5 6
20	10 1	5 2	9 4	6 1
30	11 11	5 6	11 0	6 6
50	14 6	6 2	13 9	7 0
100	18 6	7 8	18 1	8 0

upon the temperature and quantity of the mashing liquor ; secondly, upon the temperature and specific heat of the malt ; thirdly, upon the system of mashing ; fourthly, upon the position of the tun and the quality of the malt. Of course by far the most important element is the temperature of the mashing liquor ; this point will be shortly dealt with at length. The temperature and specific heat of the malt may be regarded as constant for practical purposes, and they may, therefore, be disregarded from the present considerations. The position of the tun is of more practical importance, for if, as sometimes happens, there is great loss of heat by radiation, the heat developed by the mash as the result of chemical action may be lost, and the temperature, instead of rising during the conversion, may actually fall. It is therefore necessary to make due allowance for any such loss of heat, either by increasing the temperature or the relative quantity of mashing liquor. The quality of the malt plays a not unimportant part in the temperature of conversion ; " slack " malt, for instance, will certainly give rise to no increase in temperature as the result of chemical action ; and when material of this kind is used, this fact must be taken into account.

These points are, however, of minor importance compared to the influence of the system of mashing upon the tempera-ture of conversion. For instance, suppose we mash a malt with two barrels of water to the quarter, at 160° F., in an out-side machine, our actual conversion temperature—that is the temperature of malt and water after intermixture—will be *lower* than if we were to take the same malt, the same quantity of liquor, and at the same temperature, but instead of mixing them in an outside masher as before, run the liquor into tun first, the grist afterwards, and then intermix in the tun itself. In order to obtain a given conversion temperature, or initial goods heat, on both systems of mashing, it is necessary to have the temperature of the liquor some two or three degrees higher when using an outside machine than when mashing in the

tun itself; the reason of this will be seen on considering the
essential differences between these systems. When the malt
is run into liquor lying in the mash-tun, the first portion will
come into contact with liquor at its maximum heat; but the
more malt we get into our tun, the more shall we have cooled
the temperature of the mixture; the last portion of the malt
will, as a matter of fact, be introduced into the tun-contents
at a temperature very little, if at all, higher than the goods
heat desired. Thus, suppose we mash malt for a goods heat of
150° F.; the first portion of the malt will be run into liquor at
just about the maximum point; the last portion will run into
the tun standing at or about 150° F., the intermediate portions
of grist coming into contact with mash at intermediate heats
all of which are *below* the maximum. When, however, an
outside masher is used, each lot of grist is brought into contact
with liquor at the same *maximum* heat, since the grist and
water are intermixed in these machines at the rates required
for the mash.

On this ground alone some preference should be given to
the second method, for the excess heat to which each lot of
grist is exposed is uniform; on the other system the first lot
of grist is exposed to a maximum excess heat, the last lot to
no excess heat at all. To ensure the uniform conversion of
the tun-contents we depend, on this system, upon bringing
the more active diastase of the last portions of the grist to
play upon the earlier portions of the mash, the diastase of
which will have been much enfeebled by the excess heat to
which it was exposed. For this purpose very complete intermixture of the mash is necessary; experience shows, however,
that this is very difficult to effect in practice. In the outside
system of mashing we are independent of the necessity for
complete intermixture in the mash-tun, and it is a fact that
mashes made without side machines are, as a rule, more satisfactory than those made on the alternative system.

From, Chapter II. we know that as the temperature of the

mash rises above 140° F., the diastase is increasingly checked. When, therefore, we conduct a mash at 150° F., that is, when our mixed malt and water are 150° F., we shall be restricting our diastase. But the restriction of the diastase in such a case will not only be due to the temperature of the mash exceeding 140° by 10°; there is a further restriction due to the difference in temperature between that of the mashing liquor and the conversion temperature aimed at. For the malt, or at any rate part of it, is not only exposed to the temperature of the final mixture, but also to the higher temperature of the water necessary to produce the desired mash-heat. This excess temperature, if we may so term it, is a factor of considerable importance, and must not be ignored. Its full importance is plainly seen in the application of underlets. The underlet is generally a small "piece" of liquor; in order, therefore, to raise the temperature of the mash by means of it to any appreciable degree, it is necessary to apply it at a high heat—from 175° to 190° F. are not uncommon temperatures in practice. Even at the lower limit, the diastase in that portion of the mash upon which the underlet will first impinge will be destroyed. But the underlet will cool by coming into contact with further portions of the mash, so that the final portions of the mash which it will touch, will be exposed to but a very small excess temperature. In such a case the restriction of diastase caused by the underlet would not be represented by the two or three degrees rise in temperature of the mash obtained after thorough intermixture of the underlet with the tun-contents. The restriction is far greater, being principally due to the very appreciable enfeeblement of diastase in the earlier portions of mash into which the underlet was introduced, and to the complete annihilation of diastase in the earliest portions of the mash.

The excess temperature is generally overlooked; the mash-heat is alone regarded as responsible for the type of conversion. But this is not really the case ; the type of conversion is not only dependent upon the conversion temperature but also upon the way in which that conversion temperature is attained. If, for instance, we mash ordinary cold grist with water at such a heat as will give us an initial goods heat of 150°, we shall get a wort very different in type from one made by mashing the same malt for an initial heat of 150°, where this was attained by heating the grist as well as the liquor to this temperature, and thus mashing them both while hot.

The restriction of diastase then—which we shall presently see to be requisite in mashing operations—is dependent not only upon the temperature of the mixture, but also upon the manner in which that temperature has been attained. It is desirable to repeat and emphasise this statement, for we shall subsequently see that its bearings are of very great practical importance.

THE DUE RESTRICTION OF DIASTASE IN THE MASH-TUN.

It has already been said that it is as a rule necessary to restrict the diastase in the mash-tun, and this is done by employing higher conversion temperatures than those at which diastase is most active, and also by exposing the malt to water at temperatures in excess of such conversion temperatures. The object of so restricting the diastase is to secure in the wort a sufficiency of maltodextrins. Their type will for the present be neglected : for the moment we will only consider why it is that a sufficiency of this group of bodies is necessary. It will be convenient to classify their functions, and consider each individually. They may be divided up as follows :—

    (*a*) Condition, or after-fermentation (with which is generally associated freedom from hop-resin turbidity).

    (*b*) Body, or fulness.

    (*c*) Soundness, or stability.

    (*d*) Viscosity and " head "

(*a*) It will be remembered that the normal or medium maltodextrins are unfermentable during the primary fermentation, but degrade to maltose and then ferment at the instance of the cask yeasts. The steady after-fermentation of beer in cask or other storing vessel is therefore dependent on these substances ; and if wort, and therefore beer, contains none of them, or too little of them, steady and persistent after-fermentation will be either defective or non-existent. The "condition" of beer is of course essential to success ; without it the beer, however good in other respects, would be unsaleable.

It frequently happens that beer remains turbid through suspended hop-resins. This is particularly the case where there is a lack of hot aëration of worts during cooling, but it may arise from some imperfection of the hops, or some imperfection in the yeast. The detection of the resin turbidity will be subsequently dealt with ; but it may here be briefly stated that where resin turbidity exists there is no better means for dealing with it than by getting the beers into a vigorous after-fermentation. The reproduction of yeast in the cask seems specially effective in removing the resins, and there is every reason to suppose that their removal is not an absorption of them by the yeast, but a physical attachment of the resinous matter to the surface of the yeast-cells. Since the cask fermentation—or at any rate natural cask fermentation—depends upon a due amount of maltodextrin, it is clear that a sufficiency of these bodies is necessary to provide against the contingency of hop-resin turbidity. Beer defective in maltodextrin is specially likely to suffer from resin turbidity, and where this is so, it can only be removed by artificial cask fermentation, by priming in fact. But priming is not always permissible or advisable, partly in respect of flavour, partly in respect of the possibility of an excessive cask fermentation. The steady persistent cask fermentation, such as follows when the beer is racked with a due amount of maltodextrin, is the best incentive to a removal of this form of turbidity, while the flavour and other characteristics of beer are not interfered with as they might be by artificial additions.

When beer is racked with what appears to be an appreciable amount of unfermented maltose a certain kind of after fermentation is produced. But this after-fermentation is very different from the steady natural after-fermentation wanted; in summer it is tumultuous, and constitutes fretting.

(*b*) Since the normal maltodextrins do not ferment out during the primary fermentation, they will exist in the racked beer to their full extent; and since they are only slowly degraded and fermented in cask, well-brewed beer, stored for however long, will always contain a good proportion of them. As maltose gives sweetness to the beer, and to the dextrin is due roundness, dry-fulness and viscosity, and as the flavour of a maltodextrin is that of a mixture of the constituent maltose and dextrin groups, it is clear that the presence in the beer of maltodextrins at all stages is very important in respect of fulness. The fulness of a beer may be either of a sweet or of a dry character, and either kind of fulness may be attained by controlling the relative proportions of combined maltose and dextrin locked up together as maltodextrin. That this can be done will be subsequently demonstrated; but of whatever kind the fulness is to be, it will be necessary that there should be a sufficiency of the maltodextrins.

(*c*) The stability of a beer is a subject demanding attention from two separate points of view : the nature and amount of infection, and the degree of resistance offered to that infection. For the moment the nature and amount of infection will be taken as a constant—for at present we are dealing only with the chemical constitution of the worts. That worts containing a fair abundance of maltodextrin do raise the resisting capacity of a beer, is not only borne out by practical experience in the brewery, but is capable of experimental demonstration; for if a small amount of maltodextrin (such as can be artificially prepared by acting upon starch with sulphuric acid under certain conditions) be added to one set of samples of a beer (preferably for this purpose an inferior beer), and if another set of samples of the same beer are left untreated, and if these two sets of samples are forced side by side on the same forcing-tray for twenty-one days, it will be found that the samples to which the maltodextrin has been added show more stability than those untreated in this way; in other words, they are more free from bacteria, and from the acid products of these bacteria, than are those samples which were not treated with maltodextrin. It is not, however, to be assumed that the maltodextrin possesses any direct prejudicial action on bacteria; in every probability it has no such action. But we must remember that by increasing the maltodextrin we give the cask yeasts more work to do, and the continued steady activity and growth of the yeast will serve to keep down the activity and growth of bacteria. The case is, in fact, similar to what occurs in an earlier stage of brewing. The worts, as they lie on the coolers, become infected with myriads of organisms which could not fail to break down the stability of any beer, were it not that the subsequent pitching or yeasting of the beer, and the consequent growth of that yeast in the fermenting vessel, keep down the development of the organisms which the worts have absorbed. Similarly, then, by keeping a yeast active and vigorous in cask, the bacteria are checked and partially crowded out.

In the lectures on brewing of which this work is an enlargement, a different reason was given to explain the stability induced by a due restriction of the diastase in the mash-tun. It was there suggested that the restriction of the diastase would be effected concurrently with the restriction of the ferment (the so-called "peptase") which acts upon the malt-albuminoids during mashing, forming from them the so-called peptones. According to that view, and so far as stability was concerned, the brewer in aiming at a reduction of the diastase, was really aiming at a reduction of the amount of peptones formed. These bodies were credited with the quality of putrescibility, and it was assumed that the defective stability of a beer brewed on the lines of insufficient restriction in the mash-tun conversion was directly due to the above putrescible bodies being formed in excess.

The above view was put forward with due reserve, and it was suggested as a theory, and a theory only. At the present moment we are inclined to drop the theory, although we are not prepared to say that it is altogether inaccurate. It receives some confirmation from van Laer's researches on the viscous ferment, in which these peptones were found to be particularly well adapted to nourish the ferment in question ; and it is at any rate not improbable that what applies in this connection to the viscous ferment may similarly apply to other disease-organisms of beer. It may be added, too, as suggestive, that peptones play a large part in all nutritive preparations for the culture of bacteria. But since there is no direct evidence on this point, and as no work has been done to show any direct connection between instability of beer and peptone percentage in worts, it is wiser to put aside the theory, and to accept only that which has been investigated. That the maltodextrins in a beer do keep the cask yeasts active, and that, by so doing, they do indirectly induce stability, is a matter on which there is no doubt. Whether, however, the stability of a maltodextrinous beer, is wholly due to the facts mentioned is a question which cannot be answered in the present state of our knowledge.

As van Laer's work has been mentioned in connection with the peptone theory, it may be well to add that in the same investigation results are quoted which go to show that the ropy ferment will not develop so vigorously in the presence of much carbohydrate matter, as when the carbohydrate matter is low in amount or absent. This carbohydrate matter in the case of a beer is mainly maltodextrin ; and it is clear that, with regard to this particular ferment, van Laer comes to similar conclusions to those mentioned above, although in a very different way.

(*d*) The viscosity of a beer is in great measure due to the dextrin, whether that dextrin be free, or combined as malto-dextrin. Since the free dextrin in beers varies only within very slight limits, it is clear that it is the varying and controllable combined dextrin to which we have to look for the determination of a sufficient viscosity. The viscosity of beer is the cause of its retaining in it for some time the little bubbles of carbonic acid which rise to the surface when the pressure is relieved. A sufficiently viscous beer therefore will, in the language of the practical brewer, be " beady " or " lousy," and, as every brewer knows, this is a thing to be distinctly aimed at. For when a beer is insufficiently viscous the gas bubbles will at once fly to the surface, leaving the beer without any apparent sparkle or life in it ; it will be " still," and not effervescent in appearance.

The " head " too is also in great part dependent upon the dextrin : for it is this constituent which forms the pellicle over the escaping gas bubbles which give rise to the well-known froth. Whether that froth disappears at once (as it does in champagne, for instance) or whether it is retained, as in the case of well-brewed stout, will depend upon the dextrin ; and every brewer knows that the persistency of the " head," or froth, is a thing much prized by his customers.

The degree of headiness, and the permanentness of the head, are of course primarily dependent upon the condition or after-fermentation of the beer. The after-fermentation is, as we know, due to maltodextrin ; that the result may be perfected, especially so far as the appearance of beer is concerned, is, as we now see, also due to the same group of bodies.

That the above remarks on head and headiness are not purely theoretical may be shown by any brewer who will prime two samples of the same beer, in the one case with invert-sugar syrup, in the other with maltodextrin syrup.

Each syrup will give rise to a fermentation in cask, so that in either case the gas necessary for head and headiness will be there. But while in the case of the invert primed beer, the gas bubbles will rise rapidly to the surface, and the head will be transient; in the case of the maltodextrin primed beer, the gas bubbles will only slowly rise to the surface, and the head will be relatively permanent.

The Excessive Restriction of Diastase in the Mash-Tun.

When the diastase is excessively restricted in the mash-tun we get more than a sufficiency of maltodextrins—in fact, an excess of them, coupled with an unduly high type; and although, as we have seen, a sufficiency of these bodies when of normal type is essential to success, an excess, or abnormally high type of them, may be equally harmful. The defects to which an excess of maltodextrin will lead are of course different from those following an insufficiency—in fact they are the exaggeration, to a prejudicial degree, of the benefits accompanying a sufficiency of them.

Since the normal or medium maltodextrins are unfermentable during the primary fermentation, their percentage will clearly govern the racking gravity or the attenuation of the beer; and in the case of a wort containing an excess of them, the attenuations will be stubborn—in other words the racking gravity will be insufficiently low. A beer which does not attenuate sufficiently will be defective in several respects. In the first place it will want "cleanness" or delicacy of flavour, retaining some of the mawkish sweetness of the original wort. In the next place, clarification (whether natural or artificial) will be found difficult. Owing to the excessive amount of solid matter dissolved in the finished beer (and to the viscosity of the substances constituting the solid matter) the beer will rack with an excess of yeast suspended in it. In a less viscous medium the yeast finds no difficulty in working out during the fermentation, and consequently there is no excess in the racked beer. The excess viscosity which prevents the proper working out of the yeast during the primary fermentation also prevents the ready deposition of the yeast during storage, so that within the period of a moderate storage no natural clarification can be expected; and the finings used as an artificial clarifying agent are far less readily effective in the case of an unduly viscous beer than in a normal one. The consequence of excessive maltodextrins will therefore be to delay and sometimes prevent clarification.

Such beers as these too frequently throw up a yeasty head during storage, owing to the reproduction of the yeast-cells in fermenting out the excess of maltose formed by them from the excess of maltodextrins, and this is sometimes accompanied by a tumultuous and excessive cask-fermentation known as a fret.

Frets are certainly more often due to another cause—to the presence in the beer of what appears to be unfermented maltose, but what are really somewhat fermentable low-type maltodextrins. This will be dealt with subsequently. Still when the maltodextrins are of normal type, and when they are in decided excess, there is a tendency for the beer to show a species of fret, one in which there is much yeast reproduction and consequently much turbidity. Both forms of fret are of course more likely to be met with in the summer months, when the worts absorb from the air the large amounts of wild yeasts distributed in it during the warm season. This excessive kind of cask fermentation is not by any means conducive to stability. Stability is certainly determined by a sufficient conditioning, but any tendency to violence or tumultuousness seems on the other hand to subsequently induce the development of bacteria. That such is the case is only in accord with analogous natural phenomena.[*] Al-

* Compare 'Phenomena of Symbiosis.'

though, therefore, a sufficiency of maltodextrins is necessary for stability, an excess of them is harmful in this respect.

Although we cannot lay down any hard and fast rules for the maltodextrin rate, since this will depend greatly upon the class of beer it is desired to produce, and the use or non-use of sugars, yet as a rule, the best results are obtained when the worts contain from 12 to 16 per cent. of them calculated on dry wort solids.

Conditions other than Mash Liquor Heats Affecting the Restriction of Diastase.

(a) *The increase of heat due to the chemical action of diastatic conversion.* When malts are of good quality, and where the vessel is so constructed and so placed that there shall be a minimum of loss of heat by radiation, there is always a distinct rise in the temperature of the mash commencing shortly after the intermixture of the grist and water. This rise will vary according to the materials, the structure, and the position of the tun. Under favourable circumstances it may amount to 5° F.; it is more frequently 2–3° F., and in very many cases there is no appreciable rise at all. When the rise is appreciable, and especially when it is rapid, it becomes a factor for consideration; and upon the actual goods heats prevailing during, and subsequent to, the rise, will the type of conversion in great part depend. This fact is sufficiently obvious and requires no further explanation. It will also be clear that the extent to which the heat-increase will affect the conversion will depend upon the rapidity with which the rise commences and with which the maximum heat is attained. This is manifest from the considerations dealt with shortly in this chapter (p. 225) and in Chapter II. (p. 93). When in a given tun the rise in heat is in excess of what it usually is in that tun, we get some indication of an excessive diastatic conversion; and this should be met at once by the employment of a higher conversion-temperature, by which we shall keep the excessive vigour of the diastase in check.

(*b*) *Influence of concentration.* The concentration of a mash has been found to exercise a not inappreciable influence upon the type of wort produced ; the more liquid the mash, other things being equal, the more active the diastase. The brewer, however, cannot hope to exercise any very marked control over the type of his conversion by alterations in the liquor rate, for he is compelled, for practical reasons, to mash with not less than $1\frac{1}{2}$ barrels per quarter and not over 3 barrels per quarter. Although these limits are somewhat narrow from a theoretical standpoint, they are yet sufficiently wide, to allow of an extra diastatic conversion at the higher, as compared to the lower, rate. By this, we mean that any given malt mashed with three barrels will give a wort richer in maltose, and poorer in maltodextrins, than when worked with half that quantity of liquor, the conversion-temperatures remaining the same in each case. Whether this is due to any direct chemical influence of the increased bulk of liquor upon the diastase is doubtful ; it seems to us more probable that the increased diastatic activity prevailing in the presence of much water is due rather to the more complete extraction of such diastase as may be present under the influence of the greater volume of extracting medium. In other words, we are inclined to regard the matter as mechanical rather than chemical. That it should be so, would be only on a line with observations on all other extractive processes. Whenever we desire to extract the whole, or the maximum, of a soluble substance from a mixture with an insoluble substance, we use a large volume of water. In addition to the increase of water mechanically assisting the rapid extraction of diastase, it doubtless assists the conversion as a whole by wetting the starch more rapidly, and thus bringing it into a condition to be more rapidly acted upon by the diastase. It might be argued on the other side, that the mashing process is not quite analogous to an ordinary extracting process, in so far as such diastase as would not be dissolved out with $1\frac{1}{2}$ barrel per quarter, would certainly be extracted during the sparging operations. This is so, in a way; but it must be remembered that considerations of extract (as will subsequently be shown) compel the brewer to use hotter liquor for the spargings than for mashings, and although the bulk of sparge liquors might be adequate for the complete extraction of the residual diastase, its temperature would be such as to render it partially inert for purposes of saccharification.

(*c*) *Influence of time.* The researches quoted in Chapter II. (see p. 106) demonstrate that the composition of a wort is to all intents and purposes settled by the conditions prevailing during the first thirty minutes of mashing; and this amounts to saying that if we abstract from a mashing a sample of wort at the end of this time, and at the end of, say, two hours, the percentages of the maltose, maltodextrins, and dextrin, will be practically the same in each case, when calculated upon the dry wort solids. We are aware that this statement, which we base upon analyses of worts brewed on the practical scale, as well as upon the researches already referred to, is in direct contradiction to some experiments published by Graham.[*] Graham's experiments went to show that as a wort stands, so does the maltose contained in it increase; in other words, that the longer we stand, the more work do we get out of our diastase. It is not difficult to understand, however, when the methods adopted by Graham are considered, that the figures in question must not be accepted as representative of the influence of time upon diastatic action; for in the investigations two points have been ignored, and it is due to them that the results gave rise to an erroneous deduction. Graham has omitted to take into account (1) the ready-formed sugars in the malt, (2) the changes occurring in these

[*] Society Chemical Industry; Proceedings, Annual General Meeting, 1881, p. 17.

sugars during mashing. He calculates his maltose from the gross reducing action of the wort samples, and his dextrin is obtained by converting the maltose and dextrin into dextrose by boiling with acid, estimating the dextrose by Fehling's solution, subtracting the dextrose due to the maltose, and referring back the balance of dextrose to dextrin. We are not concerned for the moment to criticise the above method of determining dextrin; this is treated of in the analytical section of the work (p. 456), and for the moment we will assume that the above method does give us the dextrin with sufficient accuracy. It will be clear, however, that the two points above referred to are of substantial importance.

In estimating the maltose from the total reducing power of the wort, Graham includes the invert-sugar present in the malt. In converting the dextrin into dextrose for the estimation of the latter he will include, besides the dextrose due to the conversion of the maltose, the invert-sugar formed from the cane-sugar in the malt by the action of the boiling acid. The maltose, therefore, includes the invert-sugar; the dextrin includes the cane-sugar. If these sugars suffered no change during the mashing process, the above method, although inherently faulty, would not necessarily detract from the deductions which might be formed from the results in respect to maltose and dextrin proportions. But they do alter, for the cane-sugar will gradually become invert-sugar as the wort stands in the mashing vessel (see p. 155). The effects of this, according to Graham's method, will be clearly to raise the *apparent* maltose, and to decrease the *apparent* dextrin during the prolongation of the mashing process. Suppose only 1 per cent. of cane-sugar be inverted, this will show in the relative proportions of maltose and dextrin, as estimated by Graham's method, as follows: if there were, previous to that inversion, 43 per cent. (apparent) maltose and 14 per cent. (apparent) dextrin, the figures after inversion will stand at maltose 45 per cent., dextrin 13 per cent.

In view of these facts, we do not feel inclined to regard Graham's figures as necessitating any qualification of the statement we have made, that is, that, so far as time is concerned, with respect to the quality or type of wort produced; the wort after standing for thirty minutes, and for two hours, will be to all intents and purposes the same.

Having regard to the fact that the type of wort is practically settled within the first half hour of the intermixture of the grist and the liquor, it will follow that in the employment of " underlets," the time at which the underlet is applied will exercise a very appreciable influence. If, for instance, the underlet is added immediately after the mash is in, the rise of temperature at this stage will be seen to restrict the diastatic activity very appreciably, and, as a consequence, we shall find in our wort less maltose and more maltodextrin than had the underlet not been applied. But if, on the other hand, the underlet is applied, say, thirty-five minutes after mashing, its influence in regard to diastatic action will be inappreciable, for by the time it is applied, the type of wort is practically already settled. The earlier we apply an underlet, therefore, the more restriction shall we get from the rise in heat which its application effects; and after a certain time we may regard its application as influencing only the type of extract which is yielded by the malt during later mashing and during sparging. So far as that portion of the extract is concerned which is yielded by the malt at once, or at any rate quite early in the process—and in good material this constitutes nearly the whole of the total extract—its influence will be quite immaterial.

It may occur to readers to inquire at this point why, if this be so, the worts should not be taken at the expiration of the time when the chemical composition of the wort has arrived at a point of equilibrium; why, in fact, the brew should stand two hours or more, when thirty minutes would seem from the above consideration to suffice? The answer to this is

Q 3

that although the quality of the extract may be the same in either case, the quantity will be different, and it this question of quantity of extract, that compels the brewer to stand his mash for a longer time than that which determines the relative chemical composition of the wort. This will be shortly treated of.

(*d*) *Influence of saline water constituents on diastatic restriction.* This has been dealt with in Chapter I., to which reference should here be made (p. 19).

RESTRICTION OF DIASTASE IN MASH-TUN COMPARED TO RESTRICTION OF DIASTASE ON THE KILN FLOORS.

Although hitherto only lightly touched upon, it will be remembered that the higher our heats in our mash-tun the more restricted will be the action of the diastase—in other words, the more rich in maltodextrins will be our worts, and the higher will be their type. That statement will apply equally to any given quantity of malt, of whatever quality or kind.

It will be remembered, from considerations given in detail in Chapter II., that the higher we dry a malt on the kiln floors, the less diastase will there be in that malt. Hence, if we now regard the mashing system as a constant, and the malt as variable, we may say that the higher dried our malt is, the more restricted will be the action of the diastase, and the more rich in maltodextrins will be our worts. Now it is quite possible to conceive that we may obtain an equal amount of restriction by mashing on the one hand a high-dried malt at a low temperature and a low-dried malt at a high temperature. That such a thing is not only possible but practically attainable must be clear. Let us suppose now that we have so adjusted our two mashing systems to our two malts that in each case the diastase has been working under an equal burden of restriction. The point arises, will the two worts so produced be identical in chemical composition, or will they differ? This point is one of great importance in our opinion, and we shall have to deal with it at length.

Two worts prepared in the manner above described, if the adjustment has been perfect, will polarise identically. This will be clear from the fact that the diastase restriction has been identical in each case. Now the polarimetric value of a wort depends upon the relative amounts of total maltose (i. e., free maltose *plus* maltose in maltodextrin) and of total dextrin (i. e., free dextrin *plus* dextrin in maltodextrin). The two worts, then, contain equal amounts of total maltose and total dextrin. It might at first sight be assumed from this that the worts would be identical; but they would not be so; and there will be this important difference. If we subject the two worts to fermentation we shall find that the one (that made from the lightly-dried malt mashed high) will ferment to a lower point than the wort yielded by the high-dried malt when mashed low. If, after the fermentation, we determine the reducing matter present by Fehling's solution, we shall find a further difference: the wort yielded by the low-dried malt when mashed high containing after fermentation far less reducing matter than the other wort. Now the reducing action in a fermented wort is due to the maltose in the maltodextrin*; consequently the wort yielded by the high-dried malt when mashed low will—although containing the same amount of *total maltose* as the other wort—contain a larger amount of maltose locked up in the maltodextrin. In other words, the total maltose in the two worts is the same, although it is differently apportioned; the wort from the high-dried malt mashed low containing more combined maltose (or

---

* We will for the moment disregard the unfermentable and undegradable substances formed in malts, which also reduce Fehling's solution; for the moment these may be assumed to be the same in each case, and the same allowance be made for them.

maltodextrin maltose) than the other ; and since the totals
are the same it will follow that the wort from the high-dried
malt when mashed low will contain less free maltose.   That
such is the case explains the less attenuation obtained in the
case of that wort when fermented.

From these facts it follows that two worts may be prepared as above described which give identical percentages of
total maltose and total dextrin, and which yet behave
differently during fermentation, and which will be different in
other respects, and further that such differences are due to the
differences in type of maltodextrins yielded in either case.

To make the matter clearer we will take a hypothetic
example in which, for the sake of simplicity, the figures are
assumed to come out as integers.

Wort A is from low-dried malt mashed high ; wort B is
from high-dried malt mashed low.

	Wort A (calculated on 100 parts starch conversion products).	Wort B (calculated on 100 parts starch conversion products).
Free dextrin	20	20
Dextrin in maltodextrin	10 }	10 }
Maltose in maltodextrin	5 }	15 }
Free maltose	65	55
	—	—
	100	100
In the above,		
Total dextrin .. .. .: ..	30	30
„ maltose .ʼ	70	70
„ maltodextrin	15	25
Type of maltodextrin	{ 1 maltose 2 dextrin	{ 1 maltose 0·66 dextrin

The above figures show clearly then how two worts,
although containing the same total maltose and total dextrin,

may contain different quantities of free maltose and of maltodextrins, and the type of maltodextrin may be also different. It is clear, too, that the wort A would ferment to a considerably lower point than wort B; and it is similarly clear that the reducing action of wort B, after fermentation, would be much higher than that of wort A.

We do not in the least commit ourselves to any theory attempting to explain why, when the reduction of diastase is mainly in the mash-tun, we get a higher type of maltodextrins than when the reduction is mainly on the kiln floors. All that we can say on this point is that, in the analysis of many hundreds of worts, we have found the type to be higher when low-dried malt is mashed high than when high-dried malt is mashed low, even although in both cases the worts may contain identical quantities of total maltose and total dextrin.

These considerations open up fresh ground—the control of the type of maltodextrins, as well as their percentage, by the selection of malts and the application of certain systems of mashing to such malts. To put it briefly, we may say that when we get our main restriction of diastase in the malt itself, we get a maltodextrin richer in maltose than when we aim for that restriction in the mash-tun. In other words, the dry heat of the kiln floor tends to give us worts containing maltodextrins richer in maltose than the wet heat of the mash-tun. Which type of maltodextrin is the better, we shall now proceed to discuss. As a matter of fact, it will be seen that for certain beers we require the one kind; for other beers the other. Both are equally desirable in their respective places.

THE INFLUENCE OF THE TYPE OF MALTODEXTRIN IN
WORT UPON THE CHARACTER OF THE BEER.

It will be convenient to discuss this question under
distinct heads, which we will take in this order.

(*a*) Conditioning, or secondary fermentation.

(*b*) Soundness.

(*c*) Flavour.

(*d*) Brightness.

(*e*) " Head."

(*a*) *Conditioning.* It will be remembered that the natural
conditioning of beer depends in normal circumstances upon
the degradation of maltodextrins in cask, by means of the
diastase-like ferment secreted by these secondary yeasts, and
upon the consequent fermentation of the maltose so produced.
Now although it is true that all maltodextrins, of whatever
type, are completely broken down by diastase in a relatively
short time into maltose, when the diastase operates at temperatures and under conditions which are most favourable to
its activity ; yet, when the degrading ferment is not diastase,
but only the diastase-like ferment which is secreted by the
secondary yeasts, and when the temperatures and conditions
are those which prevail in the cask, and which of course are
not those most favouring its activity, the duration of degradation is by no means short ; in fact, beers stored for very long
periods, are found to still contain an appreciable amount of
maltodextrins.   Thus an old beer, brewed by Messrs. Worth-
ington and Co., of Burton, about the year 1790, was proved
by Brown and Morris to contain 17·57 per cent. of maltodextrins, calculated on the wort solids.   The degradation of
the maltodextrins in cask, and the elimination of the resulting
maltose by fermentation, is therefore a slow process, and it is
in regard to its duration and the time which has to elapse
before its initiation, that the type of the contained maltodextrin plays an important part.

If the same beer is analysed at different intervals during storage, it will be found that the type of maltodextrin in it will have become lower, i. e., the proportion of maltose in the maltodextrin will have become greater. The following figures show this to be the case.

	*A.	B.
Percentage of combined maltose †	13·33	11·84
,, ,, dextrin	5·24	2·87
Total maltodextrin	18·57	14·71
Type	$\begin{cases} 2\cdot5\ \text{M} \\ 1\cdot0\ \text{D} \end{cases}$	$\begin{cases} 4\ \text{M} \\ 1\ \text{D} \end{cases}$

These results are confirmed by Brown and Morris in the paper just referred to, in which lengthy storage is clearly proved to exercise a degrading influence upon the type of maltodextrins. The authors find that in a new strong ale the type was $\begin{cases} 1\cdot7\ \text{maltose} \\ 1\ \text{dextrin,} \end{cases}$ while in an old strong ale it was $\begin{cases} 2\cdot4\ \text{maltose} \\ 1\ \text{dextrin.} \end{cases}$

It is thus manifest that the degradation of maltodextrins proceeds, so to say, in two stages: first, the lowering of the type until the maltodextrin shall contain a certain high proportion of maltose; second, the degradation of such a maltose-rich type into free maltose. Now this being so, it will follow, other things being equal, that if our beer, to start with, contains a type of maltodextrin rich in maltose, such a beer will come into condition more rapidly than a beer containing an equal amount of a dextrinous maltodextrin. In fact by starting with the maltose-rich maltodextrins we

* The interval between analysis A and analysis B was four months.
† These figures are calculated on 100 parts of dry wort-solids.

dispense to some extent with the preparatory degradation process, i. e., the degradation of the type. A maltose-rich maltodextrin will, in other words, produce maltose more rapidly than a dextrinous maltodextrin, or beers containing it will come into condition more rapidly.

It will therefore follow that if in practice we require a beer to come into rapid condition, we shall aim for a malto-dextrin in our racked beer rich in maltose. This requirement applies particularly to running mild ales, which must as a rule be delivered in condition a very few days after racking. In such beers, therefore, we must use those means which yield the desired amount of combined maltose. From what has been said, it will be remembered that a high-dried malt mashed at a low heat will best fulfil our object. Practical men have long been accustomed to brew their mild running ales on these lines; and the connection between early condition and the high pro-portion of combined maltose in the maltodextrin so obtained is one of the reasons for these lines having been successful.

In a stock beer, on the other hand, early condition is neither requisite nor desirable. In such beer we want rather a slow and persistent condition, an after fermentation which shall continue throughout the lengthy period of storage; but its actual commencement may without harm be delayed to several weeks after racking. In order to produce this result, it is obvious from what has been said that our type of malto-dextrin should be of the opposite character to that required for running ales. We want, in fact, a dextrinous maltodextrin; and such a maltodextrin is obtainable by mashing our malts high in the mash-tun.

Intermediate between running mild ales on the one hand, and stock ales on the other, come a variety of different beers which have sprung into demand with the recent variation in public taste. A light family mild ale, such for instance as forms the staple product of many breweries at the present day, must come into fairly rapid condition—much more rapidly than stock ales, and not necessarily so rapidly as running mild ales. For ales of this character we should therefore want a maltodextrin of intermediate character.

The following determinations of maltodextrin in the worts of typical ales will serve as examples.

—	Family Bitter (19 lb.).	Stock Bitter (24 lb.).	Running Ale (18 lb.).	Stout (27 lb.).
Maltose in maltodextrin ..	4·19	3·75	4·70	10·00
Dextrin in maltodextrin ..	7·24	7·44	4·66	8·30
Total maltodextrin	11·43	11·19	9·36	18·30
Type	{1 M. {1·7 D.	{1 M. {2 D.	{1 M. {1 D.	{1 M. {0·8 D.

(*b*) *Soundness.* As has already been explained, the soundness of a beer will depend in a great measure upon the continued activity of the cask yeasts. For a beer, therefore, which it is desired to stock, we shall require a sufficient amount of maltodextrin, and it must be of such a type as will yield a long and persistent cask-fermentation rather than a rapid and relatively transient one. For stock ales, therefore, we shall want a dextrinous maltodextrin, one in fact obtainable by high mashing heats. When, however, marked stability is of secondary importance, as in mild running ales, the type need not necessarily be dextrinous; and if a relatively low type maltodextrin is more suitable on other grounds for beer of this sort, there will be no need, on the ground of stability, to aim at altering the type. Now, so far as conditioning is concerned, we have seen that a maltose-rich maltodextrin is most suited to running ales; and we shall afterwards see that it is similarly the case in respect of flavour. That being so, and beer of this sort not requiring particular stability, it follows that the low heat in mash-tun (combined with high-dried material) which we know to produce the required type of maltodextrin for mild ales, will on the whole best meet the general requirements of ales of this class. For stock ales, on the other hand, where soundness is of paramount importance, a dextrinous maltodextrin is a necessity, and to obtain it we must needs use high mashing temperatures.

(*c*) *Flavour.* The flavour of the maltodextrins is (like their reducing power and their opticity) the same as that of a *mixture* of the maltose and dextrin contained in them. This is a point upon which we have satisfied ourselves by tasting the qualities of maltodextrin made on the practical scale by the limited conversion of starch by means of acidulated water at the boiling point. Now since the maltodextrin constitutes by far the largest proportion of the solid matter of a beer as racked, it is clear that the flavour imparted by that malto-dextrin will in any case be a prime factor in determining the flavour of the beer itself. It is, of course, the maltose con-stituents of the maltodextrin which will confer sweetness and that palate fulness which English brewers always associate with sweetness. The dextrin constituent will confer a certain dry body, but certainly no fulness in the sense of sweet fulness. It will follow, then, that if our maltodextrin is one rich in maltose, and granting a sufficiency of it, the flavour of the beer will be sweet and full, such a flavour, in fact, as we demand in a successfully brewed mild ale. When, on the other hand, the maltodextrin is dextrinous, we shall have no great sweet-ness, but a clean, dry and delicate flavour : such a flavour as is demanded of a successfully brewed pale ale, say of the Burton type.

Now it has been shown on other grounds that the condi-tions required for the production of dextrinous malto-dextrin are those suited to the production of pale ales ; and it is now seen that also in point of flavour, such conditions tally at once with our practical requirements. Fairly low-dried malts, then, mashed high, will, by producing dextrinous maltodextrins, give the right flavour for pale ales. Analogously, the determination of a sweet flavour by means of a low-mashed high-dried malt, will tally with the practical requirements of a mild ale.

(*d*) *Brightness.* Brightness has been already touched upon in connection with the maltodextrins, but rather from the view of percentage amount than of type. It will be remembered that in the event of our having an excess of maltodextrin, the viscosity of the wort or beer made is such as to seriously impede the natural subsidence of yeasts, or their removal by finings ; and it will also be remembered that when these bodies are markedly deficient, the lack of condition may leave hop-resins uneliminated, which, in their turn, will make the beer dim and thick. We will now, however, assume that the actual amount of maltodextrin is fixed, and that it is the type only which varies. So far as the first point is concerned (viscosity) it is clear that this will depend upon the dextrin constituent of the maltodextrin, i. e., the more dextrinous it is, the more viscosity will any given amount of maltodextrin confer. It is therefore clear that if we desire early brilliancy, we must aim rather at a maltose-rich maltodextrin than a dextrinous one ; in mild ales, therefore, we shall do well to aim for such a type. Consequently, in such beers, we shall do well on this ground, as well as on the other grounds previously considered, to obtain a maltose-rich maltodextrin by means of a low-mashing heat and a high-dried malt.

In the case of pale ales, such early brilliancy is not a requisite ; for beers of this sort are invariably stored for some time before consumption, during which time the amount of maltodextrin will have become reduced, and the combined dextrin will be reduced not only proportionately to the reduction of the total maltodextrin but to a greater extent on account of the degradation of the type. From these considerations, then, and having regard to the longer storage of pale ales, we can safely aim for a dextrinous maltodextrin in the case of pale ales of the ordinary kind. When by chance, however, pale ales are required to come into brilliant condition particularly early, it might be well to reduce the dextrin constituent of the maltodextrin by mashing lower, and compensating for this extra play of diastase in the mash-tun by slightly raising the heats on the kiln floors.

So far as the question of hop-resin turbidity is concerned, the considerations applying to them will be the same as those applying to cask fermentation. Consequently, for mild ales where early brilliancy is requisite, we must aim for a maltodextrin rich in maltose, such a type we know being particularly easily degradable and fermentable, and therefore giving rise to after-fermentation with particular ease. On this ground, then, we shall brew these beers at low heats and from high-dried material. But, pale ales, not requiring to fine so rapidly, may be allowed to come into condition more slowly, and hence a dextrinous maltodextrin will be permissible to such beers.

(*e*) *Head.* This point has been already touched upon in connection with the percentage of maltodextrin, to which it bears, perhaps, a closer relationship than with the type. But it is a fair presumption that it is the dextrin in the maltodextrin which gives rise to the permanency of the head. This appearance would probably be of advantage in the case of mild as well as pale ales. But in mild ales it is hardly expected, and it is certainly not worth while altering the type of maltodextrin which we know to be most suitable for mild ales to secure a property which is not by any means an absolute requirement. In the case of fine pale ales, however, the permanency of head is a thing looked for and demanded; and the high rate of combined dextrin which the conditions of pale ale brewing provide, will serve to secure this property to such ales.

THE DRAWING-OFF OF WORT.

The requisite amount of maltose and maltodextrin and the required type of the latter having been obtained by the accurate adjustment of mashing-heats and concentration, and a sufficient time having elapsed for the conversion of that portion of the starch which refuses to gelatinise during the more active conversion, the taps are opened and the wort drawn off. The time required for standing the mash with normal material is about two hours. With badly vegetated malts, which contain starch, requiring a longer time for its solution, two and a half hours may be given; with well-germinated malts, on the other hand, one and a half hours will frequently suffice; taking matters generally, however, two hours is the time usually allowed by practical men, and it is found to answer well in the majority of cases.

The taps being opened, the wort filters through the false bottom, and emerges from the taps bright; at any rate it should be so after the first ten minutes or so. Turbidity during the very early stages is not of very much account, being caused by particles of husk which have accidentally passed through the holes or slots of the false-bottom plates. If the turbidity is persistent, things are not going as they should do. Such persistent turbidity may be caused by excessive use of the rakes, or by their use too late in the operations, and in these circumstances the remedy is of course easy. But when the worts are turbid with starch (which can be readily tested for in the usual manner (p. 96), it is a certain proof that the mashing-heats are wrong. Starch in the worts at this stage is so unusual as hardly to become a point worthy of lengthy discussion; the later runnings are far more likely to be starchy, and the means to be adopted for dealing with this defect will be discussed when we consider the later stages of the mash-tun operations.

The wort as it flows from the mash-tun either pours into

a sort of collecting vessel known as the underback, or it may flow direct into the coppers. The underback is, however, a very commonly used vessel on account of its convenience. Worts have very frequently to be pumped into the copper, and where this is so, some such vessel as an underback is obviously necessary. In some breweries the mash-tun commands the copper, but this is rare; and unless the whole brewery is worked on the gravitation principle, where pumping is wholly unnecessary, it is wisest to have the pumping before the boiling stage, rather than after it. A pump is a relatively difficult thing to keep perfectly clean, and it is best that such contamination as may be picked up by wort in this way should affect it before boiling. Where, then, pumping has to be done somewhere, it is judicious to get it over before boiling, and in this case the underback is a necessary appliance.

The underback is a vessel seldom large enough to take the whole of the wort; nor is this necessary, for the spending of the wort and the pumping to copper can be carried on simultaneously. It is usual to fit this vessel with steam coils sufficiently powerful to either boil the wort, or at any rate keep it from any drop in temperature. The benefit of these coils at this stage seems to be generally admitted by practical men, although the reason generally adduced for these benefits is an erroneous one. It is stated, in fact, that the object of the coil is to prevent further diastatic conversion in the wort in underback, and with this object brewers generally aim at raising the temperature of the wort beyond the diastatic limit. But as a matter of fact there is no risk of any such action. The type of wort is settled during the first 30 minutes of mashing; and although there is an abundance of diastase present at this point, such residual diastase has no further action upon the maltodextrins in the wort, either during the later stages of mashing or sparging, or during digestion in the underback at temperatures suited to its activity. It is true that this diastase is quite competent to degrade to a certain extent starch, or starch products higher in the scale than those obtaining in the wort ; but so far as the wort constituents are concerned, there is no further action either during late mashing, sparging, or digestion in the underback. So far as the latter point is concerned, it has been definitely shown by one of us, with Levett and Swift, that worts digested for one hour at the same heats as those at which they were mashed showed no degradation whatever. The following figures prove this :—[*]

—	$K_{3.86}$ before digestion.	$K_{3.86}$ after digestion.
Malt from a London Brewery :—		
Wort A, mashed at 140° ..	50·0	50·1
Wort B, mashed at 150° ..	46·3	45·5
Wort C, mashed at 155° ..	40·0	40·4

In the face of the general consensus of opinion among practical men that underback-wort should be kept hot, it is, perhaps, unsafe to say there is no benefit in so doing. But whatever the benefit may be, it has no connection whatever with a continued diastatic conversion at this stage.

Although the diastase of the wort in the underback is without influence upon the maltodextrins contained in that wort, yet it has (as has been already stated) a very distinct and appreciable influence upon starch and upon starch-products, which are higher in the series than the maltodextrins existing in the wort containing this diastase. It is to this property of diastase as it exists after say two hours' stand that we owe the conversion of the starch which is brought into solution during sparging, and to which reference will shortly be made.

* Brewing Trade Review, 1891, p. 6.

R

We have found, in fact, that the diastase of underback-wort, though inoperative as regards the maltodextrins contained in that wort, will tend to convert starch and the higher starch-products down to the same type of extract as happens to be in the wort. That this is so, tallies with what we know as to the stages of equilibrium reached in conversions of starch-paste by diastatic infusions previously restricted by heating to definite temperatures (Chap. II., p. 113).

Some brewers, in dealing with that kind of unsatisfactory material which yields starch towards the end of the sparging process, have been benefited by retaining in the underback a portion of the first wort at about the temperature at which it spends into that vessel, mixing it afterwards with the starchy later runnings, and digesting the mixture at about 155° for from thirty to sixty minutes. By so doing the starch 'is converted, and the conversion stops short at a reasonable point ; and in this respect the method is preferable to treating the starchy runnings with diastase solutions, for these will very rapidly run down the starch to free maltose and dextrin, and there will be but little chance of retaining any malto-dextrins in them.

The wort then either passes from the mash-tun into the underback, and thence to copper, or else to copper direct. If the goods are drained as dry as is possible, they will still retain a fair amount of wort : rather under one barrel of wort to the quarter of malt. By merely opening the taps, then, we obtain only half to two-thirds of the total amount of wort which we have prepared. To extract the remainder it is necessary to wash the goods in some way or other. This is usually done by means of what is known as a sparger.

It is necessary to carefully guard against an impression that might at first be formed that sparging operations consist entirely in the mechanical washing out of the wort already prepared. In a great measure this is the case ; but something more than this is done. Were sparging merely an extraction of ready prepared wort, it would be a matter of indifference at what temperature the sparge liquor were applied, so long as there were sufficiency of it. But the temperature of the sparge liquor is only second in importance to that of the wort liquor ; and for this reason.

During the standing of the mash, some of the starch which is ungelatinised, and therefore unconverted, during the quite early stages of mashing, is gradually brought into a soluble form, and then converted. This is the object of standing our mash for about two hours, although, as we know, the type of wort is settled in about 30 minutes. But the temperatures existing in the tun at this stage, and the time allowed, are insufficient for dealing with the whole of the residual malt starch ; consequently, when the tun is set to drain at the expiration of the stand, the goods still contain a very appreciable amount of starch which up to this point has resisted gelatinisation. The amount of such residual starch will vary principally with the malts ; it being far less in well-germinated, good malt than in under-vegetated, inferior malt. That this is so, follows from the fact that in under-vegetated malts the dissolution of cellulosic tissue surrounding the starch-granules is not sufficiently complete, and, as a consequence, the starch will not entirely gelatinise at the relatively low heats prevailing during the early stages of the mash. It will also vary with the mashing conditions ; a low mashing heat clearly leaving a greater amount ungelatinised than a high mashing heat. But it may be safely asserted that even the best malts mashed on the most appropriate lines will always fail to convert entirely during early mashing, and consequently when we start sparging the sparge liquor will impinge on what we may term a starch-containing material.

Now it is obvious, this being so, that the temperature of our sparge liquor is no indifferent matter. If the temperature be low, we merely extract the wort, leaving the residual starch unaffected ; and this means a loss of total extract. If,

R 2

on the other hand, we sparge too high, we certainly gelatinise
the whole or nearly the whole of the residual starch ; but
since the diastase left in the tun at this stage will be enfeebled
by these higher heats (and perhaps to a dangerous extent),
and as the high temperature of the sparge liquor will still
further enfeeble it, it follows that sparging of this sort will
lead to the gelatinisation of the starch, and to the extraction
of that body as such with the worts. To have starch in
our worts in an unconverted condition is, it is hardly necessary
to say, bad. It will at any rate interfere with brilliancy,
on account of its viscosity.

In order then to avoid loss of extract, and in order to
prevent the introduction of starch, or of starch-products very
nearly allied to starch, into our worts, it is advisable to adopt
a medium course between high and low sparge heats. At any
rate, they must be sufficiently high to somewhat raise the
temperature of the goods in the tun. If they were not, they
would not be competent to affect the starch which refused to
gelatinise during previous mashing operations. But, for the
reason before given, they should not greatly exceed this heat,
even at the end of operations. It has been found by experience—and experience in this case tallies with what one
would suppose on logical grounds—that a slight and gradual
rise in the temperature of the goods is the course to be aimed
at. If, for instance, our goods are at 152° when taps are set,
it should be our object to raise that temperature to about
154–155° during the early stages of sparging, permitting a
gradual rise to about 160° or so towards the end. In applying
our sparge liquor, however, it must be remembered that during
the standing of the mash and the spending of the wort the
top strata of the goods will have cooled to a considerable
degree, and it is necessary on that ground to sparge on the
first ten barrels at a higher heat than the rest. The sparge
liquor will therefore go on hottest at first, and gradually drop
in heat, but the temperature of the goods (and it is this which

of course determines the nature of the changes) will slightly rise.

Sparging is generally continued until nearly the whole of the available matter has been extracted. Not quite the whole, however, for all grain, even after most exhaustive sparging, will always contain some unaffected starch. To obtain this minute residue would not, however, pay: an excessive amount of sparge liquor would have to be employed, the subsequent necessary evaporation of which would be a costly process. Besides this, towards the end of sparging operations, as generally conducted, the whole of the diastase would be washed out, and the last traces of starch in the goods would therefore pass as such into the worts.

The following analyses of worts, A from well-grown malt, B from under-vegetated malt, show the maltodextrins in the respective extracts.

		1st Copper.	2nd Copper.
A. From well-grown malt.	Combined maltose	3·75	4·22
	Combined dextrin	7·44	8·40
	Total maltodextrin	11·19	12·62
	Type	{ 1 M. { 2 D.	{ 1 M. { 2 D.
B. From badly-grown malt.	Combined maltose	2·06	4·10
	Combined dextrin	2·64	12·43
	Total maltodextrin	4·70	16·53
	Type	{ 1 M. { 1·3 D.	{ 1 M. { 3 D.

It will be seen that in the case of worts from malt A, the maltodextrins are of practically the same type in the first and second coppers. This is the result of well-made malt mashed appropriately. But where the ill-vegetated malt **B** was used, we find in second copper, the result of a very incomplete conversion of that portion of the starch (and in this case it would constitute an appreciable fraction of the whole) which refused to gelatinise until sparging operations had commenced.

In dealing with fairly good material, abnormal types and amounts of maltodextrins are not to be anticipated. But when material is under-vegetated, and when, in consequence, the starch is stubborn to gelatinise, the tendency to yield either starch, or what may be termed incompletely transformed starch, during the later stages of sparging, is a very serious difficulty, and it may be said that, to all intents and purposes, these very dextrinous maltodextrins which are closely allied to starch, are as bad as starch in practical operations. In the year 1889, brewers using the 1888 barleys, experienced an enormous amount of difficulty from this particular cause. The weather of 1888 was so unseasonable as to produce barleys, not only here, but in many countries abroad, which vegetated most unsatisfactorily. As a consequence, they contained much starch which refused to gelatinise during mashing, thus leaving an excessive residue for extraction during sparging.

In dealing with malt of this sort, it is well to mash in such a manner that the mash-tun heats shall rise gradually throughout the operations, that the maximum limit shall be a fairly low one, and that the various standing periods shall be increased. It is found to be an advantage, too, to have a second mash before starting sparging operations, in order, of course, to reduce the amount of residual starch which would otherwise fail to be gelatinised until sparging was reached. The most successful plan is to mash for about 147° as an initial goods-heat, bringing up to 150° by a "piece" 25 minutes after mashing, standing 2½ hours from time of mashing, setting tap, and drawing off one barrel per quarter, then mashing up

again with 1½ barrels per quarter, at such a heat as will raise
the temperature of goods to 154°, and standing for about forty
minutes. Taps are again set, and the sparge liquor turned on,
the heat of this being never allowed to raise the goods-heat
beyond 158°.

It will be plain, on considering the question, that such a
course as this is the one most appropriate to the difficulty.
If we mash too low, we leave a greater residue of starch to
be extracted during sparging. If we mash high, the diastase,
which will have the subsequent burden of partly converting
the sparge-extracted starch thrown upon it, will be too enfeebled during the early stages to do its work during the latter.

When starch makes its appearance in the last runnings the
brewer should slightly raise his mashing heats, either of the
initial or of the piece, and lower his sparging heats. He should
also allow the first mashes to stand longer, and, if necessary,
mash up a second time. To shut taps periodically during
sparging, and digest the goods for some little time, will be of
similar service. These suggestions will be obvious on reference
to previous considerations.

The amount of liquor poured over the goods in sparging
will depend upon various circumstances. As a rule, it will vary,
for beers of ordinary gravity, between four and five barrels per
quarter, making, with two barrels per quarter of mash liquor,
a rough total of six to seven barrels per quarter. When sugar
is used, the amount of sparge liquor will be greater than in all-
malt beers ; again, it will be greater when the worts are boiled
in two or more lengths and not in one. The effect of the
amount of sparge liquor is a curious and an obscure one.
It seems pretty certain from the experiences of practical men
that the more liquor is poured over goods, the lower will be
the subsequent attenuations. Thus, if in one case six barrels
are passed over the goods, and in the other case eight, and if
evaporation is so conducted as to produce a wort of the same
ultimate gravity in both cases, during fermentation the second

beer will attenuate more completely than the first. We personally have known instances where these points seem to have been proved in practice, and where they could not be referred to other possible causes. If the solids in the later runnings contained more maltose and less maltodextrin than those in the earlier worts, the thing would be clear enough ; but, as a rule, this is not the case, and if there is any difference between them, it is usually in the direction of the later runnings being less relatively rich in maltose. Practical men have ascribed the fermentability of the later runnings to the extraction at that stage of some peculiarly assimilable nitrogenous matter. But this suggestion, which is diametrically opposed to logical deduction, rests on no experimental basis, and it may be ignored. At the present moment the point is unexplained.

The Mashing of Stout and Porter Worts.

The considerations, which have been hitherto touched upon, and which refer to ales, pale and mild, will assume a somewhat different aspect when we turn to the mashing of black worts. A very essential difference between a black beer and a pale or mild ale is, that while in ales brilliancy is of paramount importance, in stouts and porters it need not be considered. This being so, we are at once relieved from many restrictions which it is necessary to observe in ale brewing ; thus, for instance, marked viscosity, which is so risky in ale brewing, is not only permissible but actually desirable in stouts and porters.

The principal requirements of stouts and porters are (putting aside their characteristic empyreumatic flavour) fulness, high condition, a marked head, and a certain amount of stability. It will be plain from what has been said that to obtain these properties we must clearly aim for a very high proportion of maltodextrins—such a proportion in fact, which in the case of ale would be quite inadmissible on the ground of subsequent turbidity. In order, therefore, to obtain the desired amounts of maltodextrin, it is clear that we shall want a marked diastatic restriction, either in the malt-kiln or in the mash-tun. It is usual in stout brewing to rely rather on the former than on the mash-tun restriction, and in the majority of cases this would be wise, since the full flavour which is desired is, we know, producible rather from high heats on the kiln floor, than from high heats in the mash-tun. When, however, stouts of particularly marked head, body, and viscosity are demanded, brewers do well to supplement the already great restriction in the kilns by a further restriction in the mash-tun. The disadvantage, however, of any very high heat in the mash-tun is in respect of the fermentations, the amount of malto-dextrin produced in this way being so enormous as to prevent attenuation beyond about one-half the gravity, or yeast re-production to a normal extent. Stouts brewed in this way, however, are remarkably full; they come into rapid condition, which is long sustained, and they invariably give a firm and persistent head.

By talking of the restriction of the diastase of black-beer malts on the kilns, it is not meant that the malts in question are all uniformly cured at high heats for this purpose. In point of fact, stout grists are usually compounded of mixtures of various malts, black, crystallised, brown, amber, and ordinary fairly pale material. The black, the brown, the crystallised and other similar material, is heated to such an extent as to have had its diastase entirely killed : in fact, from various analyses of these malts we are able to say that black and crystallised malts contain absolutely no diastase, brown malt containing such a trace as to be inappreciable. The diastatic capacity of amber malt is about 5 to 10 (Lintner's standard). The relatively pale malt used for stout brewing would average about 29. The mean diastatic capacity of the whole, then, would be fairly low, and we are within our right in referring to a stout grist as one in which the diastase has been very extremely checked by dry heat.

In brewing ales we had to carefully avoid the dextrinous maltodextrins and the starchy matter which is likely to be extracted towards the end of sparging operations. In stouts, however, the viscosity, the frettiness, and the "head" they will confer are to be aimed at, since the risk of thickness does not apply in this case. Brewers, therefore, sparge their stout goods at higher heats than they do their ale goods, and it is usual for them to encourage the extraction of starchy or semi-starchy matter at this stage by raising the heat of the sparge liquor gradually towards the end, instead of dropping it, as in ale brewing. The low initial mashes adopted by some successful stout brewers would probably aid this extraction of starchy matter towards the end of operations ; for a larger residue of starch would remain to be converted at the conclusion of the mash proper, and this would certainly stand no chance of conversion towards the end of sparging operations.

CHAPTER VI.

THE BOILING AND COOLING OF THE WORT.

OBJECTS ATTAINED BY BOILING.

WHEN the wort has been prepared, it is boiled for some time with hops. The objects of this process are these :—

(*a*) The concentration of the wort.

(*b*) The extraction from the hops of their soluble constituents.

(*c*) The complete sterilisation of the wort.

(*d*) The coagulation of the albuminoids.

(*e*) The destruction of the diastase, and consequent fixation of starch conversion products.

(*a*) *The concentration of the wort.* In the event of our being unable to dispel a certain proportion of the water used in sparging operations by subsequent ebullition, we should either have to greatly lessen the amount of the sparge liquor, or else add an enormous proportion of sugar, unless we were content to brew beers of very low gravity. As a rule, during copper-boiling, about 25 per cent. of the wort is dispelled; and there is in addition a slight subsequent concentration during cooling, owing to evaporation at that stage. The structure of the copper, and the employment or non-employment of various subsidiary boiling appliances, will determine a greater or less evaporation ; but about 25 per cent. may be taken as a fair average under normal conditions, and there seems no particular object in aiming at any greater degree of concentration. If we endeavoured to reduce the necessity for concentration by reducing the volume of sparge liquor, we should certainly lose extract; while any endeavour to keep up gravities by an excessive use of sugar, would certainly change the general character of the beer.

(*b*) In Chapter III. the constituents of the hops were enumerated and described. It is during boiling that we avail ourselves of the properties which they possess. In the first place we extract the tannic acid, or more strictly speaking a large portion of it; for spent hops which have been boiled twice—the total boiling period amounting to 5 hours—will yet be found to contain a certain proportion of this constituent. The function of the tannic acid is to precipitate certain nitrogenous bodies, which are uncoagulated by heat. These substances, i. e., those which are uncoagulated by heat but are precipitated by tannin, are referred to in brewing literature as peptones, though there is no absolute evidence in support of these being bodies coming strictly under this designation. Reasons have been given for assuming that these particular bodies are undesirable from the standpoint of beer stability ; their partial precipitation by the tannic acid of the hops, and their consequent separation in the hop back, would therefore be a point to the good, so far as the soundness of beer is concerned.

The coagulable albuminoids require no tannic acid for their coagulation, and these substances would appear to separate on merely heating and without the agency of tannic acid. But the uncoagulable albuminoids referred to would appear to require some such precipitant as tannic acid for their removal. The precipitation is, however, an incomplete one. If, for instance, we take a wort, boil it to separate the coagulable albuminoids, filter these off, and then boil the whole with an amount of hops equivalent to the exceedingly high rate of 30 lb. per quarter, the filtrate after removal of the so formed precipitate will still give a further precipitate on the addition of more tannic acid. There would therefore seem to be an amount of these bodies in wort in excess of that capable of being precipitated by the tannic acid contained in—or rather extracted from—such amounts of hops as are used in practical operations. That this is so is curious, and the whole matter requires investigation. For, on other grounds it would seem that finished beer, when fairly heavily hopped, contains some free tannic acid. At any rate a strongly hopped bitter ale will give a precipitate on intermixture with a heavy, but lightly hopped, mild ale, which certainly appears to be identical with the precipitate of tannic acid and degraded albuminoids. On the one hand, therefore, the degraded albuminoids would appear to be in excess; on the other hand, the tannic acid would seem to be in excess. There is no adequate explanation for these facts, and it is only possible to state barely what has been found. But in spite of many points which require clearing up, it is yet a reasonable presumption that the more hops we add to our worts, the more of these albuminous bodies do we precipitate. Granting that these substances are undesirable—and there is reason to suppose that they are so— it will follow that tannic acid, as being responsible for their removal, may be regarded (indirectly) as an antiseptic.

The chief antiseptic value of the hops, however, lies in the bitter resins which they yield to boiling wort. It will be remembered that one of them—a soft resin—is endowed with particularly active antiseptic properties; it will further be remembered, however, that on protracted boiling this soft resin is converted into a hard resin which is not endowed with antiseptic properties. That such is the case would be a reason for not boiling a wort with spent hops only; at any rate this should be guarded against in the case of ales which it is desired shall keep sound for some time. Such a practice obtains in some breweries, so far as the main bulk of wort is concerned, but these breweries are very few. It is, however, a fairly common practice to boil the second wort with the spent hops from the first wort. Such hops, however, would contain no appreciable antiseptic properties so far as the resins are concerned, and, excepting in very quick running-ales, it is, on the grounds named, more judicious to use a smaller hop rate for the first copper, reserving a portion of the fresh hops for the second copper.

Apart from the antiseptic value due to the resin, and in a less degree to the tannic acid, there is yet another hop constituent which must be regarded as having some indirect antiseptic value—we refer to the acid of the hops other than the tannic acid, and which is probably mainly composed of malic acid. From the following experiments by one of us,[*] it will be seen that worts contain about double as much acid after hopping as before, the rate of hopping in question being such as is customary in ordinary practical operations.

	Gravity of wort.	Acid as Lactic acid.
		Per cent.
Wort from mash-tun	1077·5	0·0724
Same wort (hopped) from copper	1090·15	0·1317
Wort from mash-tun	1078·6	0·0730
Same wort (hopped) from copper	1093·1	0·1338
1st copper wort (hopped)	1093·05	0·1450
2nd ,, ,, ,, ...	1051·88	0·0991

We shall shortly learn that the acidity of the wort is in great measure productive of its complete sterilisation during copper boiling; and the extra acidity derived from the hops must be credited with playing a useful part in promoting this essential end. Apart from this, however, we know on general grounds that acidity, within reasonable limits, will, while not discouraging alcoholic fermentation, distinctly check the development of bacteria; and it is probable that the hop acidity constitutes a valuable factor in the keeping properties of beers.

The hop-oils which contribute the delicate flavour and

[*] Morris, Transactions Laboratory Club, iii. (1880), p. 34.

aroma of beers, are extracted during the boiling process, but are partially expelled. The aromatic atmosphere in the neighbourhood of a brewery where wort is being boiled, is sufficiently suggestive of the loss of these volatile bodies during ebullition. It is customary in the case of beers which it is desired shall possess a delicate flavour and aroma, to reserve a portion of fresh hops until ten minutes or so before turning out the coppers. The portion so added yields the greater portion of its oil, the time allowed being insufficient for its expulsion. The practice of adding hops to the finished beers in cask—or dry hopping, as it is termed—is in great measure performed to ensure the same end.

The alkaloids will also become dissolved in the worts during boiling. These bodies, however, are, so far as we know, of no chemical interest; they are rather of importance from the physiological standpoint, and they need not be considered at this stage.

The hop leaves—that is, the insoluble portion of the hop—perform their useful filtering functions when the wort has finished boiling, and has passed to the next vessel, the hop back. But, having a certain duty to perform subsequently, it is necessary that during boiling they should not be unduly disintegrated. In that case they become much reduced in value as a filtering medium. Their condition at the conclusion of boiling, in this respect, does not depend, so far as we know, upon any chemical properties of the wort; but much depends upon the manner in which the worts are boiled. There are, for instance, certain boiling appliances—fountains and wort-heaters—which unduly agitate the wort and the hops, bringing them in the case of fountains with considerable force against the cap of that appliance. It is agitation of this sort which disintegrates the hop leaves, and which, instead of leaving the hops in the hopback in the form of a loose filtering medium, makes them a sort of compact muddy layer, through which filtration is never satisfactory. The hop leaves, again, may mechanically assist in the satisfactory coagulation of the albuminoids, which we shall shortly consider. When they are unduly disintegrated any such assistance would be lost.

(*c*) The sterilisation of the wort is perhaps the most important of the various changes occurring during copper boiling. It puts all worts on an equal footing so far as existing organisms are concerned, whether such worts are summer-brewed or winter-brewed, or prepared from pure materials or bad materials, or in clean plant or dirty plant. Subsequent deterioration will or will not take place according, firstly, to the nutrient matters (in respect of bacteria) contained in the wort, and secondly to the infection the wort and beer may suffer subsequent to the boiling process. Although in strict conformity with conclusions deducible from previous research, the question of the complete sterilisation of wort during copper boiling was one about which there was, not many years ago, considerable argument, it was alleged by those who questioned the assertion, that the researches of Tyndall and others, proving the survival of bacteria after protracted boiling in water, threw doubt upon the destruction of similar organisms during wort-boiling. This objection was answered by those who upheld the sterilisation argument that boiling in water and boiling in hopped wort were two very different things, and that such organisms as might survive ebullition in water could not live through ebullition in a slightly acid medium, like wort containing hop-extract, and boiling at a slightly higher point than water. The matter was argued with some force on either side, and was only put at rest finally by the researches of one of us.[*] It was then shown that all malts, whether good, bad, or indifferent, will yield an infusion in which, when hopped and boiled, all contained organisms are destroyed, and that, too, in a far less time than the period usually devoted to boiling operations in practice. The following experiments,

---

[*] Morris, Transactions Laboratory Club, iii. (1890), p. 33.

which are quoted from the researches in question, show this to be the case.

Malt used.	Mashed at 158° F.		Mashed at 158° F., and the filtered mash afterwards boiled for 15 mins. alone.	Mashed at 158° F., and the filtered mash afterwards boiled for 15 mins. with hops.
	After 48 hours' forcing (unboiled).	After 96 hours' forcing (unboiled).		
London ..	Groups of bacilli with long leptothrix forms, and some short active bacteria.	Growth much developed.	No trace of growth after 21 days' forcing	No trace of growth after 21 days' forcing.
Sheffield ..	Some short bacteria.	Full of short, constricted bacteria, and some short rod bacteria.	Do.	Do.
Hereford ..	Few short bacteria.	Swarming with bacilli, short rod and constricted bacteria and micrococci.	Do.	Do.
Lincolnshire, No. I.	Great many short constricted bacteria, and some micrococci.	Growth much developed.	Do.	Do.
Lincolnshire, No. II.	Very great many medium and short rod bacteria and some bacilli.	Do.	Do.	Do.
Burton (Mild Ale).	Trace only bacteria.	Full of bacilli, short rod and constricted bacteria and micrococci.	Do.	Do.
Burton, 1888 (dried at 215° F.).	Full of bacilli with leptothrix forms, rod and constricted bacteria.	..	Do.	Do.
Burton, 1889 (dried at 215° F.).	Few constricted bacteria.	Great many rod and constricted bacteria and some bacilli.	Do.	Do.

It will be observed that boiling with hops for only 15 minutes completely sterilises all these worts, made from all kinds of malts, some of them having been purposely selected on account of their very inferior quality. There is no need now to feel the minutest shadow of doubt upon the sterilisation of wort, and the opposition to it at one time put forth by

S

chemists of name and position, has now died down to the very occasional questionings of irresponsible persons.

The sterilisation of wort must not be construed, however, as removing the need of freedom from preventable infection during the stages anterior to boiling, or as removing the necessity for a pure water and good material. So sure as water is polluted, or mash-tun plant or pipes dirty, so will bacterial deterioration make itself felt in the resulting beer. For impure materials and dirty plant will convey into the wort matters of a kind suited to the subsequent development of organisms gaining access to the brewing, after the boiling process is completed, and the mere fact that bacteria are killed during boiling will mean a solution in the wort at that stage of the substance of such organisms, substances which must on every ground be peculiarly suited to the nutrition of organisms introduced at a later stage. Although, therefore, the boiling of wort must be regarded as certainly killing all bacteria existing prior to that stage, yet there is still the utmost need for complete cleanliness of plant used in processes anterior to boiling.

The fact that copper boiling effects sterilisation makes it at once clear why summer-brewed beers are less stable than those brewed in winter, even when material and system and plant are identically the same in both seasons. By placing worts, brewed at whatever season, on the same footing, as boiling does, deterioration will clearly depend upon some factor operating at a later stage. It is well known that the atmosphere in summer is enormously richer in organisms than the atmosphere in the same locality during winter ; it is therefore not surprising that, other things being equal, the wort exposed to the infected summer atmosphere during cooling operations should yield a less stable beer than the wort exposed to the relatively pure atmosphere of the winter months. If worts were not sterilised during boiling it would be hard to explain these facts : in fact it might be fairly said that without sterilisation in copper, the number of bacteria originally adherent to malt, and surviving the mashing process, would suffice, in the case of the best materials, the best system, and the cleanest plant, to ruin all chances of the stability of the resulting beer.

(*d*) So far as the coagulation of the albuminoids is concerned, this change proceeds under normal conditions readily and well. It is retarded to some extent when the hop leaves are unduly disintegrated by excessive agitation ; but its completeness or incompleteness is settled less by the conditions of boiling than by the chemical composition of the wort ; and, so far as we know, apart from the mechanical influence of the insoluble portion of the, hops its satisfactoriness is not dependent either upon the quality, the quantity, or the distribution of the hops. It will be remembered that the saline constituents of the water have an important bearing on this point ; the mashing operation, in determining the amount of coagulable albuminoids transformed into the uncoagulable albuminoids must also play a part in the total amount of nitrogenous matter separated by the boiling process. The temperature of the boiling will probably play no part in this reaction. In fact it may be that the higher the boiling point, the more do we redissolve of the albuminoids once coagulated. Judging from appearances, wort heated to 175° F. " breaks " more satisfactorily than wort boiled ; at any rate the large curds separated at the lower point break up into smaller particles during protracted boiling. Were the coagulation of the albuminoids the only object we had in view at this stage, it might be well on this ground to boil for shorter periods than are customary. But the extraction of the soluble hop constituents necessitates our passing over the possible advantages of the more satisfactory coagulation which we obtain just before boiling, or during quite the early stages of boiling.

(*e*) That the diastase is absolutely killed before boiling requires no further comment.

PRACTICAL CONSIDERATIONS OF BOILING AND COPPER CONSTRUCTION.

Having sketched the changes which the brewer desires to produce during boiling, it will be necessary to briefly consider the practical means of effecting them. In the first place, the dimensions of the copper must be considered.

In the case of deep coppers we shall raise the boiling-point, owing to the pressure exerted by the greater depth of liquid. The ordinary boiling-point of worts of average gravity boiled in coppers of ordinary dimensions has been found to vary between 214° and 216°. With a very deep copper, the boiling temperature might be sufficiently elevated to produce a distinct caramelisation of the carbo-hydrates. Even in ordinary boiling there would appear to be a very slight caramelisation; but any excessive change of this sort would be inadmissible, since the paleness of pale beers would become affected. The elevation of the boiling point would extract from the hops some of those rank bitter substances (probably resinous) which under ordinary conditions are only yielded to the wort if the boiling is unduly prolonged, or if the water contains such solvent substances as sodium or potassium carbonates. In the case of all pale and other delicate beers the acquisition of this harsh, bitter flavour would be fatal to the value of the beer.

If, on the other hand, the copper is shallow, we expose a very large surface to the air, and there will consequently be an undue loss of the volatilisable hop oils. It is unnecessary to say that the excessive loss of these bodies is at once uneconomical and damaging to the quality of the beer. It has been suggested by practical men that in shallow coppers the boiling point is so lowered as to prevent the extraction from the hops of the tannin and the resins. This however, requires confirmation, and is probably inaccurate. As a rule, brewers seem inclined now to adopt shallow rather than deep coppers, using the latter for such beers (stout, for instance) where an increase in colour is immaterial, and where there is a sufficiency of body in the beer to mask any rankness of hop-bitter. Reverting to the case of shallow vessels, it is unquestionable that the shallow boiling backs now infrequently found, but still existent, are very unsatisfactory vessels. These vessels are made of wood or iron, and heated by steam coils. If *all* the wort could be got to boil, the temperature would probably suffice for the sterilisation of the wort and the extraction of the desired hop constituents. But in vessels of this sort, there is no guarantee that the wort underneath the coils boils at all. It is to this fact, probably, that the unsatisfactory results obtained from vessels of this sort are due, and not to the somewhat lower boiling temperatures of the wort. In some of these boiling vessels the wort cannot be got to boil at all, it merely simmers. In such cases no satisfactory results can possibly be anticipated.

The usual form of boiling vessel is the copper, and it is to the copper that we refer when discussing the ways and means of boiling. The copper may be heated either by steam or by fire. In the former case, it is usually jacketed to about one half its height; in the case of fire, it is heated by the direct action of the furnace placed beneath it. Practical men seem generally to prefer fire coppers, although it is generally admitted that steam coppers are capable of giving good results. There is no experimental evidence as to the relative merits of the two systems; we have merely practical experience to go upon, and conclusions based upon practical experience are always liable to be obscured by side issues. So far as our own experience goes, we should always advise a brewer, when building a new brewery, or remodelling his old plant, to heat his copper by fire. A well-set fire copper requires no subsidiary appliances—such as fountains, domes,

&c.; it boils from the middle, and the natural circulation effects the entire boiling of the whole of the wort. In a steam copper it is certainly necessary that the steam should be under sufficient pressure (at least 40 lbs. at the copper itself). Even then, however, the wort will boil rather from the jacketed side than from the centre; and in order to ensure the sufficient degree of circulation, and to prevent fobbing or boiling over, it is almost absolutely necessary to adopt a fountain, or a dome, or some such appliance.

SUBSIDIARY BOILING APPLIANCES.

The fountain we regard as unsatisfactory. We admit that it promotes circulation, and that it guarantees the boiling of the whole of the wort. But there is the objection, once before referred to, that the hops are much disintegrated when boiled with this apparatus, and their value as a filter-bed in the hop-back is thus greatly discounted. The disintegration of the hops leads to their yielding the coarse, harsh bitters, which are only extracted, under ordinary conditions, when boiling is unduly protracted, or when the boiling-point is unduly elevated, or when the water is alkaline.

The movable dome is a better appliance. There the wort rises out of the aperture, and flows gently down the sides of the dome. There is, therefore, no violent agitation of wort, and no disintegration of hops, and the dome is quite as effective in promoting circulation as the fountain. The dome is useful, too, in collecting and retaining some of the valuable hop oils, the loss of which in a shallow copper is always likely to be great.

The advantages claimed for these appliances, apart from the circulation which they do undoubtedly effect, is that they promote the aëration of the wort. We are inclined to regard this claim as one unsubstantiated by fact. It is difficult to see how a wort boiling rather over 212° F. can possibly absorb air under any circumstances. So far as air is concerned, the action would surely be the contrary one to absorption, for any air contained in the wort prior to ebullition would assuredly be very rapidly expelled. Aëration of wort can only commence when the wort has left the copper.

THE DURATION OF BOILING—SINGLE COPPERS AND DOUBLE COPPERS.

Worts are either boiled in what brewers call one or two lengths; that is, either the whole of the wort is boiled in one portion, or it is subdivided into two portions of about the same bulk, and boiled separately. In some cases, the wort is boild in three, or even a greater number of lengths, but this is unusual except in very large concerns, and involves no difference in principle to boiling in two lengths. The size of the copper will be a prime consideration in the brewer's decision as to whether he boils in one or more lengths. The mashing and sparging liquor amounts to about six to seven barrels per quarter, and in many breweries the coppers are insufficiently capacious to accommodate this amount of boiling wort. In such cases, the brewer is obliged to boil in two or more lengths; but even where his coppers are sufficiently large to take the whole of the wort, there are certain considerations which may lead him to select a double or treble boiling. By boiling in one length, it is not practicable to evaporate the same amount of water as in a double or treble boiling; for if a single boiling were as protracted as the total period covered by a double or treble boiling, the hops would suffer disintegration, and would be made to yield the coarse, harsh bitters already referred to. In fact, it may be said that hops will barely stand more than 2½ hours' boiling if the resulting beer is to possess a delicate hop flavour. On these grounds, then, a single brewing cannot be regarded as producing so great a concentration as a double or treble boiling. This being so, when the worts are boiled in one length, the amount of sparge liquor has to be restricted as compared to the amount permissible when the worts are boiled in two or more lengths. The restriction of sparge liquor may not be a matter of much importance with malts which convert readily and completely; but in the case of inferior grain, in which there is more starch, only difficultly gelatinisable, than there should be, there would be some chance of losing extract. Putting the question of capacity of copper aside, it may be said that the double-copper system is that preferable in the brewing of running ales in which a more or less inferior malt is employed. When, however, high class materials are used, such as it is customary to employ for pale ales, the single-copper system is technically as good, and is more economical in point of time and fuel.

Worts boiled in one length are generally boiled from 2–3 hours; 2½ being a fair average. Although the sterilisation of the wort is completed in a far shorter period, we require about 2½ hours to extract from the hop its desired principles, and to sufficiently concentrate the wort.

When the wort is boiled in two lengths, it is usual to boil the first about two, and the second two-and-a-half hours. There seems a general consensus of opinion in favour of boiling the second copper longer than the first. Practical experience certainly would seem to favour this plan. It may well be that the second wort will contain, relatively to its concentration, a greater amount of those degraded albuminoids which are formed during mash-tun operations from the coagulable albuminoids, and these in their greater amounts would demand as complete an extraction as possible of such tannin as might be contained in the second copper hops. The wort, too, being weaker in gravity, weaker in acid, and weaker in actively antiseptic resin, might well require a longer period for complete sterilisation, than would the first wort; it must be admitted, however, that a considerably shorter period of boiling than that given would suffice for the sterilisation of even this second wort. A third or fourth wort would be boiled for about the same time, or for rather longer than the second wort.

THE DISTRIBUTION OF HOPS.

When boiling in a single length, all the hops would be added directly the wort boils, or very shortly after, in the case of running ales, and of such beers as are relatively lightly hopped, and in which the delicacy of hop flavour is of secondary importance. In the case of pale ales, however, as has been already stated, it is well to retain about one-tenth to one-sixth of the hops for addition about a quarter of an hour before turning out. These latter hops—which should be the most delicate of those used—will contribute their full amount of hop oil, the period of boiling being insufficient to expel any appreciable fraction of it. The spent hops from a single boiling, are either sparged in hop-backs, or pressed, to wash or extract from them the wort which they hold, and the bittering principles which they are yet capable of yielding. The pressing of hops, however, is not advisable, except in the case of the very finest. Inferior hops yield under this system a coarse and harsh bitter flavour, which is distinctly disagreeable.

In the double copper system, it is advisable, in higher class ales, to split up the hops between the first and second coppers, generally in the proportions of two-thirds to the first, and one-third to the other. When a large proportion of hops is not used, a certain portion of the spent hops from the first copper can be returned to the second, and boiled together with the fresh hops reserved for that purpose. It is also advisable, and for the grounds previously stated, to reserve about one-tenth to one-sixth of the fresh hops destined for the first copper, for addition shortly before turning out. As before, the residual wort and bittering substances retained by the hops can be extracted by hop-sparging or hop-pressing.

In the case of lightly hopped running ales, it is usual to put the whole of the fresh hops into the first copper, using for the second or other coppers the spent hops from the first.   It is better, however, even in cases of this kind, to reserve a certain portion of fresh hops for the second copper, say one-fourth or one-fifth; for in spent hops from the first copper, the resinous matter will have become hardened and rendered antiseptically inactive, and the tannin will have been nearly all extracted.   When this is done, or when the whole of the hops are added to the first copper, it is customary in running ales to return the whole of the first copper hops to the second. In these beers, delicacy of flavour is no great consideration, and any slight bitter harshness will be covered by the greater body of these beers.   It is therefore well to obtain as much extract from the hops as they are capable of yielding.

It is usual to add any brewing sugars that may be required to the copper during boiling.   This addition is, of course, no essential portion of the boiling process; but it is well when using sugars to avail ourselves of the purifying action of boiling for sterilising them, should they require any such process.   The sugar is added in the case of double boilings, sometimes to the first, sometimes to the second copper, and sometimes it is distributed between the two—so far as we know, however, there is no appreciable difference in the two methods.   It may perhaps be preferable to add the sugar to the weaker worts of the second copper, the equalisation of gravities can do no harm, and may possibly do good ; the second wort, by being made stronger, would probably sterilise more quickly, and the volatisable portion of the oils might be better retained by the wort, rendered somewhat more viscid in this fashion.

After the completion of boiling, the worts are generally run into a shallow vessel provided with false bottom plates.   The plates retain the hops, the wort penetrating first through the hops, and thence through the plates.   The vessel is provided with a tap to let the clear wort on to the coolers.   The hop-back is frequently provided with a sparge arm for the sparging of the hops; when the hop-back is circular, the ordinary automatically revolving sparge arm is a convenient appliance. When it is rectangular the hops have to be sparged through a rose-fitted hose.

The filtration of the wort through the hop-bed is an important matter. It will be remembered, that the saline constituents of water influence the form assumed by the coagulated albuminoids ; and these will be either retained or not retained in the hop-back according to their nature. Upon the state of the hops, we are similarly dependent for a good or an unsatisfactory filtration ; and when they are disintegrated through excessive boiling, or through undesirable boiling appliances, their filtering power is lost. Similarly, when hops are too old, they are apt to form, not a loose filtering bed, but a muddy compact mass incapable of efficiently filtering wort. Besides the coagulated albuminoids, we depend upon the hop-layer to filter off the nitrogenous substances precipitated by tannin. Upon the efficiency of this filtration we are in great measure dependent in respect of brightness, and in some measure in respect of stability (see p. 15).

In the hop-back, the worts begin to absorb oxygen and to deposit those glutinous nitrogenous bodies which are soluble in boiling wort, but separate during cooling. These changes it will, however, be far more convenient to study when we are considering the wort as it is being cooled and refrigerated. At the same time, it is undeniable that the aëration of hot wort does commence in the hop-back, and some brewers go so far as to force air by pumps into the wort at this stage. Where cooling room is deficient, this plan is not without its advantages. It might, however, be advisable to leave worts to cool somewhat in the copper before turning out, and then inject air into them there, instead of in the hop-back. The injection of air into the hop-back may disturb the hop-bed, and interfere with

its filtering capacity. This objection is got over, and the aëration obtained equally well, by injecting air into the copper wort
after this has been permitted to cool somewhat, say to 190° **F.**

THE COOLING AND REFRIGERATION OF WORT.

The changes taking place during this stage are :—

(*a*) The lowering of the temperature to the point requisite
for a healthy fermentation.

(*b*) The aëration of the hot and of the cold wort.

(*c*) The deposition of the glutens and other amorphous
matters.

(*d*) The infection of the wort.

(*a*) The primary object of exposing the wort on coolers
and passing it over refrigerators is to bring it down from the
temperature at which it stands in the hop-back to about 60° F.,
the average temperature at which the wort is seeded or pitched
with yeast. The considerations leading to the adoption of
60° F. for the initiation of fermentation, do not require to be
dealt with here ; at this stage we have only to discuss the
ways and means of bringing the wort down to this heat, and
the changes incident to this process. The old-fashioned
brewers endeavoured to effect this object by exposing wort
in shallow coolers until the desired loss of temperature was
attained. This plan was, however, a lengthy one, especially
in summer time, and the resulting infection of the wort during
this protracted exposure to the myriads of organisms contained
in the air during the summer months, must have been a fatal
blow to the stability of the resulting beers. It was doubtless
chiefly on this ground that brewers used to confine their brewing operations entirely to the winter months. Certainly with
gravities cut down, and hop rates as low as they now so
frequently are, such a system as this would be impossible,
and the need for some system of cooling which should diminish
the infection brought forth the invention of machines known

as refrigerators, in which the worts are rapidly cooled by the agency of cold water applied in the manner now familiar to every brewer.

When refrigerators were first introduced to the notice of practical men, the ease and rapidity with which large volumes of wort could be cooled down to the requisite point, and the very small space required for them, induced many firms to either totally abandon the old-fashioned cooler, or to very much curtail its area. But this innovation was not one which stood the test of time, and it was soon found that in spite of its many conveniences the system was one which gave beers defective in point of brilliancy. That such was so was to be anticipated, for, as we shall presently see, the exposure of the wort for relatively long periods while it is yet hot, is of the greatest importance if brilliancy of beer is to be aimed at. In the absence of coolers, or where the cooling area was very small, in fact where refrigerators did the whole or the bulk of the work, the cooling process was too rapid to allow of this necessarily prolonged exposure. The practical result was that the great majority of brewers who had abandoned and curtailed these coolers, reverted to their use ; and at the present day the most successfully conducted concerns avail themselves both of the cooler and the refrigerator. This combination is certainly that which gives the best results. On the cooler we get the slow aëration of the hot wort, while the employment of the rapidly acting refrigerator greatly reduces the infection of the wort, and also the time and space necessary when the whole of the cooling is done on coolers.

In most breweries convenience alone dictates the point at which the wort runs from the cooler to the refrigerator, nor would there appear to be any need for any fixed temperature at which the transference should take place. So long as the wort leaves the cooler above 130° F. and below 150° F. the results, so far as aëration and infection are concerned, will be satisfactory.

So far as the construction of coolers is concerned, it is clear that the greater area of wort exposed the more successfully they will accomplish their object. It is well, therefore, to have as shallow a cooler as possible, one in fact in which there shall not be more than four inches of wort. Exigencies of space frequently prevent the cooler from being as shallow as this; where this is so it is well to have two or more, one immediately above the other. Head-room is not frequently wanting, even in the most cramped brewery, and there is generally space for such an arrangement as this. Metal coolers (copper of course being the best) are preferable to wooden ones. The worts are exposed to great risks of infection on the coolers, and the more cleanable the vessel the better. A rotten or unsound wooden cooler is one of those dangers which no care or skill in other points can cope with. It is true that wood coolers, when properly looked after, are capable of being kept sound and clean for long periods; but there is always a risk attaching to them, and it is a risk well worth avoiding.

The coolers should be placed in a room by themselves. As will be subsequently stated, every chance of infection of all kinds should be avoided at this stage; and when the cooler, as is frequently the case, is not shut off from the rest of the brewery, there is always a great danger of infection by barley-dust, malt-dust, and other bacteria-laden dust, which cannot be prevented from distributing itself over the air of all parts of the brewery. The roof and walls, too, require attention. Frequently the roofing consists of rough wooden beams and joists, supporting old tiled roofs encrusted with the dust of many generations. This is very bad. The dust is a source of perpetual danger, and may help to infect wort brewed even during the winter time, while the roughness of the wooden beams and tiles prevents their being properly cleansed. Where this kind of arrangement is in force, it is well to have an inner shell constructed either of smooth match boarding or metal; either of these surfaces can be kept scrupulously clean. Walls, too, should be smooth and clean. The necessity for isolating the cooler from the rest of the brewery applies with equally great force, if not greater force, to the refrigerators; as do also the remarks concerning the need for absolute cleanliness.

It is unnecessary to say much as to the construction of the refrigerator, for most of the modern appliances answer their purposes admirably. The old-fashioned refrigerator, where the wort ran inside and the water out, should of course be discarded wherever it is still in use. For apart from its obvious lack of aëration, there is the greatest possible difficulty in thoroughly cleansing the wort pipes. Wort at this stage forced through dirty piping cannot under any circumstances be depended upon to yield stable beer.

It is probably on the refrigerators that the worts suffer most infection. During this process the whole of the wort is exposed to the atmosphere in a very thin film, so that the area exposed to infection is much greater than on the coolers. In addition, the lower temperature of wort during refrigeration would render the infection more dangerous.

The time will doubtless come when brewers will refrigerate their wort out of contact with air-borne organisms. Air at this stage is necessary, but the air should be filtered free from its contained organisms. Attempts at some such system have already been made, and with success, and there is no doubt in our minds that the time is not far distant when infection at this stage will be greatly reduced, or prevented, by some suitable appliances. In that case one of the chief dangers to the soundness of beers will be removed.

(*b*) *Hot and Cold Aëration.*—For our knowledge of the process of wort-aëration and its effects we are entirely indebted to Pasteur.[*] Pasteur found that air absorbed by cold wort plays a different part to that absorbed by hot

[*] 'Studies on Fermentation,' p. 353 (English edition).

wort. The oxygen taken up by cool worts is in fact merely
mechanically dissolved, and it will be dissolved to an extent
dependent, within limits, upon the lowness of the temperature.
Oxygen taken up by hot wort plays a different part. It is
not mechanically dissolved, but is chemically fixed, entering
(as Pasteur shows) into some form of combination with the
hop resins. It is this form of aëration that plays so important a part in the natural clarification of beer, or in its
ready clarification by isinglass finings. When the resins are
modified by the chemically fixed oxygen, they conglomerate
into particles of greater density; these sink easily to the
bottom of the storing vessel, forming a compact sediment,
and leaving the supernatant beer bright. When, however,
for some reason or other, the aëration of the hot wort is incomplete, the resinous substances, instead of conglomerating
and acquiring the density necessary for their rapid deposition,
remain suspended in the finished beer in a very fine state,
and in that condition they are equally unready to deposit
naturally or to yield to the action of finings. The exact
temperature at which oxygen is most rapidly absorbed by
wort has not been determined, but it would seem that the
higher the temperature, up to within say 20 degrees of the
boiling point, the greater the relative amount of oxygen
chemically fixed in the manner required. It is clearly on
the coolers that we get this chemical fixation of oxygen. The
relatively protracted exposure of the hot wort to air at this
stage provides the requisite condition for its oxygenation.
It is clear then that we should endeavour to expose as large
a surface as possible to the air, for oxygen (as Pasteur has
shown) will not penetrate into wort at a distance appreciably
below its surface. It is also clear that when conditions of
space or convenience make a large cooling area impossible,
artificial devices for the oxygenation of the hot wort should
be resorted to.

The oxygen mechanically dissolved by wort plays an

entirely different part to that chemically fixed. It would appear to have no influence upon clarification, except perhaps indirectly. Its function is eventually to act as what Pasteur calls a *primum movens* of life and nutrition to the yeast during the first few hours in which it is in contact with the wort. In other words it is an oxygen supply for the yeast during the earliest stages, during which the ferment is unable to obtain its supplies by the breaking down of the carbohydrates. As a rule, the whole of the mechanically dissolved oxygen is removed during the first ten to twelve hours of fermentation. The yeast having utilised it, will by that time have acquired sufficient vitality to forage for its own supplies of oxygen. The oxygen chemically fixed may to some extent assist the vigour of the yeast, but not unless accompanied by mechanically dissolved oxygen. At any rate the main function of the chemically fixed oxygen is not to nourish yeast, but so to modify the hop resins as to prevent their remaining in suspension, and persistently clouding the finished beer.

The mechanical solution of oxygen takes place principally, as has been said, while the wort is cool ; and in practical operations it is clear that it is mainly during refrigeration, and to some extent when it runs from the refrigerators into the fermenting tuns that the wort becomes aërated in this way. It is obvious then, on all grounds, that if we desire the benefits derivable from chemically fixed and from mechanically dissolved oxygen, we should cool our worts by means of coolers and refrigerators.

(c) While the wort lies on the coolers it throws down a greyish deposit known by brewers as the cooler grounds. This deposit consists of a mixture of substances which come out of solution during the cooling of the wort, and the oxygen absorbed would appear to have some influence in securing a more complete and rapid deposition of the grounds. The grounds, too, will also separate out the more readily the

T

shallower the vessel ; for the shallower the vessel, the more of its surface is in contact with wort, and surface exercises an attracting influence upon minute particles of suspended bodies.

The predominant constituent of the grounds would appear to be a nitrogenous substance derived from the malt, soluble in hot wort, but insoluble in cold, and similar in many respects to the class of albuminoids known as gluten-caseins. But besides this glutinous matter, the grounds contain also an appreciable amount of bitter resinous matter from the hops, with small amounts of mineral bodies.

Practical men aim at securing a maximum deposition and separation of these grounds, and it may well be that any want of completeness in respect to their separation at this stage may operate detrimentally on the brilliancy, and perhaps on the stability of the beer, although with regard to stability direct evidence as to any connection of this sort is wanting. But there is one form of beer turbidity known as " greyness " which practical men invariably attribute to an incomplete separation of the cooler grounds. Certainly grey beer does seem to be rendered turbid by the kind of substances usually ·deposited on the coolers. Direct evidence on this point is wanting ; but it may well be that while the complete separation of the grounds during cooling cannot do harm, and may be very necessary, yet their incomplete separation may do harm and cannot presumably be of use. It is well, therefore, to aim at a sufficient separation of these suspended particles, and this will be adequately accomplished when the worts are exposed on shallow coolers for the normal period. It is said that worts containing any excess of acid retain in solution a portion of the glutinous matters which, under normal circumstances, are deposited during cooling, and the fact that some malts containing rather more than usual percentages of acid yield " grey " beers is attributed to this supposed fact.

This may or may not be right; so far as we are aware, there is no evidence to substantiate it.

(*d*) The infection of the wort during cooling is an inevitable and unfortunate incident in the English process of brewing. To avert it various systems of " pure air " brewing have been introduced, in which aëration is claimed to be effected without exposure to a germ-laden atmosphere. These inventions, however, starting with Pasteur's, and including others since advocated, may be regarded as yet in their infancy ; but we feel no doubt that as a demand will surely arise for some such system, suitable means will be forthcoming. In most English breweries, as they are constructed nowadays, the due degree of oxygenation of wort, and the proper separation of suspended particles, can only be effected on the usual coolers and refrigerators, that is, at the expense of infecting the wort with the organisms contained in the air to which the wort is exposed. The worts, of whatever kind or sort, having been completely sterilised during copper boiling, it may be said that they all start on the same footing when the copper taps are opened. Whether or no beers from these worts will stand storage or not, will depend, firstly upon the amount of subsequent infection (and, of subsequent infection, infection on refrigerators is by far the most dangerous) ; and secondly upon whether the wort contains those substances that afford to the organisms which gain access to the wort, such nutriment as will enable them to survive the bactericidal influence of a vigorous primary and secondary fermentation.

So far as the infection goes, beer brewed during the summer months, when the air teems with micro-organisms, must necessarily have a greater predisposition towards deterioration, than beer brewed in the winter months, when the air to which the worts are exposed is relatively speaking poor in them. The difference between summer air and winter air, taken at the same place (a roof in South Kensington), will

T 2

show roughly the relation between summer and winter air in respect of organisms.*

	Average number of colonies per 10 litres of air.		Average number of colonies per 10 litres of air.
January	4	August	105
March..	26	September.. ..	43
May	31	October	35
June	54	November	13
July	63	December	20

It is on this ground that brewers brew their stock beers (i.e., beers which are stored for some time before consumption) during the winter months; while during the summer months they brew only running mild beers, consumed sufficiently quickly after manufacture to avoid deterioration.

But apart from the difference between summer air and winter air in any one locality, there will be a similar difference between the air of different localities at any given season. The experiments of Miquel, Frankland, and many others have shown (what, in fact, would be anticipated) that the air of densely populated districts is far richer in micro-organisms than the air of rural, thinly populated districts; and the relation between town and country will apply equally to summer as well as to winter air. The following few data show the amount of organisms in air in different places.†

Place.	Number of Organisms found in 10 litres of air.
Sea air	0
Reigate Hill (May)	13
Garden, Reigate (May)	25
Hyde Park (May)..	43
Exhibition Road	254
Natural History Museum	280
Burlington House	326

It will be clear that, so far as soundness of beer is concerned, the country brewer stands at an enormous advan-

* P. Frankland, Journal of the Society of Arts, 1887, p. 489.   † Ibid.

tage over the town brewer; in fact, those breweries situated in densely packed manufacturing districts have, during the summer time, very great natural difficulties to cope with. In some country districts, on the other hand, especially those near the sea, the air in winter may be considered to all intents and purposes, germ-free, while even in summer time it is not so germ laden as to prevent the successful brewing of semi-stock pale ales during these months. It is principally due to this variation in the kind of air to which worts are exposed that makes brewers in different districts adopt different materials and devices. Many brewers have found, for instance, that with identical, or nearly identical water supplies, they can do with safety in one brewery, what cannot be done in another. A brewer may obviously employ a low-grade malt, for instance, in a brewery situated in a pure atmo-sphere, when the same malt, mashed on the same basis in a town brewery, would end in disaster. Similarly, brewers have frequently found that while they can employ bisulphite of lime or other somewhat unstable antiseptics in one brewery, they cannot do so in another. In the latter case the bi-sulphite will be decomposed at the instance of excessive bacteria infecting the cooling worts.

The micro-organisms in the air, to which reference has been made, are of two main kinds. These will be more particularly dealt with elsewhere; for the present consider-ation it will suffice to class them as bacteria and wild yeasts. What has been said with respect to the greater abundance of organisms in summer, as compared to winter air, covers both these classes, so that undue summer infection will mean deterioration in the direction of acidity by bacteria and of fretiness and other diseases caused by wild yeasts. The relation between town air and country air in respect of wild yeasts, must, however, be somewhat qualified; for in the country, in breweries which are surrounded by fruit trees—especially plum, apple, and pear trees—there will always be a particular tendency to wild yeast contamination, with the resulting predisposition towards unduly low attenuations in some cases, and frettiness in others, according to the predominant variety of wild yeast.

The ideal brewery would be situated on an eminence in a bare, open country, and by preference in such a place that sea air (which is, of all air, the most germ-free) could sweep clean through it. The coolers and the refrigerating room should (as has already been said) be isolated from the rest of the brewery, the walls and roofing scrupulously clean, and presenting smooth surfaces capable of being kept so. There should be no access to the cooling room of malt-dust, barley-dust, hay-dust, straw-dust ; grain heaps, and spent hop heaps, if allowed to dry, are also a source of danger, for in that condition adherent bacteria are easily taken up by passing breezes and wafted into the cooling room. For similar reasons, fruit trees should not be in great numbers near a brewery, the flowers and fruit being covered with wild yeasts which are thence transferred by winds into the brewery. Cask refuse, spilt beer and yeast, dregs from casks, and cask washings, are also a source of danger. Slaughter-houses are bad sources of contamination, as would probably also be wool-sorters' factories, tanneries, mills, and any concerns in which an abundance of fine organic dust is inevitable. Further enumeration of sources of contamination is unnecessary. It will suffice to say that although scrupulous cleanliness and the avoidance of all unnecessary infection should be strenuously enforced in all parts of the brewery, it is particularly on the coolers and refrigerators that attention in this respect should be unceasing. Organisms infecting brewings in the early stages are at any rate killed during boiling, and their direct deteriorating influence thereby checked ; infection after the wort is pitched is not so dangerous as during cooling, for organisms and wild yeasts are checked and crowded out by the actively developing pitching yeast. But during cooling and refrigeration, the wort is in its most critical condition, and as this infection is to some extent inevitable during that process, every means should be taken to reduce it to its very lowest amount.

But however careful we may be, we cannot altogether avoid infection during cooling and refrigeration, and assuming that by every available means all evitable infection has been excluded, there will yet be many organisms to deal with ; and whether this irreducible minimum of organisms is to be of harm or not will depend upon two factors, the composition of the wort, and the vigour of the subsequent fermentation. The conditions under which a vigorous fermentation, so far as the yeast is concerned, is attainable, will be considered subsequently. So far as the relation of fermentation to the proportions of the carbohydrates, minerals, and other con-stituents is concerned, this has already been touched upon in the chapter on mashing, and will be treated again under fer-mentation. We will therefore merely consider the first factor named above—the composition of the wort. In this connec-tion it is of principal importance to consider the nature of the substances which particularly afford sustenance to those organisms that infect the cooling worts. Of these rank first and foremost those putrescible matters which are yielded to the wort by uncleanly plant and by impurities in the materials. Impure water, mouldy and unsound malt, will inevitably give to the wort the very substances upon which bacteria subse-quently introduced can thrive. Bacteria introduced before the boiling stage, although then killed, will yield similarly putrescible matter. Dirty plant will yield analogous bodies.

But even when plant is clean and materials good, faults in manipulation of it may be almost equally productive of harm ; for, as has been explained, we require for the nutrition of yeast certain bodies, and these, when present in correct proportions, will, by making yeast vigorous, check or crowd out bacteria. Even good materials under faulty manipulation may fail to yield substances necessary for the nutrition of yeast, and under these circumstances disease organisms will freely develop; while again, they may even yield substances which, although not so putrescible as those due to uncleanly plant or other impurities, are yet capable of forming a suitable aliment for bacteria.

Taking our English system as it is, therefore, and regarding, as we do regard, a certain amount of infection during cooling and refrigeration as inevitable, the only hope for success lies in keeping worts free from the putrescible bodies which afford nutriment to the infecting organisms. In other words, safety lies in the use of sound materials, in manipulating them correctly, and in the most scrupulous and ever watchful care in regard to cleanliness of plant.

The following figures* show the amount of infection contained in the last few barrels of wort upon the coolers. The results are given in organisms per barrel of wort.

(A) Deep cooler   ..   163,548,000 colonies per barrel.
(B) Shallow cooler   1,406,512,800   „   „

* Morris, Transactions of Laboratory Club, vol. iii. p. 15.

CHAPTER VII.

FERMENTATION.

INTRODUCTORY.

In the preceding chapters we have described the preparation of worts, their hopping, boiling, cooling, and aëration. In this chapter we have to deal with the conversion of the wort into beer by fermentation. To effect this change we employ yeast, the main function of which is to convert the maltose and other fermentable sugars of the wort, as well as any fermentable sugar added as a malt substitute, into alcohol and carbonic acid gas. Concurrently with this change the yeast reproduces itself, and in so doing assimilates from the wort nitrogenous and mineral matters and a little carbohydrate, these substances being necessary for the building up of the new yeast-cells. Heat is disengaged, and the temperature rises during fermentation. The fermentation being over, and the bulk of the reproduced yeast separated, the beer is allowed to settle, in order that the residuary yeast may deposit as far as possible and leave the beer clear; it is then stored in cask for varying periods, and when brilliant is ready for consumption.

The above is a bare outline of an everyday fermentation and its objects. To understand it, however, we must study a series of phenomena the complexity of which is hardly to be equalled in technical processes. The act of fermentation has been the subject of philosophic speculation from the infancy of civilisation, but it is only within comparatively recent years that our ideas on the subject have taken a definite shape. Each investigation has, in a sense, been the sequel of previous investigations, and in that sense the theories of Pasteur may be regarded as a sequel to the doctrines of Cagniard de Latour, Schwann, and Turpin. Yet it must be freely and openly confessed that to Pasteur do we mainly owe a debt of gratitude for throwing light upon the complex problems of fermentation, and for placing them on a sound and scientific footing. There may be points in Pasteur's teachings which time has proved, and may yet prove, to be inaccurate ; and there can be but one opinion that the work of Hansen has supplemented that of Pasteur, and that in a manner worthy of the latter. Yet for every discovery and for every great reform there must be a pioneer, and Pasteur undoubtedly holds that position with regard to alcoholic fermentation. We shall have much to say on Hansen's work, and much to say on the new era for brewers which his brilliant researches foreshadow. In expressing to Hansen the gratitude brewers owe to him, we imply no slur on the work of Pasteur ; and we make this remark, in that there is a decided tendency on the part of certain of Hansen's disciples and others to forget the great work which Pasteur did, and to magnify its errors and minimise its merits.

HISTORICAL.

The phenomenon of fermentation has been known from the earliest ages, and Osiris and Bacchus are credited with having taught the Egyptians and Greeks respectively the art of making wine. The alchemists of the thirteenth to the fifteenth centuries used the terms fermentation and ferments, and we find descriptions of the former in the writings of Basil Valentine, Libavius, Von Helmont, and other celebrated alchemists, each of whom enunciated theories to explain the various actions which were all classified as fermentations. Stahl, about the year 1697, was the first to advance a scientific theory of fermentation. He maintained that fermentation and putrefaction were analogous processes, and that the former was a particular case of the latter. According to him, a ferment is a substance possessing a peculiar motion, which it is capable of transmitting to the fermentable substance. He mentions sugar, milk, and flour as fermentable bodies, and considers spirits of wine to be a "product of fermentation." Boerhave, a contemporary of Stahl, also considered fermentation to depend on an internal motion, and he was the first to suggest that true fermentation only takes place with vegetable substances, and putrefaction with animal substances. In 1766 Cavendish proved that sugar, when submitted to alcoholic fermentation, yielded 57 per cent. of fixed air (carbonic acid gas). From the time of Stahl no great advance was made in the study of fermentation until the French chemist Lavoisier revolutionised not only the general ideas on chemistry, but also those on fermentation. He showed the relation of sugar to the products of fermentation, and considered that sugar, like an oxide, was separable into two parts, the oxygen combining with a portion of the carbon to form carbonic acid, and the other portion of the carbon combining with hydrogen to form a combustible substance, alcohol, so that were it possible to again combine the carbonic acid and alcohol, sugar would be regenerated.

Accurate analyses of sugar and alcohol were performed by Thénard, Gay-Lussac, and De Saussure about 1810, and the results supported the conclusions of Lavoisier. Thus Gay-Lussac wrote: "If it be now supposed that the products furnished by the ferments can be neglected, as far as relates to the alcohol and carbonic acid, which are the only sensible results of fermentation, it will be found that, given 100 parts of sugar, 51·34 of them will be converted during fermentation into alcohol, and 48·66 into carbonic acid."

This reaction he expressed in the following equation—

$$C_{12}H_{24}O_{12} = 4C_2H_6O + 4CO_2$$
Cane-sugar   Alcohol.   Carbon dioxide.
(according to Gay-Lussac).

These numbers were afterwards found to be incorrect, since Gay-Lussac was unacquainted with the other products of alcoholic fermentation—glycerine and succinic acid, and he was also unaware that sugar assimilated the elements of a molecule of water before it fermented. He was, however, acquainted with the fact that substances which had been boiled and kept out of contact with the air neither putrefied nor fermented, and from this he assumed that oxygen was necessary for fermentation.

So far, however, no advance had been made in the knowledge of the cause of fermentation. That the foam formed by fermentation, was able to again excite fermentation was known to the ancients ; and we find that Pliny states that the foam or froth produced during the fermentation of the drink made from grain was used for again starting fermentation. The nature of yeast was, however, entirely unknown. For a long time it was supposed that all decaying nitrogenous vegetable and animal matter was capable of setting up fermentation.

Leuwenhoeck, in 1680, discovered the spherical forms of yeast. He did not, however, consider them to be living organisms, but compared them with starch-granules. In 1787 even, Fabroni considered yeast to be a vegeto-animal substance of a glutinous nature. The knowledge on this subject appearing very insufficient, the French Institute offered, in 1800, a prize for the elucidation of the subject of ferments and fermentable substances. Three years later, in 1803, Thénard published his memoir on alcoholic fermentation ; he recognised yeast as the cause of fermentation, but considered it to be of animal nature, since it contained nitrogen and yielded ammonia on distillation. During the fermentation it lost its nitrogen little by little, and partly disappeared, since it became converted into soluble products. Exleben, in 1818, expressed his opinion that yeast consisted of organisms, which by growing promoted fermentation. Desmazières, in 1825, regarded the small unicellular organisms in fermenting liquids as infusoria, and later they were considered to be fungi, to which Persoon gave the name *Mycoderma*, and Meyen the name *Saccharomyces*, whilst Kützing classed them with the Algæ.

The connection between the microscopic organism, yeast, and the formation of alcohol and carbonic acid, was established in 1836 by Cagniard de Latour, who unearthed the forgotten microscopical observations of Leuwenhoeck. He noticed that yeast consists of a mass of unicellular organisms belonging to the vegetable kingdom, and capable of reproducing themselves by buds. He considered that the liberation of carbonic acid from saccharine solutions, and the conversion of the latter into alcoholic liquids, was connected with this process of vegetation.

Almost at the same time as the discovery of Cagniard de Latour, similar results were arrived at quite independently by Schwann, who found that fermentation was connected with the presence of yeast. No further fermentation took place in a fermenting liquid in which he had killed the yeast by boiling, and which he protected from the entrance of new yeast germs. When he again allowed the air to have access to the liquid, the fermentation recommenced ; and thus Schwann proved that germs capable of producing fermentation were present in the atmosphere. This was a great step in advance, and we may consider that the work of these two naturalists, Cagniard de Latour and Schwann, laid the foundations from which later workers were able to advance step by step, and gradually evolve the theories which explain the many-sided phenomenon of fermentation.

Two years later Turpin, as a consequence of his researches, formulated the theory : " Fermentation as effect, and vegetation as cause, are two things inseparable in the act of decomposition of sugar." This theory, which is known as the vitalist theory, was founded on the results obtained by the three last-mentioned men, and is, indeed, the one which was afterwards supported and amplified by Pasteur. It had, however, as early as 1839, most powerful opponents in Liebig, Berzelius, and others.

The theory of Liebig, enunciated in 1839, was merely a revival of that of Stahl, yet it held its ground, and was the generally accepted theory until Pasteur brought forward his classical researches in 1857. Liebig, without any regard to the investigations of Cagniard de Latour, Schwann, and many others, considered yeast to be a lifeless, albuminous substance in a state of decomposition. This nitrogenous substance has its composition altered by the presence of oxygen, and the equilibrium of the power of attraction, which holds its constituents together, is disturbed and allows progressive changes, accompanied by the formation of new substances; a motion of the atoms of the nitrogenous substance follows, and this motion is transmitted from one atom to the next, and thus the gradual decomposition or fermentation of the nitrogenous substance takes place. When the body comes in contact with another substance—sugar for instance—the motion of the atoms can be transmitted to the atoms of this substance, and thus the breaking up of this, and the consequent formation of new compounds, is brought about. Portions of the yeast are, according to Liebig, in a state of motion, which results in a breaking down of the yeast into simpler compounds, that in the absence of air undergo no further change; the motion then ceases, and a state of equilibrium is established. This theory of Liebig, which is essentially a chemical-physical theory, is known as the mechanical theory.

Berzelius, as we mentioned above, also opposed the vital theory of fermentation. He explained the phenomenon by supposing that, when sugar is fermented, the nitrogenous substances in the solution are converted by the action of oxygen into a ferment, and that this brings about the splitting up of sugar into alcohol and carbonic acid by means of a catalytic action. The "apparent organised form" of the yeast was considered by Berzelius to be the natural state of an amorphous precipitate. Both Mitscherlich and Berthelot acknowledged the organised nature of yeast, but shared the views of Berzelius with regard to its action.

There were thus, at the time Pasteur took up the question of fermentation, three rival theories in the field. The first, the vitalist theory, maintained by Cagniard de Latour, Schwann, Turpin, &c., and later adopted and extended by Pasteur; the second, the mechanical theory of Stahl and Liebig; and the third, the theory of catalytic forces and of acts of contact, the old theory revived and maintained by Berzelius and Mitscherlich. The theory most generally accepted was that of Liebig, since the philosophical explanation given by him was capable not only of application to alcoholic fermentation, but also to other phenomena of the same nature, such as the transformation of saccharine solutions into lactic and butyric acids, in which actions no organised ferment had at that time been observed, and which apparently resulted from the action of a substance, in course of decomposition, on a fermentable body.

In 1857, Pasteur published the first of those researches which have made his name so familiar to brewers and to students of fermentation generally. Starting from the almost forgotten results of Cagniard de Latour, Schwann, and Turpin, he laid the groundwork of all that has been done since that time in questions relating to fermentation. He proved, by experiments which admitted of no question, that the yeast which produced fermentation was no dead mass, as assumed by Liebig, but consisted of living organisms capable of growth and multiplication; that germs of yeast are present in the air; and that fermentation is intimately connected with the presence of these organisms in the fermenting liquid. We cannot do better than quote Pasteur's own words, as given in his first memoir, to show the decision with which he at once advocated the physiological theory:

" My decided opinion," he says, " on the nature of alcoholic fermentation is the following : The chemical act of fermentation is essentially a correlative phenomenon of a vital act, beginning and ending with it. I think that there is never any alcoholic fermentation without there being, at the same time, organisation, development, and multiplication of globules, or the continued consecutive life of globules already formed."

He also proved that yeast could grow both in pure sugar solutions, and in sugar solutions containing various nitrogenous substances ; but that in the former case the nitrogen which was at first present in a soluble state in the parent cells, became converted into insoluble nitrogenous substances in the young cells ; whilst in the latter case the young cells were found to contain soluble nitrogen, which was available for the development of a further quantity of yeast-cells when the former were sown in a sugar solution. That is to say, yeast, when grown in pure sugar solutions exists at the expense of the sugar and its own nitrogenous and mineral substances, and during alcoholic fermentation a portion of the sugar is assimilated by the yeast, forming cellulose and fat.

Later, in 1861, Pasteur gives the results of his investigations on the part which oxygen plays in the development of yeast during fermentation. He found that yeast could grow in a solution containing sugar and albuminous substances even in the entire absence of oxygen. The amount of yeast formed is, however, very small, and the fermentation progresses very slowly, although the amount of sugar which disappears is very considerable, as much as 60 to 80 parts of sugar being fermented by one part of yeast. When the air has access to a large surface of the fermenting liquid, fermentation goes on much more quickly, and a large amount of yeast is formed in proportion to the sugar which disappears. In this case oxygen is taken up by the yeast, which grows quickly, but apparently has not so great a fermentative power, since only 4 to 10 parts of sugar disappear for one part of yeast. The same yeast, however, acts energetically when introduced into a solution from which oxygen is excluded. From these facts Pasteur concludes that yeast which causes a fermentation in the absence of air, withdraws oxygen from the sugar, and upon this action its power as a ferment depends. In yeast-water,* for instance, the cells also grow, but somewhat sparingly, even if the liquid does not contain any sugar whatever, provided there is a sufficient supply of oxygen. If the air is excluded no growth takes place, although the solution may contain, in addition to the nitrogenous substances, an unfermentable sugar, such as milk-sugar.

After the publication of these researches of Pasteur, Liebig so far modified his views as to give up his former opinion that yeast was a lifeless mass, and he admitted the connection between the growth of yeast-cells and fermentation, but he still maintained the mechanical theory of fermentation up to the time of his death. In a paper published in the Journal which bears his name, he points out that the amount of growth of the yeast-cells in a solution of pure sugar is quite inadequate to account for the amount of sugar converted at the same time into alcohol and carbonic acid, if the growth of the yeast be regarded as an essential condition of the transformation. In proof of this he quotes one of Pasteur's experiments, in which 9899 milligrams of sugar were completely decomposed by a very small quantity of yeast, and at the end of the fermentation the quantity of yeast produced was found to be only 152 milligrams, the cellulose in which amounted to less than 30 milligrams. Liebig then asks: "Are we to regard the assimilation of 30 milligrams of cellulose as the cause of the conversion of 9899 milligrams of sugar into carbonic acid and alcohol ?"

* Yeast-water is prepared by boiling pressed yeast with water, and filtering. The term applies to the bright filtrate.

Starting from the assumption that since yeast contains a nitrogenous substance soluble in water and capable of converting cane-sugar into invert-sugar, there was also present a soluble-ferment similar to invertase, but which was able to convert dextrose into alcohol and carbonic acid, Liebig says : " Yeast consists of vegetable cells which develop and multiply in a solution containing sugar, an albuminate, or a susbtance resulting from an albuminate. The greater part of the cell-contents consists of a compound of a substance containing nitrogen and sulphur with a carbohydrate or sugar. From the moment that the yeast is fully developed, and is left to itself, a molecular movement takes place in pure water, and shows itself in the decomposition of the constituents of the cell-contents ; the carbohydrate (or sugar) contained in the cell breaks up into carbonic acid and alcohol, and a small part of the constituents containing sulphur and nitrogen becomes soluble, and retains its inherent molecular motion in the liquid ; in consequence of this these bodies have the power of converting cane-sugar into dextrose. When cane-sugar is added to a mixture of yeast and water, this change into dextrose at once takes place, and the particles of sugar penetrating through the cell-wall of the yeast behave in the cell exactly as the sugar or carbohydrate, which forms a constituent of the cell-contents, and in consequence of the activity brought to bear upon them they break up into alcohol and carbonic acid (or succinic acid, glycerine, and carbonic acid) ; or, as is commonly said, fermentation of the sugar takes place. Up to the present no well authenticated case is known in which yeast is formed without sugar, or in which sugar is converted into alcohol and carbonic acid without the presence of yeast-cells."

We see from this that Liebig regards the rearrangement of the atoms of the sugar molecule as the result of a vibration or molecular motion caused by the chemical changes which take place in some unstable substance secreted by the yeast cells. According to this idea the growth of the yeast is only indirectly connected with fermentation. He says further: "It is possible that the physiological process stands in no other relation to the process of fermentation, than that, by means of it, a substance is formed in the living cell, which, by an action peculiar to itself—resembling that of emulsine on salicin or amygdaline—determines the decomposition of sugar and other organic molecules. In such a case the physiological action would be necessary for the production of this substance, but would be otherwise unconnected with the fermentation properly so called." And again he says: "In the process of fermentation there takes place, so to say, an action from without upon substances which break up into products which are of no further use to the living organism. The vital process and the chemical decomposition are two phenomena which must be explained independently of each other."

We see from the foregoing that the views of Pasteur and Liebig agree on one point—namely, that the process of fermentation is intimately connected with the presence of yeast in the fermenting liquid, but their explanations of the mode of action of the yeast are entirely different. Liebig considers that fermentation takes place in consequence of a molecular movement or vibration which is communicated to the fermentable substance through the medium of a soluble-ferment secreted by the yeast-cells; whilst, on the other hand, Pasteur regards the action as a consequence of the life and growth of the alcoholic ferment.

The controversy between these illustrious *savants* continued for some years, Liebig, as before stated, maintaining his mechanical theory to the last, and Pasteur defending his views on alcoholic and acetic fermentation, and maintaining the accuracy of the experiments on which they are based. The formation of acetic acid from alcohol in the ordinary process of vinegar making offered another point for discussion as bearing upon the general question of fermentation.

Pasteur, in accordance with his view of the phenomenon of fermentation, regarded the conversion of alcohol into acetic acid as being directly dependent on the vital action of an organised ferment, *Mycoderma aceti ;* whilst Liebig maintained that the action was entirely one of oxidation brought about by the oxygen of the air. De Saussure and Schönbein had shown that certain organic substances have the power of absorbing oxygen from the air in a similar manner to finely divided platinum ; the oxygen thus taken up being transferable to other substances and causing their oxidation. Liebig adopted this view of the action of " mother of vinegar," or the pellicle of *Mycoderma aceti*, and asserted that its action was not a physiological one, but simply that of a carrier of oxygen to the liquid, thus causing the oxidation of alcohol to acetic acid.

As additional evidence of the correctness of his theory, Pasteur, however, cites the following experiment : If a fermentable saccharine liquid be sterilised in a suitable flask and protected from outside contamination, and then a trace of pure *Mycoderma vini* be sown in it, the surface of the liquid will in a few days be covered with a pellicle of the organism, which can be shown to grow at the expense of the air, absorbing its oxygen and giving out nearly the same volume of carbonic acid, and producing no alcohol in the liquid. If now the pellicle be submerged in the liquid, bubbles of carbonic acid soon begin to rise and alcohol is formed ; at the same time its cells swell up and cease to multiply, and the internal structure of their plasma becomes very greatly modified. Thus the same cells acquire or lose the power of acting as a ferment according as they are deprived of air or exposed to its action.

Yeast and other ferments therefore differ from the other lower organisms in that they possess the power of living and multiplying regularly and continuously without contact with the air. Instead of requiring free oxygen to burn the materials which serve for their nutrition, they obtain the heat necessary to their existence by living upon oxygenated bodies like sugar, which can furnish heat by their decomposition.

When fermentation is viewed in this light, it is evident that all living cells may become ferments under certain conditions. That this is so has been shown by more than one observer. Bohm has found that plants when immersed in an indifferent atmosphere entirely free from oxygen, form carbonic acid in consequence of internal respiration. The plants are not suffocated, but obtain the necessary force for maintaining their existence by the combustion of a portion of their bodies. Lechartier, Bellamy, and Pasteur observed that a large number of fruits, when excluded from the air, form alcohol and carbonic acid without the intervention of a ferment. Pasteur has also detected the presence of alcohol in green plums and rhubarb leaves when submerged in water. An action takes place in the living cell just as in the cells of yeast, but it is less energetic. Instead of the respiration of oxygen in the ordinary atmosphere, decomposition processes take place when the respiration is prevented, and amongst the products, carbonic acid, and in many cases alcohol also, can be detected, without the slightest trace of yeast or other organised ferment being present.

In connection with the preceding, it may be well to give a short account of Adolf Meyer's views on the formation of carbonic acid and alcohol during fermentation. He considers that yeast, like every other organism, must have chemical tension at its disposal in order to perform its vital functions, and that such tension is transformed into the form of heat or mechanical motion. The breaking up of sugar by yeast into alcohol and carbonic acid approaches very nearly a combustion phenomenon, the yeast-cells calling into existence, independently of the presence of free oxygen, the force necessary for the performance of their vital functions by "internal combustion" by the decomposition of the sugar, just as the life-processes of all organisms are exerted at the expense of forces which are furnished by the oxidation of organic material. The breaking up of sugar is connected with a loss of chemical tension; the alcohol formed having a smaller heat of combustion than that corresponding with the amount of sugar from which it has been obtained by the fermentation. The sugar, according to Meyer, enters the yeast-cell by a simple osmotic process, and is partly employed for the formation of the plasma of the cell, and is partly split up.

The theory of the osmose of the sugar through the cell-wall and its decomposition within the cell has very recently received indirectly strong confirmation from the results of experiments undertaken by Bourquelot to throw light on the vexed question of selective fermentation. In these experiments he proved that the rate of fermentation of a mixed solution of carbohydrates was exactly proportional to the rate at which the sugars diffused through a membrane; in a mixture of levulose and dextrose, for instance, the sugars fermented in the same ratio as they would dialyse through a porous membrane; thus showing that in the fermenting liquid the cell-walls of the yeast acted as a membrane through which the sugars dialysed.

The physiological theory of the action of yeast was now thoroughly established, but the precise part played by oxygen in the growth of the ferment and in its fermentative action was still a subject of much controversy.

Oscar Brefeld investigated these questions, using methods which enabled him to employ the microscope. The three questions which Brefeld proposed to solve were—

(1) How does yeast behave in a nutritive solution in the absence of oxygen? can it, as stated by Pasteur, grow without free oxygen?

(2) Is it the non-growing yeast which causes fermentation?

(3) Does increase of the yeast take place without fermentation?

As the result of his experiments, Brefeld arrived at the conclusion that the first question must be answered in the negative: that yeast cannot grow without free oxygen. The experiments showed, however, that a very small amount of oxygen—1 part in 6000—was sufficient to cause growth; but that the cells ceased to multiply when the whole of this was used up. The results of these experiments are opposed to Pasteur's theory, which assumed that yeast could obtain the necessary supply of oxygen by breaking down compounds containing oxygen.

The experiments made to answer the second question proved that fermentation can take place vigorously long after all growth of the yeast-cells has ceased. This again is contradictory to Pasteur's theory, which assumes that fermentation should cease when the multiplication of yeast came to an end.

The third problem was approached by adding a very minute quantity of yeast to a large volume of sterilised and bright wort, and placing the mixture under conditions which allowed the multiplication of the yeast, but prevented all fermentation. The results showed that the former could take place without the latter.

Brefeld's experiments still left undecided the question whether growth and fermentation can take place in the same cell at one and the same time, and the general conclusion which Brefeld drew was that fermentation is a pathological phenonemon, which commences at the moment when the yeast, in a favourable nutritive solution, can no longer grow, and which ceases with the death of the yeast-cell. He considered also that the yeast which caused fermentation in sugar solutions gradually lost a portion of the cell-contents and that this consisted of nitrogenous substances which might again serve as food for the yeast. He regarded the theory that the substance which separated from the yeast-cells was the actual promoter of fermentation as being very probable.   In this respect his views approached those of Liebig.

The experiments and conclusions of Brefeld were opposed by M. Traube, F. Mohr, and A. Meyer, and also by Pasteur, who continued to maintain his view that yeast can grow in the absence of oxygen.   He ascribes the failure of the preceding workers to obtain this phenomenon to the use of old yeast in Brefeld's case, and to the employment of impure yeast in Traube's case; the further growth of the yeast would be thus checked by the foreign organisms.

As the result of later researches, Brefeld modified his views, and brought them more in accordance with Pasteur's. He found that growth was brought about in oxygen-free media by means of sugar, and growth could take place without free oxygen, since the sugar served as the source of the oxygen, and gave the energy which even in uncombined oxygen is necessary for the development.   That is to say, Brefeld held that by fermentation sugar is not only split up in order to liberate heat by this decomposition, but that the process must be regarded as both one of reduction and one of oxidation, whereby on one hand alcohol, on the other hand carbonic acid, is formed.   The first process probably uses up energy, and sets no heat free, whilst in the second, heat is liberated, and this is noticed in the fermenting liquid by a proportionally less energetic growth of the yeast-cells.   This liberation of heat is probably more considerable than in other processes of vegetation which are known in the vegetable kingdom.

Brefeld, however, still differs from Pasteur on some points. The latter, it will remembered, considers that fermentation cannot take place without simultaneous growth, development, and multiplication of the yeast, whilst Brefeld stills maintains that in oxygen-free media both fermentation with growth, as well as fermentation without growth, can take place. Brefeld, in fact, distinguishes between fermentation with growth, fermentation without growth and loss of substance, and fermentation without growth but with loss of substance continuing to exhaustion of the substance.

Brefeld does not doubt that in a pure fermentation—i. e., "growth during fermentation without pathological appearance of death "—no other product is formed besides alcohol. The by-products, glycerine, succinic acid, &c., arise, he concludes, from the death of the yeast-cells, and their presence and amount probably varies with the presence and amount of dead cells.

We now come to the "molecular-physical" theory of Carl v. Nægeli. This is really a modification of Liebig's mechanical theory, of which we have already treated. Nægeli, however, in a brief review of the rival theories, considers that the work done on this subject, both by himself and others, leaves the theory of Liebig entirely without experimental basis. He holds that if there is present in the yeast-cell a substance which gives rise to fermentation by its decomposition, it should be capable of extraction, like invertase. There is also no analogous case to that suggested by Liebig, in which a chemical vibration, especially the decomposition of a substance brought about simply by the decomposition of another substance in its presence, takes place without being accompanied by a corresponding physical vibration. The nearest analogy—namely, that in which an inorganic or organic substance is able to set up a chemical decomposition simply by contact action—is not applicable, since the catalytic substance itself undergoes no chemical vibration, but remains unaltered.

Nægeli also advances important objections to the different theories of fermentation which have been advanced by various chemists—from Traube to Hoppe-Seyler—and at which we have glanced in the preceding pages. The hypothetical substance supposed by these workers to cause fermentation has not been yet prepared—in fact, there are chemists who consider this substance to be incapable of separation from the living cell, and it must therefore differ from the other known chemical ferments. This difference is sufficient to condemn the soluble-ferment theory. Nægeli emphasises the fact that fermentation is connected with the plasma of the living cell, and only takes place under the immediate influence of this. If an organism acts at a distance, it separates a ferment. Ferments convert non-nutritive or feebly nutritive compounds into extremely nutritive compounds ; whilst fermentation destroys the most nutritive substances and produces substances which are, without exception, unfavourable to the organism. Sugar, for instance, is decomposed, forming alcohol, lactic acid, &c. The products of fermentation are not favourable to the organisms of fermentation. By the action of soluble-ferments or enzymes, the organic compound is decomposed, molecule for molecule, into its components (starch into maltose and dextrin, cane-sugar into invert-sugar, and so forth) ; by fermentation, a number of products are always formed, and the relative amounts of these vary under different conditions. In fermentation carbonic acid occurs as a by-product, but is never formed by the action of enzymes. The soluble-ferments can also be replaced by other substances—for instance, by acids, alkalis, even by water at a high temperature ; but fermentation can only be produced by yeast or other organisms. Finally, Nægeli states that by fermentation heat is liberated, whereas by the action of soluble-ferments heat is absorbed. The latter statement is, however, obviously incorrect.

Nægeli thus distinguishes between yeast and soluble-ferments ; the action of yeast he calls yeast—or fermentative —action, in opposition to the action of enzymes. He also considers that the classification of ferments into "organised" and "unorganised" is not admissible.

In the course of a series of experiments to test Pasteur's theories, Nægeli found that free oxygen is favourable to fermentation, and that, therefore, Pasteur's theory, according to which fermentation follows the want of free oxygen, is not correct.

According to Nægeli, the numerous processes of fermentation must be divided into two groups—namely, fermentations which take place in the absence of oxygen, and fermentations which require an unlimited supply of oxygen. To the first belong all fermentations of the sugars and similar substances, as well as of the peptones; these can only be brought about by yeast. To the second group belong the acid fermentations, the fermentation of wine, &c., and the " oxidation fermentations."

Nægeli considers the question whether it is not possible to form a conception of the process of fermentation which shall embrace all the observed phenomena, and also be in accord with the views of molecular physics. He commences with the discussion of the action of soluble-ferments (diastase, invertase, &c.), and points out that these can only act when they are in contact with the substance to be acted on ; then, adopting the explanations of Bunsen and of Hüfner regarding catalytic action, Nægeli holds that the contact substance (soluble-ferment) is active not only on account of attraction and repulsion, but principally on account of the condition of vibration of its atoms and molecules.

According to a fundamental conception of molecular physics, molecules possess a certain internal vibratory movement which is also possessed by each single atom and group of atoms in the molecule. When a rise of temperature takes place, a portion of the heat goes to render the internal vibrations of the molecule and its atoms and atomic groups more active, and at last a point is reached when the internal vibrations of the molecule are so intense that it breaks up and forms new compounds. This result may be brought about by agencies other than heat, or by heat in conjunction with other forces ; thus if two solutions are mixed, the temperature of neither of which is sufficient when alone to bring about a decomposition, then the unequal molecular movements of the two substances tend to bring about a state whereby the former equilibrium of each is disturbed. If the disturbance is great enough, decomposition takes place, if it is not, a new equilibrium is established.

Although the actions of soluble-ferments and yeast are different in many respects, yet they agree on one point—namely, as coming within the molecular-physical law. Like the contact action of the former, the action of yeast is also governed by molecular vibrations, which disturb the former equilibrium of the fermentable substances and bring about a decomposition. Nægeli says, " Whilst, however, the soluble-ferment acts as a simple chemical compound, the yeast-cell acts by the combined molecular vibrations of many compounds, of which the living plasma in certain conditions consists."

Nægeli therefore defines fermentation to be "a transfer of the state of vibration of the molecules, atomic-groups, and atoms composing the different compounds of the living plasma (which remains chemically unchanged) to the fermentable substance, whereby the equilibrium of its molecules is disturbed, and their decomposition brought about."

It will be seen that the "molecular-physical" theory of fermentation of Nægeli resembles both the mechanical and soluble-ferment theories. Liebig, however, considered fermentation to be brought about by chemical vibrations or decomposition, and not by purely molecular vibrations. It will be remembered that Liebig considered alcoholic fermentation to be brought about by the decomposition of the albuminoids of the living yeast-cells. The soluble-ferment theory differs from Nægeli's in that it requires the presence of a different ferment for each fermentation, or, to use Nægeli's words, "that a special chemical cause must exist for the special chemical process, whilst the molecular-physical theory allows the different fermentations to result from the living plasma, which can give rise to various chemical actions corrresponding with the different organisation and mixture, both as to nutritive substance and to fermentability."

This theory allows of the explanation of the following points, which could not formerly be explained :—

(1) That the fermentation process takes place only in the cells or in the immediate neighbourhood of yeast-cells, and that it cannot be separated from these.

(2) That whilst a symmetrical splitting up takes place by the action of soluble-ferments, yet various unsymmetrical decompositions take place by fermentation.

(3) That according to the individual differences of the yeast-cells, the quantitative proportions of the products of decomposition are different.

(4) That each specific organised cell causes special combinations of decomposition, which possess nothing in common but the liberation of carbonic acid in each case.

(5) That the fermentative action of yeast-cells as a whole has not yet been able to be imitated by artificial means.

Nægeli also discusses the important question—Does fermentation take place in the interior or exterior of the yeast-cell? He carried out a number of experiments on this point. He found that under favourable conditions, yeast split up in one hour 1·67 times its dry weight of sugar, and formed 0·85 times its weight of alcohol. The most active cells must therefore, if the fermentation took place only in the interior, have taken up in the hour 3·34 times their weight of sugar, and separated 1·7 times their weight of alcohol. Enormous as this performance appears, it is not impossible, according to Nægeli's calculations, and therefore the question of osmose receives no answer from these experiments.

Nægeli then endeavoured to find an answer from analogy by observing the behaviour of other vegetable cells. Experiments in this direction pointed to the phenomenon taking place outside the cell, but the results were not sufficiently exact to warrant a definite conclusion. However, Nægeli finally came to the conclusion that fermentation takes place outside the cell, and that the action extends to a distance of one-fortieth to one-fiftieth of a millimetre from the cell-wall. He explains the action as follows : "According to the molecular-physical theory, during fermentation the vibrations of the molecules, the atomic-groups and atoms of the plasma are communicated to the fermentable material. The communication takes place in the same manner as in all analogous cases, as in the propagation of the vibrations of light, sound, heat, and electricity.   The vibrations of one molecule give rise to similar vibrations in the nearest molecule, this conveys it the next, and so on."

The principal grounds on which Nægeli founds his conclusion that the action is extended beyond the cell are, that when certain fruits are placed in grape-juice, the flesh of the fruit shows a considerable alcoholic fermentation before the slightest trace of fermentation is visible in the juice itself. If, however, the fruit is skinned before being immersed in the liquid no fermentation of the flesh of the fruit takes place. Nægeli distinguishes this from the self-fermentation of living cells observed by Pasteur, Brefeld, &c., and to which we have previously referred.   He regards the fermentation in the first case as due to the vibrations of the yeast-cells, existing on the grape-skin, being transmitted through this to the fermentable material within ; when the skin is removed, the yeast-cells are of course also removed, and the fermentation no longer takes place. Another important reason is the formation of acetic ether in solutions in which the alcoholic and acetic fermentations are simultaneously proceeding. Alcohol and acetic acid combine only in the nascent state to form acetic ether, and if these substances are set free in the interior of their respective ferment-cell, yeast and the acetic acid bacterium, it appears impossible for the combination to take place, whereas if they are formed outside the respective cells, which may be in the immediate neighbourhood of each other, the formation of the ether is easily explained.

Nægeli sums up the question thus: "The causes of fermentation are found in the living plasma in the interior of the cell, but they also act for a certain distance (at least 1-50th m.m.) outside the cell. The decomposition of sugar takes place to a small extent in the interior of the yeast-cell, but to the greatest extent exterior to the cell. This holds good for both the chief products of alcoholic fermentation, alcohol and carbonic acid ; the bye-products, glycerine and succinic acid, are probably formed in the interior of the cell."

This statement regarding the alcoholic fermentation applies equally to all other fermentations. Lactic acid, butyric acid, acetic acid, &c., all resulting in part from decompositions, which take place outside the cells of the respective organisms.

PRESENT VIEWS ON FERMENTATION.

It must be confessed that at the present time we are without any theory of fermentation which embraces the whole of the known facts. Nægeli's theory had many adherents when it was first enunciated, but we believe that at the present time it finds very little support among those interested in the study of fermentation. This is due to several reasons which will be apparent when we consider more recent work. The theory of Pasteur is perhaps the one most generally adopted, but time has shown even it to be incorrect in several important features. In his book 'Studies on Fermentation,' published in 1879, Pasteur replies to the objections of Brefeld, Traube, and others mentioned above, and reiterates his earlier statements, which may be concisely summed up by saying that "fermentation is a result of life without air." Nægeli, however, in his experiments found that free oxygen was necessary for certain fermentations, but, as we have seen, he did not include the alcoholic fermentation in this class, although his experiments showed that free oxygen acted beneficially upon the fermentative power of the yeast. Later, Buchner showed the favourable influence which free oxygen exercises upon the butyric fermentation ; he did not, however, investigate the alcoholic fermentation.

The recent work of Bourquelot [*] on the so-called selective fermentation appears to support the theory that the change of sugar into alcohol and carbonic acid takes place in the interior of the cells, for he found that any given fermentable carbohydrate will be fermented, as compared with another carbohydrate, entirely according to its relative capacity for diffusing through a membrane : i. e., if two sugars—say levulose and maltose—are fermented simultaneously, the levulose will, in a given time, be fermented more rapidly than the maltose, just in the same manner as a mixture of levulose and maltose will diffuse at different rates through a moist membrane ; the levulose diffusing the more rapidly. Bourquelot's researches appear to show a very intimate connection between the fermentability of a sugar and its diffusibility through the cell-wall into the interior of the cell. The cell-wall of the yeast is nothing more nor less than a membrane, and this is strong evidence in favour of Pasteur's theory that the actual chemical change of fermentation takes place in the interior of the cell.

Considerable doubt has, however, been cast upon the correctness of Bourquelot's conclusions by the more recent work of Gayon and Dubourg,[†] who found that certain species of yeast had the power of fermenting levulose more quickly than dextrose, whilst, with other varieties which followed Bourquelot's law, and fermented dextrose the more quickly, the rate at which this selective fermentation took place varied within very wide limits.

We have also been informed by Adrian J. Brown

---

[*] Ann. Chim. Phys. ix. p. 245.   [†] Compt. rendu, cx. p. 16.

that when solutions of dextrose, varying from 5 to 20 per cent., are seeded with the same number of yeast-cells, and fermented for a given time, the amount of sugar which is fermented in each solution is practically the same ; that is, concentration has no influence upon the amount of sugar fermented. Now if fermentation depended upon the dialysis of the sugar through the cell-membrane, we should expect to find that in the stronger solution considerably more sugar would be fermented than in the weaker, for we know that concentration has a great influence upon the rate of dialysis.

In an interesting paper lately published by A. J. Brown[*] it was shown that yeast-cells decompose or ferment sugar more vigorously in the early stages of their growth than during the later stages, when the multiplication is less or has practically ceased ; thus, during the first 12 hours of an experiment, the proportion of sugar fermented per cell was 0·237 gram of sugar, during the second 12 hours this decreased to 0·1513 gram, and in the third 12 hours only 0·0760 gram was fermented. These results were confirmed by experiments with beer wort, and are directly opposed to the generally accepted belief on this point.

Some recent experiments by Adrian J. Brown, communicated to us by him, cast grave doubts on the accuracy of Pasteur's statement that fermentation is a result of life without air. We have already mentioned the fact that Nægeli, Buchner, and others found that free oxygen was favourable to certain fermentations ; A. J. Brown has repeated these experiments with yeast and sugar solutions, under such conditions that no *multiplication whatever of the yeast takes place*. He finds that if two such fermentations are started, and a continuous current of oxygen is passed through one, whilst a current of carbonic acid is passed through the second, a slightly greater amount of fermentation takes place in the presence of oxygen than in its absence. These experiments appear to

[*] Transactions Lab. Club, iii. p. 73.

leave no doubt that fermentation does not result from the yeast-cells being deprived of oxygen. It is true that Pasteur obtained a much increased fermentation when he employed shallow open vessels for his experiments, but he ascribes this result (1) to the larger increase of yeast, and (2) to the fact that the cells, as they were formed, fell to the bottom of the dish, and being there cut off from a supply of oxygen by the vigorous budding yeast in the upper strata of the liquid, they started a decomposition and consequent fermentation of the carbohydrate, in order to obtain their supply of oxygen.

The experiments of A. J. Brown were, as we have said, made under such conditions that no cell-increase took place, and they were therefore free from the disturbing factors which such increase would cause. The cells were also bathed, so to speak, in oxygen the whole time the experiments were proceeding, so that no explanation similar to Pasteur's is possible. We are therefore forced to the conclusion, from these and other known experiments, that Pasteur's theory is incorrect on some most important points, and inadequate to explain others.

There is, however, every probability that fermentation takes place in the interior of the cell, and A. J. Brown suggests to us that the protoplasm of the yeast-cell communicates a certain vibratory motion to the sugar molecule, which thereupon splits up into simpler molecules, at the same time liberating heat. Heat being necessary for the continuance of all cell-life, that liberated by the decomposition of the sugar goes, in part, to maintain the life of the yeast. In all probability, it is for the purpose of obtaining heat, that is to say, the necessary amount of energy, that yeast is endowed with the power of fermenting sugar. The decomposition is naturally dependent upon the life of the cell, and ceases when the cell dies.

THE ORGANISMS OF FERMENTATION.

Having now shortly examined the theories of fermentation generally, it is necessary to go rather more closely into the subject of the ferments themselves. It will be remembered that we have already dealt with bodies to which the term "ferments" is sometimes applied, such as diastase, invertase, &c. These soluble-ferments or enzymes are secreted by living cells, and are incapable of reproducing themselves. The ferments we have now to describe are living organisms, capable of growth and reproduction, mostly secreting soluble or chemical ferments, and capable of effecting certain chemical changes in substances, correlative with their reproduction—these chemical changes constituting the so-called fermentations and putrefactions of the substance. The ferments we have to consider are unicellular plants allied to the fungi, and containing no chlorophyll.

The ferments may be divided for our purpose into two great classes. The first class are the budding fungi, the second the fission fungi. While the first reproduce their species by the formation of one or more buds or smaller cells, which then grow to maturity, and in their turn bud again, and also sometimes by endogenous spore formation ; the second multiply by repeated subdivision in one, two, or three dimensions of space, also reproducing themselves, under certain conditions, by spores formed endogenously. To the first class belong the yeasts, or Saccharomycetes ; and to the second the bacteria, or Schizomycetes.

Roughly this division tallies with the division into alcoholic ferments and bacteroids or putrefactive ferments. The latter division is, however, an arbitrary division, and is by no means sharply demarcated ; for while the capacity for producing alcohol is not strictly confined to the Saccharomycetes, it being possessed to some extent by certain moulds and bac-

X 2

teria, there are certain so-called yeasts which cannot be classed among the true Saccharomycetes, as we shall presently see.

In addition to these two classes, we shall have to consider certain members of a third class, the Hyphomycetes, or moulds (see p. 373). These organisms are productive of much trouble on the malting floors, and are also apt to invade brewery premises and plant, when rigid cleanliness is not exercised.

I. THE SACCHAROMYCETES, OR YEASTS.

We have seen that Meyer first gave the name Saccharomyces to the organisms which occurred in all cases of alcoholic fermentation. The knowledge of the true nature of these organisms slowly advanced, and with the increasing attention paid by naturalists to the different cases of alcoholic fermentation, a differentiation of the Saccharomyces into various species took place. This differentiation was based mainly on the microscopic examination of the organisms, and had, as its basis, the different form, shape, and size of the individual cells. The classification thus developed is the one still in use at the present time, although our more advanced thinkers are very doubtful whether forms which are held to be species under this classification should still be so considered. The pleomorphism of micro-organisms has only comparatively recently been established by Nægeli, Zopf, Hansen, and others, and this has to a considerable extent upset the earlier idea that all the various morphologically or physiologically distinct forms belong to different species. The greater part of the discussion on this question has borne upon the Schizomycetes ; and the question has engaged the attention of all the authorities on micro-organisms, including Cohn, Koch, Van Tieghem, Ray Lankester, Lister, Klein, and many others, including those mentioned above. Dr. E. C. Hansen has suggested the same idea with regard to the Saccharomycetes, and there can be no doubt that his experiments have conclusively shown the pleomorphism of various forms of Saccharomyces. On this account the whole classification of the Saccharomycetes will probably have to undergo revision at no distant period ; in the meantime, we must provisionally accept the existing classification, which, in the main, is that of Reess,* who is one of the chief upholders of the view that the forms described below are independent species.

The origin of the Saccharomycetes is still involved in obscurity, and there are some who maintain that they are not autonomous fungi at all, but merely stages of development of species belonging to other classes. Many observers, including Hoffmann,† maintain that they are derived from the moulds : *Penicillium* and *Mucor Mucedo* being generally considered as the forms from which they originated. Others, again, have considered *Oïdium* as the source of the yeast-fungi. Oscar Brefeld,† has recently elaborated a new line of work in connection with the origin of the Saccharomycetes. He considers them to be identical with budding conidia of various species of *Usitagineæ*, or Smuts. A short *résumé* of this theory—which is somewhat too special to be considered here—will be found in Grove's 'Synopsis of the Bacteria and Yeast Fungi.'

The multiplication of yeast by budding was of course known from the time when the cells of Saccharomyces were recognised as living organisms, but we owe to Reess‡ the discovery of the reproduction of yeast and other ferments by means of endogenous spores. The formation of these is favoured by depriving the cells of all nutriment, and exposing them to a constantly damp atmosphere. Engel§ afterwards confirmed this observation of Reess, and proposed the method

* Botanische Untersuchungen über die Alkoholgährungspilze. Max Reess. Leipzig, 1870.
† Botanische Untersuchungen, 1883, Heft 5.
‡ Botanische Zeitung, 1869 ; and *loc. cit.*
§ 'Les Ferments Alcooliques : Études Morphologiques.' Par L. Engel. Paris, 1872.

at present in use for obtaining the spores—namely, by the cultivation of yeast upon plaster blocks partially immersed in water and kept in a damp atmosphere. As we shall see later, Hansen has utilised the formation of spores as a means of distinguishing between different varieties of yeast.

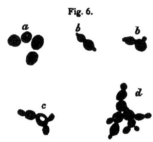

Fig. 6.

Sacch. cerevisiæ, showing various stages of budding. (After De Bary.)

Reess' classification of the Saccharomycetes is as follows :—

The Saccharomyces are simple Ascomycetes without a true *mycelium.* They are unicellular plants, multiplying themselves by the formation of buds which sooner or later separate themselves from the mother-cell (Fig. 6). A portion of the cells produced by budding directly develop to spore-forming asci. Number of spores in ascus, usually 1–4. The germinating spores multiply themselves directly by budding.

a. Species thoroughly examined :—

> *Sacch. cerevisiæ, ellipsoideus, conglomeratus, exiguus,* and *Pastorianus.*

b. Species incompletely known :—

> *Sacch. Mycoderma,* and *S. apiculatus.*

1. *Sacch. cerevisiæ,* Meyen (Fig. 6). The cells mostly round or oval, 8–9 μ ‡ long, isolated or united in small colonies. Spore-forming cells (asci) isolated, 11–14 μ long; spores mostly three or four together in each mother-cell, 4–5 μ in diameter. The alcoholic ferment of beer.

There are two varieties of this species, the so-called " top "

---

* μ is the most common abbreviation for the micro-millimetre, which is the name given to the unit of microscopical measurement, and which is equal to one thousandth of a millimetre.

or "high," and "bottom" or "low" yeast. The bottom-yeast sets up a fermentation in beer-wort at temperatures from 40° to 50° F. ; the fermentation continues for eight to ten days, and the temperature during this period rises 2·5° to 4·5° F. During the period of the fermentation only a very slight scum is formed on the surface of the fermenting liquid, and this scum contains only a very small proportion of yeast-cells. Both the old as well as the newly-formed yeast-cells settle on the bottom on the fermenting vessel as a dense sediment, sharply separated from the liquid. Bottom-yeast consists of round or slightly oval cells with a diameter of 8–9 μ, they are said to be slightly smaller and more oval than the cells of the top-yeast.

Top-yeast requires a higher temperature for its action than bottom-yeast—namely, 54° to 77° F. ; and the fermentation is usually complete in two to three days. The newly-formed cells rise to the surface of the liquid, and there form a large foam-like head. The ferment is distinguished from bottom-yeast in that its budding cells form more ramified clusters than the budding cells of the latter, which usually occur singly or in pairs. The cells also bud more freely, and the clusters of cells are carried to the surface by the carbonic acid evolved.

The earlier observers, Caignard de Latour, Turpin, and Mitscherlich, considered top-yeast to be morphologically different from bottom-yeast, and this was generally accepted until Pasteur in his earlier memoirs, concluded that the two forms were identical, and that the observed differences were simply the result of different temperatures and conditions, and that top-yeast could easily be converted into bottom-yeast, or *vice versâ*, by merely changing the conditions of growth. The results of Reess' experiments, however, threw considerable doubt upon this, and it did not appear to be such an easy thing to bring about the change. In these experiments, certainly, top-yeast and bottom-yeast approached each other more nearly in their behaviour under the conditions employed, since at a higher temperature and in a top-fermentation wort, bottom-yeast showed a more vigorous budding, but Reess found it impossible, during the duration of his experiments, to cause the differences to entirely disappear. Reess, however, considered that top-yeast and bottom-yeast have had the same origin, but that they have been so modified by numberless generations under the same conditions, that the differences now existing could not be altered by short culture. Pasteur, in his 'Études sur la bière,' adopts the same view, having thus completely changed his former opinion. He says that the top-yeast and bottom-yeast which are employed in breweries, and have. been propagated by continued culture, are two morphologically different varieties which cannot be converted the one into the other, and which possess very marked distinct properties. In cases where it has been supposed that the one has been changed into the other, he concludes that the yeast taken for the experiments was impure, and contained both top- and bottom-yeast, and therefore the former developed at the higher temperature, and the latter at the lower.

All the evidence goes to prove that these two varieties of yeast are quite as distinct as the great differences which are observed in the varieties and sub-species of higher plants and animals. They have undoubtedly been derived from some common form by selection and by the conditions of environment, just as all the species of Saccharomyces have had some common origin, perhaps from one of the so-called wild forms which is still known to us as such. It is a very moot point where the line should be drawn, and how far the different forms should be classed as species or varieties. This will be seen more clearly when we study Hansen's work, and notice the minute but constant differences which his different races of bottom-yeast exhibit. There can be no doubt that, given sufficient time, top-yeast, which will not ferment at temperatures much below 50° F., could be modified so far as to assume the functions of bottom-yeast, and ferment at temperatures of

Plate. V.

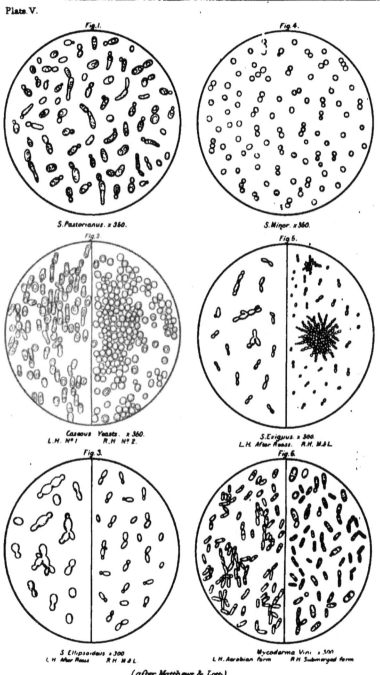

Fig. 1.

S. Pastorianus. x 360.

Fig. 4.

S. Minor. x 360.

Fig. 2.

Caseous Yeasts. x 360.
L.H. Nº 1   R.H. Nº 2.

Fig 5.

S. Exiguus. x 300.
L.H. After Reess.   R.H. M.&L.

Fig. 3.

S. Ellipsoideus x 300
L.H. After Reess   R.H. M.&L.

Fig. 6.

Mycoderma Vini x 300
L.H. Aerobian form   R.H. Submerged form

(after Matthews & Lott)

E & F N Spon, London & New York

about 35° F., by carrying on cultivations under slowly-altering conditions for a lengthened period.

2. *Sacch. ellipsoideus*, Reess (Plate V., fig. 3).—The cells are elliptical, having a length of 6 μ, and a width of 4–5 μ, and are isolated, or united by quick growth in small groups of branched cells. The spores are four, or more frequently two in each mother-cell, which are always isolated. The diameter of the spores is 3–3·5 μ. This is the ferment of wine, and is also most frequently present in spontaneous fermentations.

3. *Sacch. conglomeratus*, Reess.—The cells are almost round, of 5–6 μ diameter, and are united in clusters, which result from the formation of buds on all points of the surface of the mother-cell; the buds grow to the same size as the mother-cell, but do not separate. Spore-forming cells often occur in pairs, or combined with a vegetative cell; the spores are two to four, and on germination again give rise to clusters. The ferment occurs in wine at the beginning of the fermentation, and also on decaying grapes. Its fermentative action is doubtful.

4. *Sacch. exiguus*, Reess (Plate V., fig. 5).—The cells are conical or top-shaped, with a length of 5 μ, and a width of 2·5 μ at the widest point, and are united in sparingly branched colonies. The spores are two or three, and lie in a row in isolated mother-cells. This ferment occurs in the after-fermentation of beer, and in fermenting juice of fruit. The occurrence of this ferment in beer is to be dreaded on account of its small and light cells, which are apt to cause turbidity.

5. *Sacch. Pastorianus*, Reess (Plate V., fig. 1).—The cells are roundish-oval in slow growths; in luxuriant growths, branched colonies, with club-shaped primary members, 18–22 μ long, which throw off secondary roundish or oval cells, 5–6 μ long. Spore-forming cells, roundish or oval; the spores from two to four together, 2 μ in diameter. The ferment is a slow-acting alcoholic ferment, and occurs in the after-fermentation of wine and spontaneously-fermented beer.

This is the ferment described by Pasteur, and named *Pastorianus* by Reess in honour of that celebrated *savant*. There seems to be some resemblance between this ferment and some of the mould-fungi, notably *Dematium*, and the possible identity is discussed by Pasteur.

6. *Sacch. Mycoderma*, Reess (Plate V., fig. 6).—The cells are oval, elliptical, or cylindrical, with a mean length of 6–7 μ, and width of 2–3 μ; they form richly branching colonies. Spore-forming cells are often as much as 20 μ long; spores one to three in each mother-cell. This ferment forms a film on fermented and partially-fermented liquids, especially on wine and beer; it is known as "*fleurs de vin*," "*fleurs ou matons de la bière*." This ferment is known by several other names— *Mycoderma cerevisiæ* and *vini* (Desmazières and Pasteur), *Hormiscium vini* and *cerevisiæ* (Bonorden). Pasteur has shown that this ferment must be classed as an alcoholic ferment, since when grown submerged it produces small quantities of alcohol, but Hansen has recently stated that it does not ferment cane-sugar, maltose, dextrose, or milk-sugar, and does not invert the first named. When it grows on the surface of fermented liquids, it freely absorbs oxygen from the air, and converts the alcohol present in the liquid into carbonic acid and water.

7. *Sacch. apiculatus*, Reess.—The cells are lemon-shaped, shortly apiculate at each end, 6–8 μ long, and 2–3 μ broad, and under certain conditions slightly elongated. The cells only bud at the ends, and the daughter-cells quickly separate; the cells very rarely form united small, scarcely-branched colonies. Spores are unknown, making the position of the ferment among the Saccharomyces doubtful. It occurs sometimes in the primary fermentation of wine and other spontaneous fermentations, and also on all kinds of succulent fruits. Engel does not class it with the Saccharomyces, and gives it the name, *Carpozyma apiculatum* (see p. 337).

8. *Sacch. minor*, Engel (Plate V., fig. 4). This is a species

which is sometimes considered to be identical with *Sacch. cerevisiæ;* it is, however, somewhat smaller, and of about 3–4 μ diameter. It forms spores, and was found by Engel in the leaven of flour.

In the preceding, we have given, with the exception of *Sacch. minor*, the classification of the Saccharomyces by Reess; there are other forms described and named by various observers, the chief of which are :—

> *Sacch. sphæricus*, Saccardo.
> *Sacch. glutinis*, Fresenius, Cohn.
> *Sacch. albicans*, Robin, Reess.
> *Sacch. guttulatus*, Robin.
> *Sacch. olei*, Van Tieghem.
> *Sacch. rosaceus.*
> *Sacch. niger.*

These are, however, of no importance, and the claims of the majority of them to be included in the genus Saccharomyces are very doubtful; the two last often occur in the air, and are found on gelatin plate cultures.

In addition to the above, Pasteur describes two alcoholic ferments, to which he has given the names of "new top-yeast" (*nouvelle levûre haute*), and "caseous yeast" (*levûre caséeuse*) respectively (Plate V., fig. 2 L.H.) The former was found in a beer-wort which had been exposed to the air of his laboratory during the night; it had previously escaped notice. The cells are uniformly oval, and resemble bottom-yeast in their method of budding; it is, however, a top-yeast. It gives a beer with a very distinct, pleasant taste. Pasteur afterwards concluded that this ferment may be found in some beers.

The latter, the caseous ferment, Pasteur obtained by heating ordinary yeast at 50° C. (122° F.), in a liquid consisting of ordinary wort, solution of potassium bitartrate, and alcohol. It differs very considerably from top-yeast, and the young growth consists of jointed branches of greater or less length, which, at the junction of the segments, put forth similar cells or segments of a round, oval, cylindrical, or other shape.   Shaken up with water, the ferment sinks quickly to the bottom as a clotted sediment, and leaves the supernatant liquid almost clear.   When placed on a microscope slide and compressed by the cover-glass, it returns to its original form on removal of the pressure.

In addition to this caseous ferment, Matthews[*] has described a second yeast, which he calls caseous yeast No. 2, or *Sacch. coagulatus* (Plate V., fig. 2 R.H.).   This is a yeast . closely resembling Pasteur's in habit of growth, but with a very high degree of activity ; in fact, at temperatures above 70° F. its activity appears to be greater than that of *Sacch. cerevisiæ.* It is a bottom-yeast and yields a beer with a marked resin-bitter flavour and a high degree of acid.   Matthews considers the caseous yeasts to be of much commoner occurrence than is usually supposed.

The commercial yeast which is generally used in breweries is a mixture.   Apart from its being generally contaminated by some of the bacteroids, it is not one distinct variety of Saccharomyces, but consists of a mixture of them, the normal beer yeast largely predominating under favourable circumstances, but other varieties being abundant.   Under our English conditions, varieties other than the normal are to some extent necessary, as we shall afterwards see, but much of the trouble brewers suffer from, is due to the presence of some of the deleterious varieties of abnormal wild-yeasts, as they are usually called, or perhaps to an excess of those which in small number are necessary.

The researches of Pasteur were mainly directed to the study of the bacteria contained in yeast, and he plainly showed that when these develop they produce substances mostly of

* Transactions of the Laboratory Club, 1, p. 32.

an acid nature, which affect the potability and appearance of beer: the development of these ferments constituting the disease of beer, and the abundant formation of the products of their action the deterioration of beer. To these disease ferments Pasteur devoted his main attention, and his labours have resulted in the endeavour of the brewer to exclude these organisms from his brewings so far as is at all possible, by the employment of good materials, by the scrupulous cleansing of plant, and by skilful manipulation. To Pasteur beer disease was to all practical purposes confined to these disease ferments; and although he foreshadowed the question of yeast varieties, yet that field was left practically untouched until the great Danish naturalist Hansen put the matter of yeast variety on a firm basis.

Hansen has shown us clearly that the bacteria are not the only enemies a brewer has to contend against—that a commercial yeast, practically uncontaminated with them, is capable of producing the most serious troubles; and that the mixture of yeasts which brewers use, contains varieties each capable of imparting distinct flavours, distinct degrees of brilliancy or turbidity.

To study the properties of these yeast varieties, a new departure was necessary: the cultivation of each variety in a pure state. To Pasteur yeast was yeast, whether a mixture of varieties, or of one variety only; through the work of Hansen and his disciples we can determine what yeasts we are dealing with, their properties, and their approximate numbers as compared to normal yeasts. The method of pure cultivation is the basis upon which all this work rests; it therefore becomes necessary to describe it. Analogous methods, inculcated by Koch especially, enable us to cultivate the bacteria in a state of purity; this method, although generally applied to the pathological micro-organisms, has also yielded excellent results in regard to those which contaminate wort and beer.

(a) PURE CULTURE OF YEAST.

The method first employed by Hansen was a modification of Nægeli's "dilution" method, and, as this is still of advantage under certain circumstances, it will be advantageous to briefly describe it. The "dilution" method possesses the advantage of enabling the worker, from the moment the single cell has been sown in a flask, to carry on all subsequent operations, and to obtain a large quantity of absolutely pure yeast, without any danger of contamination from without. It is also essential to employ this method when we have a mixture of different species of yeast, some vigorous and some feeble, and wish to separate and cultivate the latter, since these will not grow in solid nutritive gelatine, but will freely do so in nutritive liquids.

Fig. 7.

Pasteur's Flask.

The method is carried out as follows: a vigorous growth of the yeast from which we desire to cultivate is promoted in a Pasteur flask (Fig. 7); the fermenting liquid is then diluted with a large, known volume of sterilised distilled water, the yeast-cells thoroughly well distributed in the diluted liquid by shaking, and the number of cells in one small drop of the liquid counted by means of a hæmatimeter. This may consist either of a microscopic cover glass, on which a number of microscopic squares have been ruled, or of a glass-slide, on which the squares are ruled in the centre of a very shallow cell. A very convenient form of the latter is made by Carl Zeiss, of Jena, in which the squares measure 1–400th of a square mm., and the cell is 0·1 mm. deep, the cubical capacity, therefore, of each square, when the cover-glass is on, being 0·00025 c.mm. The drop must not be allowed to extend beyond the squares, and its volume and cell-contents can then easily be ascertained. A similar size drop, containing, let us say 10 cells, is now withdrawn from the diluted liquid after renewed shaking, and transferred to a flask containing a known volume of sterilised water—say 20 c.c.—then the probability is that when the cells are uniformly distributed, each 2 c.c. contains only one yeast-cell. The flask is well shaken for some time, and a series of Pasteur's flasks containing sterilised wort of about 1058 gravity is inoculated, each with 1 c.c. of the liquid ; in all likelihood, therefore, every alternate flask contains one yeast-cell, and should yield a pure culture. This is, however, only a probability, and in order to determine its correctness or otherwise, the flasks are thoroughly shaken and allowed to remain perfectly still at a suitable temperature, the cell or cells then sink to the bottom of the flasks, and there remain and grow. After some days the flasks are carefully removed and examined, and the number of points of growth noted. In those flasks in which only one colony or point of growth appears, it is fair to assume that only one cell was added, and therefore the flask contains a pure culture ; those flasks which contain two or more colonies or points of growth are rejected as impure. A further check on the purity of the colonies consists in observing the total number of colonies formed in all the flasks ; this should correspond to the number of cells in the drop taken for dilution.

This method has given good results in the hands of Hansen, more particularly in cases where the species of Saccharomyces has some very marked characteristics, such as *Sacch. apiculatus* mentioned above. It is also applicable to the other cases previously referred to.

In his later investigations on the physiology and morphology of the alcoholic ferments, Hansen abandoned the "dilution" method, except in special cases, and adopted a modification of Koch's gelatine-plate method, using, however, a mixture of hopped-wort and gelatin for his cultivations, and taking more elaborate precautions to prevent contamination after the inoculated wort-gelatin is spread on the plate. He adopted this method, because by its aid a single cell can be selected under the microscope, its position marked, and its course of development followed throughout. Thus the element of chance in the method given above is eliminated, and we can actually select our single cell and keep it under observation from the beginning.

The wort-gelatine employed in this method is made by dissolving 5 to 10 per cent. of gelatin in bright hopped wort of about 1058 gravity; it should be fluid at temperatures above 30—35° C., (86—95° F.) and solid below this. In making a pure culture by this method, we start with a growth of young and vigorous cells, and dilute this with sterilised distilled water, in a small Chamberland flask (Fig. 8) of about 30 c.c. capacity, until the liquid is only slightly cloudy, then agitate in order to distribute the cells as completely as possible, and withdraw some drops with a sterilised glass-rod for microscopic examination.

Fig. 8.

Chamberland Flask.

For the purposes of this examination, and also for that to be mentioned later, it is sufficient to use a magnification which will clearly define the yeast-cells from the detritus accompanying them. A half-inch objective with B eyepiece answers this requirement very well.

This first examination tells us the number of cells in the diluted liquid and affords a guide to the next operation. We take another small Chamberland flask, of the same size as before, which has previously been half-filled with sterilised liquefied wort-gelatine, taking all proper precautions to prevent contamination, and one or two pieces of thick platinum wire about half-an-inch long. The diluted liquid in the first flask is well agitated, care being taken, both here and in the subsequent operations with the gelatin, that no froth is formed, and a piece of platinum wire, previously sterilised by heat, is dipped into the liquid and at once dropped into the fluid wort gelatin, which is then well agitated ; the temperature of the wort-gelatin should not be higher than 35° C. (95° F.) If the first microscopic examination has shown the diluted liquid to be rich in cells, the platinum wire is only dipped in it to a small extent ; if the reverse is the case, the wire is completely immersed in the liquid. The right degree of dilution at this point is most important, and can only be attained by a certain amount of experience ; roughly speaking, if we start with an ordinary barm, we shall require to dilute it in all to about one part in 1,000,000.

Having well agitated the wort-gelatin in order to obtain an equal distribution of the cells, we withdraw with a glass-rod a few drops for microscopic examination. This examination should show us only a very few cells in the mixture, and these should be well distributed throughout the preparation. If too many cells are present, the mixture must be further diluted with more wort-gelatin ; if too few, another piece of platinum wire must be dipped to the necessary extent in the first flask and added to the wort-gelatin. It is impossible to lay down any hard and fast rule as to the proper degree of dilution, since so much depends on the condition of the cells we are working with, on the nature of the species, and so forth. A little experience will soon teach the worker the proper point. As a rule, we find that more cells are present than we thought, and consequently too many colonies are subsequently obtained.

Having obtained the correct state of dilution, the next stage is to allow the cells to grow under suitable conditions. This is effected by means of a Böttcher's moist chamber (Fig. 9). This consists of a microscopic glass-slide, a little larger than ordinary, on which is cemented a hollow glass ring

Y

about 30 mm. in diameter and 5 mm. in height ; the ring is covered by a thin cover-glass. The apparatus is sterilised as usual, and then a drop or two of the wort-gelatin mixture is

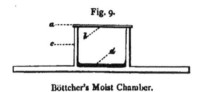

Fig. 9.

Böttcher's Moist Chamber.

rapidly spread on the under-side of the cover-glass, which is at once placed on the ring of the chamber with the layer of gelatin downwards. The joint between the upper edge of the chamber and the cover-glass is made tight by means of vaseline. It is of course necessary to keep the gelatin fluid and well agitated whilst the transfer is taking place. It is advisable to prepare two or three chambers, in case of accidents ; and the mixture in the flask should also be poured away, leaving only a thin layer on the bottom. This serves as a control on the chambers, and also as a reserve. The moist chambers are allowed to stand quietly until the wort-gelatin has set ; the joint is then made air-tight by applying a slight pressure to the cover-glass, and the latter is prevented from slipping by the application of sealing wax varnish to two or three points. The whole is then removed to the stage of the microscope and examined.

The yeast-cells should now appear well isolated and few in number, and we have to pick out one or more cells for observation, and to follow the course of their development. This may be done in several ways. For instance, permanent crosses may be made on each side of the stage of the microscope, and when we have the cell we wish to observe in the centre of the field, corresponding crosses may be made on the glass slide of the chamber, in such a way that we can always find the cell again. Or, we may use a cover-glass on which a number of microscopic squares, about one mm. square, have been etched and numbered ; the lines and numbers then serve as excellent points of reference. The most convenient method, however, is to employ the marker devised by Klönne and Müller, of Berlin ; this consists of a little arrangement which screws into the microscope in place of the objective, and by the ordinary movement of the focussing screw may be brought gently down upon the cover-glass of the moist chamber, where it marks a small coloured ring round the cell we wish to observe. These precautions serve to assure us that the small colonies which will subsequently make their appearance, and from which we shall cultivate, are absolutely pure cultures— that is to say, they are each the progeny of *one single cell.*

Having thus marked our selected cells, the moist chambers are placed in a thermostat at a temperature of 24-25° C. (75-77° F.) ; if this is not available, they may, of course, be allowed to remain at the room temperature, but in this case the colonies take longer to develop. From time to time the chambers are withdrawn from the thermostat, placed on the stage of the microscope, the marked cells found, and their progress noted. This control serves to assure us that the growing colony from each cell remains isolated, and that no coalescence of two or more colonies has taken place. Under these conditions the yeast-cells develop so rapidly that small colonies visible to the naked eye are formed at the end of two or three days.

Hansen states that this rule applies to all the Saccharomyces and all other species resembling these—*Mycoderma vini, Mycoderma cerevisiæ,* " *Torula de Pasteur,*" &c.

The colonies of Saccharomyces, and some other species, have the form and size of pins' heads, and are of a bright, yellow colour; they are sometimes waxy in appearance, and the edges may be either rounded or serrated. The same species may present all these appearances in the same culture, nothing can therefore be based on the appearance of the colonies in gelatin, as can be done with bacteria. As many as sixty colonies may be conveniently developed on a cover-glass of the size mentioned above, and if only one-half of these are available, a single moist-chamber suffices to obtain a large number of pure cultures.

The operation of transferring the colonies from the moist chamber to the flasks used in the next stage of the process, is one in which the utmost care is required, and it is most essential that the atmosphere of the room in which this operation is performed should be quiet and free from dust, and that all the vessels and apparatus employed should be thoroughly sterilised.

The apparatus required consists of a small pair of nippers, several short pieces of stout platinum wire, about half-an-inch in length, and about a dozen small Pasteur's flasks of one-eighth litre capacity. The flasks should contain bright, sterilised hopped wort of about 1058 sp. gr. In addition to the above, one or two small bell-glasses are wanted. If we are working with a single species, it suffices to take four or six flasks; if we are separating two or more species, it is advisable to take a larger number of flasks. This comparatively large number of flasks is necessary in order to enable a control to be obtained on the experiments by comparing the growths in the respective flasks, and also in order to avoid the loss of the experiment in consequence of accidents; thus we may find, in spite of all precautions, that foreign organisms have obtained an entrance into one or other of our flasks, or perhaps, on the contrary, no growth may develop in some of the flasks, owing to the wire employed having been too hot.

The moist-chambers having been examined with the microscope, and those colonies marked the growth of which from single cells has been followed from the commencement, the transfer is effected as follows: The cover-glass of one of the chambers is removed from the ring, and placed by preference on a dark background, with the gelatin surface and colonies upwards, and immediately covered by one of the small bell-glasses. The dark background throws up the whitish colonies very distinctly, and makes their removal much easier. One of the short pieces of platinum wire is then taken up with the nippers in the right hand, quickly passed through the flame of a spirit lamp or Bunsen-burner, and then dipped in one of the selected colonies, the bell-glass being removed whilst this is done, and then quickly replaced over the cover-glass and colonies. Meanwhile the india-rubber tube and glass-rod have been removed from the side tube of one of the Pasteur's flasks, and the piece of platinum wire with the adherent cells is quickly dropped into the flask, the stopper again replaced, and the side-tube and the india-rubber tube passed through the flame. This process is repeated with each of the selected and marked colonies. In place of the Pasteur's flasks, we may use either those of Chamberland, or of Salomonsen, or even an ordinary Erlenmeyer flask closed simply by two layers of sterilised filter paper. Each of these flasks has its own special use, but for general purposes Hansen considers that the Pasteur's flasks are much to be preferred. The flasks should not be shaken until all have been inoculated, in order that the atmosphere may not be disturbed. Whilst shaking the Pasteur's flasks it is necessary that the bent-up end of the tube should be made quite hot in order to sterilise the air which enters.

The flasks, after being labelled, are placed in a thermostat at about 25° C. (77° F.) In the course of one or two days traces of fermentation may usually be observed, and at the end of two or three days the fermentation is generally fairly vigorous, and a considerable quantity of yeast has formed. This rule applies to the Saccharomyces in general, and Hansen states that it also holds good for the majority of organisms capable of exciting fermentation.

The method described above has only one weak point—namely, the danger of contamination after the moist chambers are opened and during the transfer of the colonies to the flasks. Up to the time of the former operation, the control on the growth is complete, and after the colony has been safely introduced into the flask ·the further operations can be performed without fear of contamination.   When the chamber is opened there is danger of contamination from the air, from the clothing of the experimenter, &c., &c., and at this point this is the more dangerous since there is not a vigorous growth of cells to repel the intruding organisms.   For this reason, the method is inferior to the one formerly employed by Hansen, the method of dilution already described.   When using the moist-chamber method, however, the safest way is to use only one colony from each chamber.

It is possible, however, to estimate the amount of probable contamination by having a few flasks filled with sterilised wort, of the same kind as that continued in the Pasteur's flasks, standing open on the bench whilst the transfer is taking place, and at the conclusion of the experiment, closing these with sterilised filter paper, and placing them in the thermostat side by side with the other flasks.   In this manner, the nature and number of the micro-organisms capable of developing in the wort, and present in the air at the time of the experiment, may be ascertained.   Hansen states that if proper care be taken, and the work be performed in a suitable place, it is found that not more than two out of every three such flasks will contain a growth, and that this growth consists exclusively of a common mould-fungus, *Penicillium glaucum*, and never of bacteria or yeast-cells.   Our experience fully confirms this, and in no case has a bacterial or yeast contamination been found in the control flasks.

Having successfully transferred the colonies to the flasks with sterilised wort, the next question which arises is, How are we to select a culture with which to work ?   This is by no means an easy question to answer.   We have our series of small flasks, which at the end of two or three days will, as stated above, contain a fair amount of yeast in a state of absolute purity, that is to say, we shall have in each flask, provided we have carried out the cultivation with due care and precaution, the progeny of one single selected cell. The macro- and microscopic examination of the flasks will give us some information on the variety and species. The appearance of the sediment, whether it be granular or caseous, whether it be light or heavy, and so on, should be noticed, and the flasks may be often roughly divided in this way. A sample should also be withdrawn from each flask for examination under the microscope, taking all proper precautions. It is necessary, however, to remember that the differences in the microscopic appearance of the cells are, as a rule, very slight under these conditions of growth, and that it requires an experienced eye for the identification of the different varieties. We have then to fall back upon other points of difference, notably those under which ascospores are formed. Even this is often of very little use, and the only thing which can then be done is to successively submit the contents of each flask to a trial on a practical scale ; for there is no known method by which, from the mere morphological examination of a pure yeast variety, we can foretell its properties as a ferment.

Hansen introduced pure yeast on a practical scale into the brewery of Old Carlsberg in 1883, and he then cultivated the yeast on a fairly large scale in the following manner : The growth in the one-eighth litre flask is taken whilst in vigorous growth, and divided between ten one-half litre Pasteur's flasks containing bright, hopped, sterilised wort of about 1058· sp. gr. (All necessary precautions must be taken in order to avoid contamination during the transfer, which is effected by removing the indiarubber cap from the side tube of the small flask, after having previously passed the whole through the flame of a Bunsen burner ; the glass rod only is removed from the cap of the larger flask, which is also previously sterilised with a gas flame; the side tube of the smaller flask is then quickly introduced into the indiarubber tubing on the side tube of the larger flask, and one-tenth of the contents of the former decanted into the latter. The glass rod, indiarubber cap, and the side tubes are all passed through the flame before and after replacing the former. This is repeated with each of the ten flasks.) The flasks are then shaken and allowed to ferment at the room temperature for seven to eight days. When the fermentation is at an end, and the yeast has settled on the bottom of the flasks, the fermented liquid is decanted with all precautions, and the sediment from each flask divided between five one-litre flasks containing the same nutritive liquid; the transfer must be carried out in the manner just described. These flasks are then shaken and also allowed to ferment at the room temperature for seven to ten days.

This gives fifty one-litre Pasteur's flasks, all of which contain yeast derived from one cell. The yeast is again allowed to settle, the fermented liquid poured off, and the sedimentary yeast collected in a couple of the flasks. We have now about 2 lbs. of fairly thick yeast, and this is sufficient to barm a small brewery vessel of about 30 gallons capacity; this vessel should be placed in a quiet place, and be provided with a cover and lock. The yeast from this vessel is collected and passed on to a larger vessel; this is repeated until sufficient yeast is obtained to barm an ordinary fermenting vessel.

This method is now replaced in the Danish breweries and other large breweries on the Continent by an apparatus devised by Hansen and Kühle, and which enables a change of yeast to be made very frequently without the trouble and expense entailed in the above method; but in many smaller breweries this procedure is still adopted.

It will be gathered from the above that it is by no means an easy matter to prepare sufficient yeast for practical purposes, and when we add to this the difficulty of obtaining the most suitable variety of yeast to start from, it becomes no easy task to introduce pure yeast into a brewery. On the Continent a few varieties or races appear to be generally employed, thus the yeasts No. 1 and No. 2 of Carlsberg are used not only in the Copenhagen breweries, but also in a large number of German, Austrian, Russian, and other breweries, the original stock in each case having been obtained from Copenhagen. Pure yeasts suitable for Continental top-fermentation breweries have also been isolated by Jörgensen, of Copenhagen, and one or two of the Brewing Institutes on the Continent have certain selected varieties adapted to the requirements of local breweries.

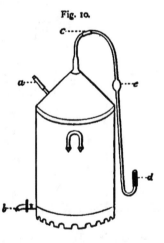

Fig. 10.

By means of the apparatus of Hansen and Kühle a large quantity of absolutely pure yeast can be obtained about every ten days for introduction into the fermenting room.

In the first place, the yeast from the small Pasteur's flasks is added to copper vessels of about 2 gallons capacity (Fig. 10); four of these vessels being used for each cultivation.

The vessels are on the principle of a Pasteur's flask, as will be seen from the illustration, and are filled with ordinary hopped wort of 1058 gravity ; they are sterilised by boiling in the usual way, and should stand for some time after sterilisation, in order that the wort may become aërated before use.

When the yeast has increased to its maximum extent in

these vessels, it is transferred to the continuous culture apparatus.

This apparatus (Fig. 11) consists of three parts, connected together by tubes; these parts are an air-pump (A) and

Fig. 11.

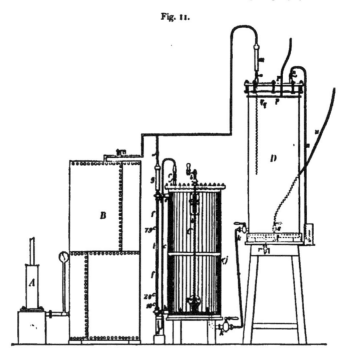

air-reservoir (B), a wort-cylinder (D), and a fermenting-cylinder (C.) The latter two parts are shown more in detail, and with the casing of C removed, in Fig. 12.

The air-pump is worked by an engine, and the air-reservoir filled with air compressed at about one to four atmospheres.

The air-cylinder is sterilised by means of superheated steam from the ordinary supply of the brewery, and then

Fig. 12.

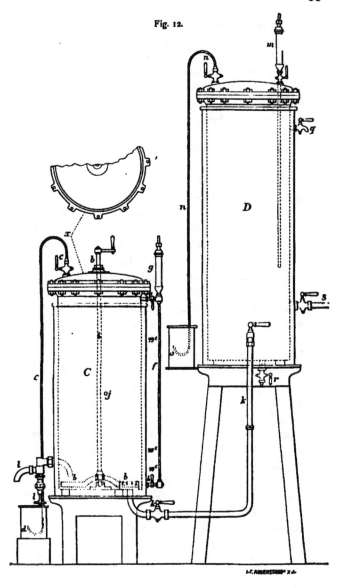

filled with sterilised air.   This air is supplied under pressure
from the air-reservoir, and is purified by being filtered through
cotton-wool contained in a metallic vessel.   The wort-cylinder
is then filled with boiling wort from the main supply from the
copper (s).   Cooling is effected by means of a system of tubes
through which cold water circulates, and the necessary
amount of air for aëration is allowed to enter through the
filter (m).

The fermenting-cylinder is sterilised in the same manner
as the wort-cylinder.   It is provided with a filter (g) similar to
that in the former vessel ; a glass-tube through which to
observe the position of the liquid (f); a suitable exit through
which the evolved carbonic acid can escape (c); a stirring
apparatus in order to mix the yeast with the liquid (b); and a
small tube through which the yeast can be introduced and
also small samples removed (l).   The tap (j) for drawing off
the fermented liquid is specially constructed in such a way
that the liquid itself provides the means of purification, and
consequently no contamination from without can take place.

The wort is transferred from the wort-cylinder to the fermenting vessel by means of the tube (k) connecting the two
vessels.   When the level of the wort reaches a certain height,
namely, just below the opening through which the yeast is
added, the supply is stopped until the yeast has been introduced, wort is then run in until 48 gallons are present.

It is only necessary to once introduce the yeast into the
fermenting cylinder, and a continuous supply of pure yeast
can then be obtained for a twelvemonth, or even longer, as
may be desired.

The fermentation in the cylinder is complete at the end
of ten days, and the fermented liquor is then drawn off and
sterilised air allowed to enter through the filter (g).   When
the greater part of the fermented liquid is drawn off, the tap
is shut, and fresh wort is run in from the wort-cylinder ; the
wort and yeast are well mixed and about 5½ gallons of the

mixture run off, this is repeated, and another 5½ gallons of the second mixture withdrawn. These 11 gallons contain sufficient yeast to barm 176 gallons of wort. The fermenting-cylinder is again filled up to 48 gallons with sterilised wort, and sufficient yeast remains in the vessel to again set up fermentation.

One fermenting-cylinder thus yields sufficient absolutely pure yeast to barm 14½ barrels of wort each month.

In working this apparatus the two chief points to be kept in mind are, that the steaming is sufficient to completely sterilise the vessels, and that during the cooling and withdrawal of the liquid an excess of pressure of sterilised air is maintained in the respective cylinders. When these two conditions are observed no contamination by the sucking back of impure air can take place.

(*b*) ADVANTAGES DERIVED FROM USE OF PURE YEAST.*

The two chief advantages to be obtained by the employment of pure cultivated yeast are, the constant properties of the resulting beer, and the freedom from wild yeast which results from the intelligent use of the method.

Let us first consider the former point. It is, of course, self-evident that it is a great advantage for a brewer to be able to turn out a beer constant in taste and aroma ; and, other things being equal, it is possible to do this by employing the same race or variety of yeast. Experiments have shown that an ordinary brewing yeast consists of a mixture of several races or varieties of yeasts, all apparently to be classed as *Sacch. cerevisiæ*, and undistinguishable the one from the other by an ordinary microscopic examination, but possessing very different properties in practice. A case in point is found in the two varieties of yeasts know as Carlsberg Nos. 1 and 2.

* See also 'Untersuchungen aus der Praxis der Gärungsindustrie,' E. C. Hansen, München, 1890.

These are two races isolated by Hansen by pure culture from ordinary yeast, and resembling each other most closely in appearance, but giving very different results in practice; thus No. 1 gives a beer well adapted for bottling, and containing less carbonic acid than No. 2. The beer has also a lower attenuation, and should remain bright in bottle for at least three weeks. This yeast is chiefly employed for home use; in fact, in Hansen's latest memoir we read that by far the larger part of the output from the breweries of Old and New Carlsberg, amounting to some 250,000 barrels, is fermented with this yeast, whilst only a very trifling amount is fermented with No. 2, which is principally cultivated for export, since it is much preferred by German brewers to No. 1; it gives a good draught beer, containing more carbonic acid than No. 1. We have quoted the foregoing in order to emphasise the necessity of each brewer selecting that variety which is most suited for his trade, and also to point out the fact that the same variety is not suited to all breweries. A yeast cultivated pure by Hansen's method from a brewery yeast will often give a beer differing much from that obtained with the original yeast, for, as mentioned above, the latter may consist of several varieties, each with a distinctive flavour and aroma, &c., and the product consists then of the average of all these differences, whilst the beer from the pure cultivated variety will have the special characteristics of the one race very pronounced. The use of the apparatus briefly described above, enables the brewer to maintain his yeast pure, and thus obtain constant results, since having once introduced the pure culture into the apparatus, a fresh supply of the same variety may be obtained for an unlimited number of times, and the brewer is therefore in a position to throw out his old yeast and introduce a fresh supply of the same variety as often as he pleases. Experience has shown that different varieties of pure cultivated yeast differ much in their resistive powers to outside contamination; thus the Carlsberg yeast No. 1, previously referred to, will remain, as a rule, pure for six to eight months, whilst the No. 2 yeast, under the same conditions, becomes contaminated at the end of only two to four months. The season of the year also exercises a considerable influence.

The second advantage obtained from the use of pure yeast is the freedom from " wild yeast " which results from the method. The diseases which Hansen has more particularly pointed out as being associated with the presence of "wild yeast" are yeast-turbidity, yeast-fret, and an unpleasant bitter taste ; but there are undoubtedly many other irregularities which must be ascribed to the presence of foreign yeast-forms ; for instance, low or excessive attenuations, and unpleasant or even offensive odours. As regards excessively low attenuations, when these are not explained by the chemical composition of the wort, we are satisfied that the excessive attenuation is associated with an abnormal species of yeast, and if many cases of offensive flavours and odours were thoroughly investigated there can be no doubt that abnormal species of yeast would be found to play an important part in regard to them.

The very universal distribution of yeast-forms in the atmosphere during certain seasons of the year is also an important factor. We are aware that during the summer and autumn months the surface of ripe succulent fruits forms the habitat of different species of "wild yeasts," and that these are carried into the air and distributed by the action of the wind and other agencies. Therefore it does not surprise us to find that the atmosphere of country places where fruit trees abound is rich in Saccharomyces forms ; but it does at first sight appear strange that these species of micro-organisms should be found in the air of large cities, as has recently been shown to be the case by Dr. Percy Frankland, who found various species of yeast among the organisms present in the air of London. Bearing all these considerations in mind, we must then not be surprised at beer and wort taking up wild yeast forms from the air during the various processes they undergo, but more especially during cooling and refrigeration. Another fruitful source of contamination is the dust from barley and malt. Hansen long ago pointed out that this dust contained various wild-yeast forms, and that wort should be carefully guarded from any possible contamination of this nature.

Before leaving this subject it may be well to mention a fact lately pointed out by Lindner—namely, that there often occur on the walls, roofs, &c., of the rooms of a brewery, numerous colonies of yeast which, from the conditions under which they exist, develop spores very freely, and these may be carried into fermented and fermentable liquids with great readiness by air currents, by the clothes of workmen, &c.

The practical application of pure yeast, has, we believe, not yet obtained a footing in England. In top-fermentation breweries on the Continent and elsewhere it has been employed, according to Jörgensen and others, with conspicuous success. And quite recently, Van Laer and Kokosinski have described its successful application to the top-fermentation breweries of Belgium and the North of France ; certain selected varieties of pure yeast are there said to have been used with the most satisfactory results. In this country, the question was taken up by Brown and Morris, who met with many difficulties in the application of this system to English brewing. These difficulties have not yet been overcome, but experiments are, we understand, still in progress in the brewery with which these gentlemen are connected. So far as we know, no other attempts have been made to adapt Hansen's system to English brewing. Top-fermentation brewing as practised in England, differs in so many respects from that in vogue on the Continent and elsewhere, that it is impossible, from the satisfactory results said to have been obtained with pure yeast on the latter systems, to draw any conclusion as to the practicability of its application to the English system. This can only be decided by experiment ; but should the difficulties alluded to above be overcome, there can be no doubt that the use of pure yeast would be attended with many advantages to the brewer.

(c) PHYSIOLOGY AND MORPHOLOGY OF THE ALCOHOLIC FERMENTS.

The work of Hansen has a scientific as well as a practical importance, and in the papers on the physiology and morphology of the alcoholic ferments, he has advanced our knowledge of these organisms to a considerable extent.*

Fig. 13.

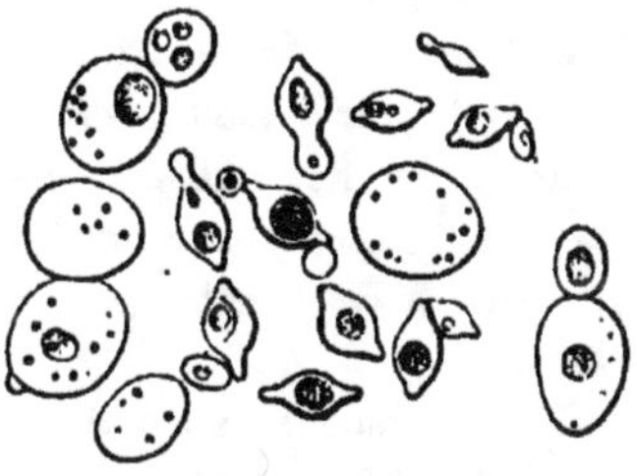

Sacch. apiculatus × 1000, compared with *Sacch. cerevisiæ.* (After Hansen.)

His first work in this direction, was that on '*Saccharomyces apiculatus* and its circulation in Nature.' In this investigation he first used the method of pure culture, which afterwards yielded such important results. This ferment has a very characteristic lemon-shaped cell (Fig. 13), and readily

* ' Recherches sur la physiologie et la morphologie des ferments alcooliques.' Par M. Emil Chr. Hansen. Résumé du compte rendu du Laboratoire de Carlsberg.

I. 'Sur le Saccharomyces apiculatus et sa circulation dans la nature.' 1er vol., 3me livraison. 1881.

II. ' Les ascospores chez le genre Saccharomyces.' 2me vol., 3me livr. 1883.

III. 'Sur les Torulas de M. Pasteur.' 2me vol., 3me livr. 1883.

IV. 'Maladies provoquées dans la bière par des ferments alcooliques.' 2me vol., 2me livr. 1883.

V. 'Méthodes pour obtenir des cultures pures de Saccharomyces et de microorganismes analogues.' 2me vol., 4me livr. 1886.

VI. ' Les voiles chez le genre Saccharomyces.' 2me vol., 4me livr. 1886.

VII. 'Action des ferments alcooliques sur les diverses espèces de sucre.' 2me vol., 5me livr. 1888.

z

lends itself to an investigation of the kind made by Hansen, who determined its exact cycle in Nature. It is found on all ripe succulent fruit, in the yeast of wine, and also in the spontaneously fermented Belgian beer. It appears to be incapable of fermenting maltose and it cannot invert cane-sugar; in dextrose solutions, however, it sets up a vigorous fermentation. Hansen's investigations with this ferment proved the very interesting fact that, although it can always be found in a healthy budding state on fresh, ripe fruits, yet it is never found on unripe fruit, or leaves or twigs, &c., and it is completely absent from the plant during winter; experiments conclusively showed that the ferment hibernates in the earth under the plant, and there remains dormant until the summer, when it is carried by the wind and insects into the air and to the ripe fruit. Recent experiments of Hansen have shown that the cells may remain dormant in the earth for several seasons.

This investigation led the way to those experiments in which Hansen separated and examined the six species to which we have previously referred. These species were obtained from different sources, and Hansen has throughout all his investigations used them as types of the differences which exist among the different varieties or species of Saccharomyces. They are briefly described as follows :—

Saccharomyces cerevisiæ I.—A vigorous top-yeast; used in an impure state in the breweries of London and Edinburgh.

Saccharomyces Pastorianus I.—A bottom-yeast; obtained from the air of a brewery in Copenhagen. This ferment causes a bitter, unpleasant taste in beer. It gives cells resembling those figured by Pasteur and Reess (Plate V., fig. 1).

Saccharomyces Pastorianus II.—A top-yeast; also found in the air of breweries; it resembles the preceding in appearance. It is apparently without action on beer.

Saccharomyces Pastorianus III.—A top-yeast; found in a beer suffering from yeast-turbidity; it is one of the causes of this disease in beer. The cells resemble those of the two preceding.

Saccharomyces ellipsoideus I.—A bottom-yeast; found on grapes; it constitutes the true ferment of wine described by Pasteur and Reess (Plate V., fig. 3.)

Saccharomyces ellipsoideus II.—A bottom-yeast; found in beer suffering from yeast-turbidity; it causes this disease in beer; the cells resemble those of *Sacch. ellipsoideus* I.

The first examination to which Hansen submitted these six species was the spore formation, which has been previously referred to. We have already mentioned the work of Reess and Engel in connection with the formation of endogenous spores in the Saccharomyces, and briefly described the method and conditions employed for obtaining the spores. This must be supplemented by the latest results of Hansen's work, which show that the most necessary conditions for this formation are the employment of young, vigorous cells, the presence of free oxygen, and a suitable temperature.

Hansen commenced his investigation of the ascospore formation with the object of ascertaining if the different species could be distinguished from each other by the appearance of spores after different periods of time at different temperatures. For this purpose he determined for each species :—

(1) The limits of temperature—*i. e.*, the highest and lowest temperature within which spores were formed.

(2) The most favourable temperature for the formation of spores—*i. e.*, the temperature at which spores appeared in the shortest time.

(3) The relation between the intermediate temperatures. The results of these determinations are stated in the following table, which gives the more important temperatures at which Hansen worked.

z 2

ASCOSPORE FORMATION.

Tempera-ture.	Sacch. cerev. I. (Hansen).	Sacch. Past. I. (Hansen).	Sacch. Past. II. (Hansen).	Sacch. Past. III. (Hansen).	Sacch. ellip. I. (Hansen).	Sacch. ellip. II. (Hansen).
Deg. C.						
37·5	none	..	..	..	..	..
36·37	29 hrs.	..	..	..	..	..
35	25 ,,	..	..	..	..	none
33·5	23 ,,	..	..	..	none	31 hrs.
31·5	..	none	..	..	36 hrs.	23 ,,
30	20 hrs.	30 hrs.	..	..	..	..
29	..	27 ,,	none	none	23 hrs.	22 hrs.
27·5	..	24 ,,	34 hrs.	35 hrs.	..	..
26·5	..	..	..	30 ,,	..	..
25	23 hrs.	..	25 hrs.	28 ,,	21 hrs.	27 hrs.
23	27 ,,	26 hrs.	27 ,,	..	..	..
22	..	..	..	29 hrs.	..	..
18	50 hrs.	35 hrs.	36 hrs.	44 ,,	33 hrs.	42 hrs.
16·5	65 ,,	..	..	53 ,,	..	..
15	..	50 hrs.	48 hrs.	..	45 hrs.	..
11·12	10 days	..	77 ,,	..	..	5·5 days
10	..	89 hrs.	..	7 days	4·5 days	..
8·5	none	5 days	..	9 ,,	..	9 days
7	..	7 ,,	7 days	..	11 days	..
3·4	..	14 ,,	17 ,,	none	none	none
0·5	..	none	none	..	..	..

It will be seen from the above, that well-marked differences occur among the six species; thus, *Sacch. cerevisiæ* I. has its limits of temperature between 10° C. (50° F.) and 37° C. (98° F.), *Sacch. Pastorianus* I. between 3° C. (37·4° F.) and 30° C. (86° F.), whilst *Sacch. Pastorianus* II. forms ascospores only between 3° C. (37·4° F.) and 27·5° C. (81·5° F.), and so on. Then, too, the time limits vary greatly with the different species, at least at the lower temperatures. Thus, at about 11° C. (52° F.) *Sacch. cerevisiæ* I. shows the formation at the end of ten days, *Sacch. Pastorianus* I. at the end of 77 hours, and *Sacch. ellipsoideus* II. at the end of 5·5 days.

Plate VII. shows the form which the ascospore formation takes in the six species in question. It also gives some idea of the appearance of cultures of the six species.

The differences shown by other yeasts, and especially those employed in practice, are in many cases much greater. Thus, the pure yeast in use at Carlsberg (No. 1), at 25° C. (77° F.) only develops ascospores at the end of five days, whilst the majority of the species given in the above table shows the formation after about one day at this temperature.

Hansen having proved that certain species of yeast have a most injurious action on wort and beer, producing yeast-turbidity, unpleasant taste, smell, &c., it is naturally most important to know when these "wild" forms are present to any extent in brewery yeast. Holm and Poulsen, in two papers lately published,[*] have thoroughly worked this subject out, employing the ascospore formation as the means of detecting the presence of wild yeast. In their first paper they used only Carlsberg No. 1 yeast, *Sacch. Pastorianus* I. (the cause of a bitter, unpleasant taste in beer), and *Sacch. Pastorianus* III. and *ellipsoideus* II. (the causes of yeast-turbidity in beer). These yeasts were obtained in pure culture, and each of the three last mixed in definite proportions with the first; the temperature of 25° C. (77° F.) was employed, and Holm and Poulson then found that it was possible to easily detect the presence of 0·5 per cent. of either of the "wild" yeasts in the Carlsberg yeast. In their second paper, they approached the question on a broader basis, and submitted a large number—twenty in all—of pure cultivated varieties of yeast, collected from different parts of the Continent, to this method. The same wild yeasts were employed for contamination as in the first paper. It was found that the yeasts examined divide themselves into two groups, one of which can be examined at 25° C. (77° F.), after forty

[*] 'Résumé du Compte-rendu du Laboratoire de Carlsberg.' 2me vol., 4me livraison, 1886. Ibid., 2me vol., 5me livraison, 1886.

hours, and the other at 15° C. (60° F.) after seventy-two hours; it was found that the pure beer-yeasts of the two groups did not show any formation of ascospores, until some days after the respective times at which 1 or even 0·5 per cent. of either of the disease-yeasts could be easily detected.

These later experiments, then, indicate that a means of checking the purity of English brewery-yeast exists, since, when the most suitable sporulating temperature for the particular yeast is ascertained it would be an easy matter to detect the presence of wild yeast. There can be no doubt that the three species named by Hansen, *Sacch. Pastorianus* I., *Sacch. Pastorianus* III., and *Sacch. ellipsoideus* II. (with probably some other species), are as prevalent in English as in foreign brewery-yeasts.

We have seen that the formation of ascospores enables us to detect 1 per cent. or even 0·5 per cent. of disease-yeast in a brewery-yeast, and when we remember that Hansen proved, by careful synthetical experiments, that these forms might be present to the extent of 2·5 per cent. of the total yeast, without bringing on a development of their particular form of disease, it is apparent that a systematic use of the method would prevent any trouble arising from this source.

Another point of difference between normal and disease-yeasts which Hansen claims to have established, is the film formation. It has long been known that fermented liquids form films or pellicles on their surface. Reess first drew attention to the occurrence of these films among the Saccharomycetes; later Pasteur dealt with the question, and in his 'Études sur la bière,' considerable space is devoted to what is termed the aërobian yeast. Hansen appears to think that the films described by himself are not identical with Pasteur's aërobian yeast; this point is, however, doubtful, and we are inclined to consider the two growths to be identical, although the explanations given by the respective authors are different. It seems probable that Pasteur was dealing with the same phenomenon, but that he did not recognise it to be a stage in the development of the Saccharomyces.

Hansen found that the formation of films by pure cultures was very common among all classes of micro-organisms; bacteria as well as moulds and yeasts being capable of giving the formation. Careful experiments, carried out on exactly the same lines as the preceding experiments on the ascospores, gave results showing great differences in this respect among the six species of yeast mentioned above. These differences occur not only in the limits of temperature between which the formation takes place, but also in the rapidity with which the film first makes its appearance at the different temperatures. It was found that the most favourable conditions for the formation of the films are complete quiet and a plentiful supply of free oxygen. For this reason, a little of a pure culture of the respective species was seeded into flasks only half-filled with sterile wort and covered with sterilised filter-paper. The flasks were then placed at the required temperature, and allowed to remain absolutely quiet and undisturbed for the required time. The differences found in this way are shown in the following table :—

FILM FORMATION.

Tempera-ture.	Sacch. cerev. I. (Hansen).	Sacch. Past. I. (Hansen).	Sacch. Past. II. (Hansen).	Sacch. Past. III. (Hansen).	Sacch. ellip. I. (Hansen).	Sacch. ellip. II. (Hansen).
Deg. C. 40	..	..	..	..	..	None
36–38	None	..	..	..	None	8–12 days
33–34	9–18 days	None	None	None	8–12 days	3–4 ,,
26–28	7–11 ,,	7-10 days	7–10 days	7–10 days	9–16 ,,	4–5 ,,
20–22	7–10 ,,	8–15 ,,	8–15 ,,	9–12 ,,	10–17 ,,	4–6 ,,
13–15	15–30 ,,	15–30 ,,	10–25 ,,	10–20 ,,	15–30 ,,	8–10 ,,
6–7	2–3 months	1–2 months	1–2 months	1–2 months	2–3 months	1–2 months
3–5	None	5–6 ,,	5–6 ,,	5–6 ,,	None	5–6 ,,
2–3	..	None	None	None	..	None

In addition to the variations in temperature and time given in the above table, the different species show a marked difference in the appearance of the cells of the film at the various stages, and in this way we are aided in distinguishing the species one from another. The most important variations in this respect are seen at 13–15° C. (55–59° F.), at which temperature the six species exhibit well-marked characteristics. It is especially to be noted that *Sacch. Pastorianus* II., and *Sacch. Pastorianus* III., which cannot be distinguished from each other when growing in the ordinary way, present noteworthy differences in film-formations at this temperature; the same remark applies to *Sacch. ellipsoideus* I. and *Sacch. ellipsoideus* II. *Sacch. cerevisiæ* I., *Sacch. Pastorianus* II., and *Sacch. ellipsoideus* II., show, at 13–15° C. (55–59° F.), only round or oval cells, whilst the other three species, *Sacch. Pastorianus* I. and III., and *Sacch. ellipsoideus* I. show branched, mycelium-like colonies. In the other films and at the higher temperatures the variations are not so marked, the films consisting, as a rule, of very large mycelium-like cells in ramified colonies.

Plate VI. illustrates the appearance of the cells of the film-formation in *Sacch. cerevisiæ* I.; 1 at 6–15° C.; 2 at 20–34° C.; and 3 in old cultures of the film.

In the course of his researches Hansen found that budding and fermentation can take place at temperatures above that at which films will form. He also confirmed Schmitz' observation of the occurrence of cell-nuclei among the Saccharomyces.

Hansen's recent work on the action of the alcoholic ferments on the different kinds of sugar deals with the action of about forty yeasts, including two new species, *Sacch. Marxianus* and *Sacch. membranæfaciens*, on four varieties of sugar—viz., cane-sugar, maltose, lactose, and dextrose. The majority of these yeasts invert cane-sugar, and ferment maltose and dextrose, but not lactose. *Sacch. membranæfaciens*

Plate.VI.

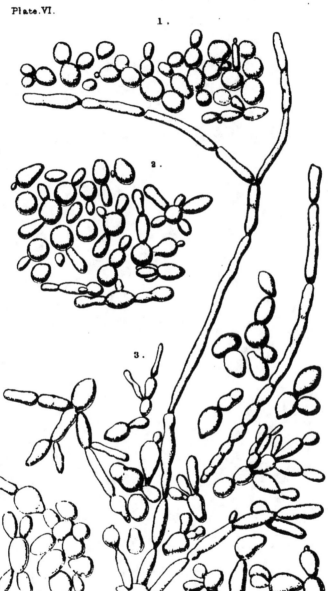

1.

2.

3.

Sacch. cerevisiæ I (Hansen). 1. Film growth at 15-6° C.
2. Film growth at 34-30° C. 3. Film growth on old
cultures. × 1000. (after Hansen)

E. & F. N. Spon, London & New York   The Roll & Son, Lith

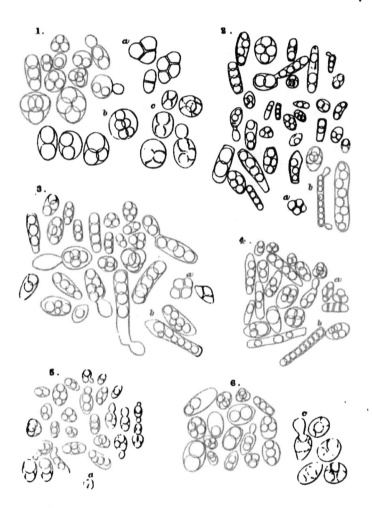

Formation of Ascospores.

1 *Sacch. cerevisiæ I*	2 *Sacch. Pastorianus I*
3 *Sacch. Pastorianus II*	4 *Sacch. Pastorianus III*
5 *Sacch. ellipsoideus I*	6 *Sacch. ellipsoideus II*
(after Hansen × 1000)	

F. & F. N. Spon, London & New York.   The Ketth Son Lith

alone is not able to ferment either of the sugars employed, and it does not secrete invertase ; it is, however, distinguished by the extreme ease with which ascospores are formed; it also liquefies gelatin. *Sacch. Marxianus* and *Sacch. exiguus* are also not able to ferment maltose. Among the other species employed the most noteworthy result was obtained with *Monilia candida* (p. 376), which is able to directly ferment cane-sugar.

Several interesting facts in connection with the products of fermentation have been obtained by means of Hansen's pure culture methods.

Thus Ordonneau published in 1886 [*] some very interesting experiments on the chemical action of different species of yeast. He found by distilling a large quantity of genuine Cognac that one hectolitre of spirit contained 218·6 grams of *normal butylic alcohol*, in addition to smaller amounts of other alcohols ; but when he submitted a spirit obtained from potatoes, beet-root, maize, &c., to the same treatment, he found the chief constituent to be *isobutylic alcohol*, and it is to this alcohol that the disagreeable odour of potato, &c., spirit is due. Further experiments established the fact that the formation of isobutylic alcohol is due to the use of ordinary brewer's yeast. When a potato mash was fermented with pure *Sacch. ellipsoideus*, the true wine yeast, normal butylic alcohol, was formed as in the spirit from the grape. This result is especially important with reference to the recent discussion regarding the presence of fusel oil (isobutylic alcohol) in beer.

The recent analyses of Carlsberg pure yeast beer by Borgmann [†] are also worthy of mention. The beer employed for these analyses was the ordinary bottom fermentation beer fermented partly with No. 1 yeast and partly with No. 2 yeast. It was stored for some months, as usual,

[*] Compt. rendu, cii. p. 217.
[†] Zeitsch. f. Anal. Chemie, xxv. p. 532.

and bottled. Several analyses of each beer gave the following mean results :—

100 grs. contained	No. 1 yeast.	No. 2 yeast.
Alcohol	4·13 grams	4·23 grams
Extract	5·35 ,,	5·84 ,,
Ash	0·20 ,,	0·25 ,,
Free acid (as lactic).. ..	0·086 ,,	0·144 ,,
Glycerine..	0·109 ,,	0·137 ,,
Phosphoric acid	0·0775 ,,	0·0828 ,,
Nitrogen	0·0710 ,,	0·0719 ,,

It will be seen from these numbers that the yeast exercises some influence on the composition of the resulting beer.   The proportion of alcohol to glycerine in these beers is also very different from that found by the author in a large number of beers fermented with ordinary yeast.   In the former case the ratio is—

No. 1 yeast.	No. 2 yeast.
Alcohol 100. Glycerine 2·63.	Alcohol 100. Glycerine 3·24.

And in the latter the proportion varied between a maximum of, alcohol 100, glycerine 5·497, and a minimum of, alcohol 100, glycerine 4·140.

These experiments open up a most interesting field of work, and show us the important part which the yeast plays in determining the nature of the resulting product.

These results have been confirmed and amplified by Amthor,[*] who examined the beers produced from eight different pure yeasts.   The alcohol-glycerine proportion was low in all the beers, in one case it was as low as 100 : 1·65.   The percentage of alcohol also varied considerably.

Quite recently Gronlund [†] has published a paper which confirms Hansen's work in a remarkable degree.   He had to investigate the case of a beer which possessed a most

[*] Zeitschr. f. physiologische Chemie, xii., p. 64.
[†] Zeitschr. f. d. gesammte Brauwesen, 1887, p. 469.

unpleasant odour and bitter taste, and by means of careful analytical and synthetical experiments he proved conclusively that the cause of this was the presence of a particular yeast form. This yeast he was able to identify with Hansen's *Sacch. Pastorianus* I., which, it will be remembered, is said by this observer to bring about this disease in beer.

II. THE SCHIZOMYCETES, OR BACTERIA.

No yeast or beer is entirely free from Schizomycetes, or bacteria, for, as we have explained, they, with the wild yeasts, are abundantly prevalent in the air, and have thus every opportunity of affecting the wort, especially during cooling and refrigeration. The plant itself is never entirely free from them, and wood-plant especially requires the utmost care, if we wish to keep down the organisms which would otherwise breed most plentifully in its pores. Unclean mains, pipes, pumps, &c., are also a very frequent source of contamination. The yeast itself introduces them, being seldom free from bacteria. The yeast and the bacteria are, in a way, antagonistic, for a vigorous yeast checks their development, and a poor one encourages it, so that under normal conditions their presence in a yeast, when not excessive, is not necessarily productive of the harm which might at the first glance be supposed to be the case.

The Schizomycetes, or bacteria, are unicellular plants which occur in round, cylindrical, rod, or spindle-shaped forms, or united in threads and masses. The individual cells are always very small, the transverse diameter ranging from $0 \cdot 0002$ mm. (2μ) in the round forms, to $0 \cdot 0001$ mm. and even less, in the rod-like cells. The cell-contents consists of a protoplasmic substance, which in the different species shows different reactions with staining solutions, and thus serves as an aid to the differentiation of the species. Some bacteria contain a starch-like substance, coloured blue by iodine, in their cell-contents; others, *Beggiatoa,* for instance, contain granules of pure sulphur; and many exhibit well-defined colours when seen in quantity. These colours, which may be yellow, red, green, violet, blue, brown, &c., depending on the species, are due to colouring matters either in the cell-contents or in the cell-membrane.

The cells are surrounded by a membrane or cell-wall, which, according to De Bary, may be regarded as the inner-most and comparatively firm layer of a gelatinous envelope surrounding the protoplasmic body. The possession of this gelatinous membrane is common to bacteria and some other of the lower organisms; examination of several forms of bacteria has shown this membrane to consist of a carbo-hydrate closely related to cellulose. With certain species and under certain conditions the gelatinous membrane is developed to a very large extent, and we then find countless numbers of cells enclosed in a gelatinous network. Such a gelatinous colony is termed a zoogloea; it often precedes the formation of spores.

In liquids, many bacteria show a peculiar rotatory or oscil-lating movement. This movement has been usually attributed to the presence of cilia or flagella at the extremities of the bacteria. These appendages have actually been found in certain species of rod bacteria, and have been considered to be analagous to the cilia of the cells and spores of many Algæ. These latter, however, are extensions from the surface of the protoplasmic body, and when the protoplasm is sur-rounded by a membrane, pass through openings in that membrane. In the case of bacteria the cilia are thread-like extensions of the cell-membrane, as has been shown by their behaviour with reagents, and therefore they have nothing in common with the cilia of the swarm-spores of the Algæ.

The multiplication of bacterium-cells takes place in two ways. The usual way is the vegetative multiplication by successive cell-division. At a certain stage in the life of a cell, a fine transverse line makes its appearance, and divides the cell into two equal parts. The line then swells, and the cell becomes divided into two daughter cells. The newly formed cells may either become detached or remain combined and form threads. The successive divisions may take place either all in the same direction or in two or three directions in space, when colonies are formed in the form of plates or cubes (*Sarcina*).

The second method of multiplication is by the formation of spores, which appear to resemble, morphologically and physiologically, the "resting spores" of Algæ and Fungi. They take their rise in the interior of the cell, in the form of a small spherule, which gradually becomes larger whilst the protoplasm disappears little by little. After a time the highly refractive, dark-coloured spore occupies nearly the whole of the interior of the mother-cell, the membrane of which is then either dissolved or ruptured, and the spore escapes; it can then multiply by cell-division. The conditions under which spore formation takes place among the bacteria are not definitely known, but it appears to occur when the substratum has, for any reason, become unsuited for the nutrition of the organism. Spores of bacteria are much more resistive to adverse influences than are the bacterium-cells themselves. Tyndall, amongst others, showed that a temperature which destroys the fully-developed cells is without action on the spores of the same species; the latter may even survive boiling for some minutes. It is for this reason that it is necessary to repeat at intervals of some hours the boiling of nutritive solutions in order to completely sterilise them. The same remark as to resistibility also applies to the action of antiseptics; solutions, for instance, of corrosive sublimate which kill adult organisms are without action on spores. Spores, too, may remain for months, and even years, in a state of quiescence before they germinate, and in this respect they show their analogy to the "resting-spores" mentioned above. When germination does take place, the spore loses its refractive power and swells slightly, the containing-wall then bursts, and the new organism emerges from the interior.

We have already briefly touched on the zooglœa formation, produced by the grouping together of a vast number of bacterium-cells due to the extension of the gelatinous substance of the cell-wall. These zoogloea vary much in appearance and character, depending on the species, but occur to a greater or less extent with nearly every species. In some species the zoogloea forms thick, almost solid gelatinous masses; a good instance of this is the so-called frog-spawn bacterium, *Leuconostoc mesenterioides*, which occurs in sugar factories, and may fill entire vats of sugar solution with a substance resembling frogs' spawn. Other species only form loose, flocculent colonies of cells, without any sharp limiting surface. Such zoogloea are usually termed swarms.

The question of species in the Schizomycetes has been the subject of much discussion among bacteriologists, and various endeavours have been made to group these organisms in good characteristic species, and thus render their study more easy. Cohn takes morphological differences as the ground of his classification, and groups the bacteria according as they form single spheres, short rods, long rods, or spirals. He divides them as follows:—

Tribe I. Sphœrobacteria,	Spheres	Genus 1. Micrococcus,
,, II. Microbacteria,	Short rods	,, 2. Bacterium,
,, III. Desmobacteria,	Threads	,, 3. Bacillus,
		,, 4. Vibrio,
,, IV. Spirobacteria,	Spirals	,, 5. Spirillum,
		,, 6. Spirochæta,

and each genus contains a great many species. But it is found that many bacteria are pleomorphic; that is, they alter the form of their cells during the different stages of development, and therefore classification by form alone is more or less unreliable. Since morphological distinctions between the species cannot be relied upon, we have to fall back upon physiological distinctions, and we may employ differences in chemical action for defining the species. Bacteria excite peculiar and distinct chemical changes in the substrata in which they are cultivated. These chemical changes may be divided into three classes, and it is possible, therefore, to distinguish between—

(1) Chromogenic species—that is, those forming and secreting colouring matter.

(2) Zymogenic species, those exciting the various fermentations.

(3) Pathogenic species, those producing diseases in man and animals.

Naegeli and other observers contend that only a few species of bacteria exist, and that all bacteria without reference to their form or mode of development should be combined together in a single genus or in very few genera. The physiological and morphological differences observed are ascribed solely to different conditions of nutriment and temperature. De Bary contends that this theory is a wrong one. He defines species "to mean the sum total of the separate individuals and generations which, during the time afforded for observation, exhibit the same periodically repeated course of development within certain empirically determined limits of variation. We judge of the course of development by the forms which make their appearance in it one after another. These are the marks by which we recognise and distinguish species." To effect this determination we must use the methods which have lately come to the front. Pure cultures of the organism must be followed with precision under the microscope, by means of moist chambers with a suitable substratum.

De Bary and Heuppe divide the bacteria into two large groups—namely :—

(1) Bacteria which form endogenous spores.

(2) Bacteria which form arthrospores, including those bacteria whose fructification is unknown.

Zopf classes the bacteria into four groups—Coccaceæ, Bacteriaceæ, Leptotricheæ, and Cladotricheæ. This classification is based not only on the form of the cells, but also on the mode of growth and connection between the forms, and fully recognises the pleomorphism which exists among the organisms.

For the purpose of the brewer these rival schemes of classification are of little value; it is here only important to know what changes are produced in beer-wort and beer by certain definite pure-cultivated species under certain conditions, and what morphological changes the organisms themselves undergo during the different stages of growth.

The conditions of existence of the bacteria vary greatly; the nutritive material, the presence or absence of oxygen, the temperature, the acidity or alkalinity of the material and so forth, all exercise a varying influence on the different species of bacteria, and we shall consider these factors as we examine those species of especial interest to brewers.*

(a) Bacillus subtilis.

Bacillus subtilis, as the name implies, is an organism shaped like a thin rod (Plate VIII., fig. 5). *Bacillus subtilis* is also known as the hay bacillus, on account of the prevalence of this organism on hay, but any organic infusion containing nitrogenous and mineral matters, when exposed

---

* The following works may be consulted for more detailed information on the points touched upon above :—

'Lectures on Bacteria.' By A. de Bary. Translated by Garnsey and Balfour. Clarendon Press.

'Manual of Bacteriology.' By E. M. Crookshank.

'The Bacteria and Yeast Fungi.' By W. B. Grove.

'Die Spaltpilze,' By W. Zopf.

to the atmosphere, is speedily infected by it. The ferment is also known as *Vibrio subtilis* (Ehrenberg).

This bacillus, as we generally see it, is in the form of thin rods, sometimes rigid, sometimes sinuous. The length ranges from 2–6 μ, the width about 2 μ. Cohn says that, at 21° C. (70° F.) division into two takes one hour and a quarter, at 35° C. (95° F.), only twenty minutes. It will thus be seen how important is temperature in its effect on the reproduction of this organism—both during fermentation and in the storage of beer and yeast.

Bacillus subtilis readily forms spores, which are remarkably tenacious of existence; but boiling for thirty minutes is sufficient to kill them. The vegetative rods are killed by merely raising the liquid containing them to the boiling temperature. The vegetative rods are provided with flagella or cilia, which cause their active movement.

We know of no work which connects *B. subtilis* with the formation of any predominant specific chemical compound. It is an agent of putrefaction, rendering nauseous any fluid in which it develops, evolving sulphuretted hydrogen and other volatile sulphur and phosphorus substances, and splitting up the nitrogenous matters into simpler nitrogen-containing substances, and the carbonaceous matter partly into organic acids of widely different kinds. We do not think that it can be regarded as in any way distinctly leading to the formation of the acids which occur in beer, when certain organisms are allowed to develop.

Bacillus subtilis is one of the species of bacteria said to possess a peptonising influence upon proteid substances. According to various investigators, certain albuminoids are, under the influence of this and other species, broken down into peptones, and then into simpler products of decomposition, among which leucin, tyrosin, and ultimately ammonia, have been found. Among the species which we have reason to regard as of interest to brewers, this action is possessed by

2 A

Bacillus subtilis, Bacillus amylobacter, and *Bacterium termo.* It is well shown, when these species are cultivated in nutritive gelatin, the gelatin immediately surrounding the colonies becomes liquefied, due to the formation of simpler decomposition products in the manner described above. These organisms commonly occur in contaminated waters, and no doubt, by their action, render the putrescible matter more assimilable for the later growths of micro-organisms.

(*b*) *The Butyric Acid Ferment.*

Similar morphologically to *Bacillus subtilis* is *Bacillus amylobacter,* also known as *Clostridium butyricum,* or *Vibrio butyricus* (Plate VIII., fig. 6). It is distinguished from *B. subtilis* by containing starch in its cells, easily recognisable by injecting a solution of iodine, when the starch is stained blue. This ferment is the *Vibrio butyricus* of Pasteur, shortly known as *Vibrio,* or butyric ferment ; the last term is open to the objection of implying that it alone produces butyric acid, which is certainly not the case. This ferment is morphologically identical with *B. subtilis. B. amylobacter* is, as shown by Pasteur, capable of development without air. It forms butyric acid in solutions of starch, dextrin and sugar. In brewing, it is frequently met with both in yeasts and beers. It is abundant in the air, and would therefore infect the wort on the coolers, but we have generally found it to be connected with dirty plant, whether this be the wood, or the pipes and mains and pumps. To the latter, air would have but little access, so that this organism, which can exist without air, might develop in places where other organisms would not survive. The presence of this organism in yeast and beer should direct immediate attention to the state of the plant.

It is plentifully abundant on the exterior of barley, and

freely grows on the germinating malt. In bruised corns attacked by mould, there is sometimes a putrid flavour, probably due to the products of this ferment. We know that it attacks starch readily, and even cellulose is not proof against it. It is an interesting fact that this ferment is abundant in the stomach and intestines of herbivorous animals, and assists in the digestion of the cellulose-containing food of these animals.

(c) Bacterium termo.

Bacterium termo is a common ferment (Plate VIII., fig. 4). It is not associated with the production of any predominant chemical compound, but is abundant in the air, and promotes general decay. It is largely present in brewing waters polluted by sewage, and in common with other species is found on the exterior of barley and hops. *Bacterium termo* being provided with a flagellum or whip, is capable of movement, but this power of movement is lost when free oxygen is absent.

This species, which recent work has shown to be a very indefinite one, has always been associated with putrefactive changes, and has been regarded as one of the chief agents in the peptonisation of albuminoids. It readily liquefies gelatin when grown in this medium.

Caution is necessary lest the presence of bacteria and other minute ferments in yeast, beer, &c., be too hastily assumed. The caution is the less necessary in the case of the organisms we have considered, since their configuration in the true bacillus form is distinctive. But minute fragments of amorphous matter under the microscope are frequently mistaken for Bacteria, Sarcinæ, Micrococci, &c. In the minute form in which the amorphous matter (from the hops, &c.) exists, it possesses a certain tremulous movement known as the Brownian movement; a movement likely to be misconstrued into life, and the amorphous substances are thus frequently regarded as bacteria. In all cases, therefore, it is well to treat the deposit with potash or ammonia solution. This dissolves the amorphous matter, but does not affect the bacteria.

(d) *The Lactic Acid Ferments.*

In the case of the ferments we have hitherto considered, we have had to deal with tolerably well-defined organisms, whose life-history and fermentative action has been experimentally determined, but when we come to the micro-organisms producing lactic acid as one of the products of their action, we find the question of the number of the species, the nature of the action, and many other points involved in the greatest obscurity.

Pasteur, in his 'Études sur la bière,' describes the ferments which produce lactic acid in wort and beer as being short bacteria, small, thin, and slightly contracted in the middle, forming dumb-bell shaped organisms which are generally detached, but sometimes occur in short chains. Their length is given as about 2 μ, and they are shown in Plate IX., fig. 2. Pasteur was the first to show that, in the old method of bringing about lactic acid fermentation in different kinds of sugar, a particular organism, or perhaps more than one, was the exciting agent. This method consisted in adding sour milk or cheese to the sugar solution, and keeping it exposed to the air at a temperature of 40–50° C. (104–122° F.). Calcium carbonate or zinc oxide was added from time to time to neutralise the acid formed, because when a certain amount of acid is formed, the fermentative action and growth of the organism is stopped. That this was the true reason for the addition of the carbonate was not, of course, known until Pasteur had proved that an organism was introduced with the sour milk, and that it multiplied in the liquid and acted as a ferment. It appears in the form of small cylindrical cells, which multiply by .

division, and at once separate from each other; immediately after division they are barely half as long again as they are broad, and average 0·5 μ in thickness. It is only very rarely that the cells remain united, and form short chains. This species somewhat resembles *Bacterium aceti*, and was named *Micrococcus lacticus* by Van Tieghem. Later this organism was examined by Hueppe, who claimed to have noted the formation of endogenous spores, and, consequently, if this is correct, it must be transferred to the bacillus division (the members of which, it will be remembered, are characterised by the formation of endogenous spores); Hueppe named it *Bacillus acidi lactici.*

The lactic acid fermentation, which has as its chief product lactic acid, can be represented by the equation—

$$C_6H_{12}O_6 = 2(C_3H_6O_3).$$
$$\text{Dextrose.} \qquad \text{Lactic acid.}$$

whereby one molecule of carbohydrate splits up into two molecules of lactic acid. It is of very common occurrence; the most familiar example is the souring of milk, attended by the coagulation of the casein and the formation of butter-milk. The cause of this is the formation of lactic acid from the sugar of the milk, by the air-sown ferment, in sufficient quantity to bring about the coagulation. Another familiar instance is the production of "sauerkraut," the acid of which is mainly lactic acid. In the lactic acid fermentation the reaction does not take place absolutely in accordance with the above equation, since there are formed, in varying quantities, a number of other products—namely, alcohol, butyric acid, carbonic acid, mannite, and other insufficiently investigated substances. It was formerly supposed that oxygen was not necessary for the lactic acid fermentation, and in fact, that it retarded the fermentation, but it has now been shown that oxygen is requisite for the continued action of the ferment, and that it certainly does not retard the action.

The bacillus or micrococcus, described above, is the most widely distributed lactic ferment, but it is by no means the only organism which produces lactic acid. Hueppe alone mentions five species, and others have been described by Lindner and other workers. The five species described by Hueppe are all micrococci. One of these is the celebrated red *Micrococcus prodigiosus*, the appearance of which in various places gave rise to many superstitions in the past. Two others were found by Hueppe to be the exciting cause of the lactic acid which occurs in the mouth.

The only work on the lactic acid fermentation of any importance to brewing, and published during the last few years, has emanated from the Berlin Brewing School. Delbrück and Hayduck have both carried out a series of experiments on this fermentation, in which they show that the most favourable temperature for the action is 40° C. (122° F.). In order to obtain a growth of the ferment free from other organisms, Delbrück states that it suffices to take air-dried malt, and, after grinding, to mash it for half-an-hour at 60° C. (140° F.), and then allow the filtered liquid to stand at 40° C. (122° F.). The lactic acid ferment, which is almost invariably contained on the exterior of malt, will, under these circumstances, start an active fermentation in a few hours, and produce a considerable quantity of lactic acid. It is the custom in the German distilleries and pressed-yeast factories to add a small quantity of sulphuric acid to the mash for the double purpose of checking the development of bacteria and degrading the nitrogenous constituents of the mash in order to render them assimilable by the yeast. Hayduck determined the amount of sulphuric acid necessary to check the lactic ferment, and found that the addition of 0·04 per cent. was sufficient to completely stop the lactic fermentation, and it was previously known that 0·05 per cent. was the maximum allowable limit, in order that the growth and fermentative power of the yeast might not be influenced. Lactic acid itself has a marked deterrent effect on the development of the lactic acid bacterium ; thus Hayduck found that the addition of 0·15 per cent. of acid to a mash was sufficient to prevent any lactic fermentation taking place, although, when the fermentation has once started, it will continue until ten times as much acid as this is formed. This is accounted for by the fact that the presence of the acid in the first instance checks not only the fermentative power of the bacterium, but also its multiplication, and the small number of organisms present are not able to produce any determinable amount of lactic acid.

As a result of the above and similar experiments, it is now recommended that lactic acid be added to the mash instead of sulphuric acid as formerly, since in addition to its power of keeping down the growth of bacteria, it possesses, in a much higher degree than sulphuric acid, the power of converting the albuminoids in the mash into substances which serve as food-stuff for the yeast.

Alcohol has a very marked influence on the lactic acid fermentation. Under the conditions mentioned above, Hayduck found that 4 per cent. of alcohol exercised a very considerable retarding action, whilst 6 per cent completely stopped the fermentation. This observation, which, of course, only applies to the species Hayduck was working with, seems to point to but slight danger existing of beer being attacked by the lactic acid ferment during the secondary fermentation or during storage, as there is usually sufficient alcohol present to prevent the growth of the organism. But we know that in some cases a considerable increase of acidity, due to fixed acid, takes place in beer on storage, and this acid we are accustomed to regard as lactic acid. We are therefore forced to the conclusion that the organism producing this acid in beer is a different species to that experimented with by Hayduck.

Direct experiments on the beer organism are still wanting, although Lindner has examined a number of the so-called sarcina-forms from beer, and found them all more or less capable of producing lactic acid. One, indeed, appeared to be so powerful a ferment that he named it *Pediococcus acidi lactici* (Plate IX., fig. 1), and considered it to be identical with the spherical bacterium observed by Hayduck and Delbrück. It closely resembles in appearance the *Pediococcus cerevisiæ* to be described later but differs from it in many important particulars. It occurs in *diplococci* and *tetrads,* but only in two-dimensional growth; the diameter of the single cells is 0·6–1·0 μ. *Pediococcus acidi lactici* forms a vigorous growth in malt-extract in twenty hours, and the liquid has a strong acid reaction. The same result was obtained when the organism was grown in hay infusion. On examination it was found that the acid consisted entirely of lactic acid. The most favourable temperature for its growth appeared to be about 40° C. (104° F.). The cells are completely killed at a temperature of 62° C. (144° F.). It occurs on malt, and, although the organism can grow in the presence of air, yet it appears to develop more freely in the absence of oxygen.

One very remarkable point in connection with this ferment is that—as was the case with *Pediococcus cerevisiæ*—it was found impossible to inoculate it into sterilised hopped wort or hopped wort-gelatin.

(e) *The Acetic Acid Ferments.*

The acetic fermentation is, perhaps, one of the commonest of the fermentations which are brought about by the action of bacteria, and certainly the organisms which are the cause of the production of acetic acid have been more completely studied, and their life-history more thoroughly determined, than other micro-organisms of the same class. It is now well established that there are at least three species of bacteria which are capable of exciting acetic fermentation; these have been exhaustively examined by Cohn and Hansen from a physiological and morphological standpoint, and by Pasteur and A. J. Brown from the chemical side.

The connection between the formation of acetic acid in an alcoholic fluid and a specific organism, was first established by Pasteur in 1864, when he published his 'Mémoire sur la fermentation acétique.' In this memoir he shows that the change which an alcoholic liquor undergoes when it stands exposed to the air at a moderate temperature, whereby it gradually assumes a characteristic acid smell and taste and the alcohol slowly disappears, is due to the intervention of the acetic acid ferment, *Mycoderma aceti*—or, as it is now more usually called, *Bacterium aceti*—which forms a film on the surface of the liquid and slowly brings about an oxidation of the alcohol by the oxygen of the air, according to the following equation :—

$$C_2H_4OH \quad + \quad O_2 \quad = \quad CH_3COOH \quad + \quad OH_2$$
alcohol.   oxygen.   acetic acid.   water.

130 parts of acetic acid are thus formed from 100 parts of alcohol. He also shows that if the fermentation is allowed to proceed after the whole of the alcohol has been converted into acetic acid, the organism then brings about an oxidation of the latter substance, resolving it into a mixture of carbonic acid and water. Under certain conditions, whereby the action of the ferment is enfeebled, Pasteur shows that aldehyde is formed in the liquid.

Pasteur recognised only one species of organism as exciting this fermentation, but, as stated above, it has since been shown that at least three species possess this power. The discovery of this fact is due to the employment of the methods of pure culture which are now universally practised.

Hansen[*] was the first to examine the acetic acid ferments by the methods of pure culture ; he obtained the organisms from a film growing on beer at a temperature of 30–34° C. (86–93° F.),

---

* Résumé du compte rendu du Laboratoire de Carlsberg, i. (1879), p. 96.

and on examination found that they belonged to two species, both of which produced the same action in the liquid—that is, caused the formation of acetic acid, whilst they were distinguished from each other by their behaviour with iodine solution. One, which Hansen called *Mycoderma aceti*, is coloured yellow by iodine; the other, named *Mycoderma Pasteurianum* by Hansen, is coloured blue by the same reagent. Morphologically, the cells of the two species cannot be distinguished one from another (Plate VIII., fig. 3.) The morphology of these two species cannot be better described than in the words of A. J. Brown:* "*Bacterium aceti* (and *Bact. Pasteurianum*), in its normal state, freshly growing as a pellicle on a surface of a liquid, appears under the microscope as a mass of cells 2 μ in length, and slightly contracted in the middle, giving them a sort of figure of 8 appearance. These cells are united into chains of variable length, which are easily broken up by pressure of the cover-glass. Frequently the cells are quite divided in the middle, thus producing strings of micrococcus-like forms; both forms being sometimes found in the same chain. The above two forms are those usually present when the ferment is growing vigorously at the surface of a liquid. But in the liquid below the surface-film, and on the bottom of the containing vessel, abnormal forms are often found differing very much from the ordinary surface growth; this is more especially the case in old cultivations. These forms often attain the length of 10–15 μ, or even more; in some cases their form is that of leptothrix threads, of even thickness throughout their length; in others, the long cells are swollen out in two or three places along their length, giving them a most irregular appearance. These cells are generally of a dark grey colour. At their ends a short chain of short rods or micrococci is sometimes observed. The other forms most frequently seen are short

* Journ. Chem. Soc, 1886, pp. 172–187; and 1887, pp. 638–642.

rods, about 3 μ in length, and micrococci about 1 μ in diameter floating freely in the culture liquid."

Both these species are strictly aërobic, and grow only in liquids containing not more than 10 per cent. of alcohol; they require both nitrogenous and mineral nutritive substances. A. J. Brown has fully confirmed Pasteur's statement that the acetic ferments split up the acetic acid itself when all the alcohol has been oxidised. The limit of acid formation lies between 1 and 2 per cent.; when this is reached, the further action of the ferment on the alcohol is checked. The most favourable temperature for the growth of the acetic ferment is, according to Hansen, 33–35° C. (92–94° F.), and under 18° C. (64° F.) the action is very slow, and decreases directly in proportion to the temperature; therefore beer stored at low temperatures runs little risk of being attacked by this ferment. Sulphurous acid is very destructive to the acetic ferment.

The chemical action of *Bact. aceti* has been very thoroughly examined by A. J. Brown, and the results described in the papers previously mentioned. He first showed quantitatively that pure dilute alcohol (ethylic alcohol) is oxidised to acetic acid by the ferment (a trace of fixed acid was also found, thus confirming Pasteur's statement that a little succinic acid is formed); and that propylic alcohol is similarly converted into propionic acid; other alcohols were unacted on, although the ferment grows freely in a 1 per cent. solution of methylic alcohol in yeast-water. The action of *Bact. aceti* on the carbohydrates is most interesting and important. Dextrose in yeast-water and in the presence of calcium carbonate is converted into gluconic acid, which forms the calcium salt with the calcium carbonate; the ferment is unable to break up or alter cane-sugar; mannite is completely converted by *Bact. aceti* into levulose, which is itself untouched by the ferment. This reaction is of great theoretical interest, since by its aid we are able to convert dextrose into levulose. A. J. Brown has proved this completely.   He first converted dextrose into mannite by acting upon it with sodium amalgam, and then converting the mannite so obtained into levulose by the action of *Bact. aceti.*   Glycol is converted into glycolic acid by the ferment, and glycerine is completely decomposed into carbonic acid and water.   Milk-sugar, starch, and maltose are unacted upon by *Bact. aceti.*

We have mentioned that there are at least three species of bacteria known to have the power of converting alcoholic liquids into acetic acid, or vinegar.   It has long been known that a peculiar gelatinous mass, known as the "vinegar-plant," or "mother of vinegar," was sometimes formed on vinegar solutions, and was formerly much used in country districts for the manufacture of home-made vinegar.   This mass is mentioned by more than one writer—Pasteur, Zopf, &c.—but it was commonly regarded as a zoogloea form of *Bact. aceti.* A. J. Brown met with this ferment during his work on *Bact. aceti* mentioned above, and as it appeared to be a distinct organism, he examined it more closely.*   It was purified by a combination of the "fractional" and "dilution" methods and also by cultivation in beer-wort and gelatin.   Grown in a favourable liquid it first forms a jelly-like translucent mass on the surface of the liquid, which gradually covers the entire surface with a gelatinous membrane, and may attain a thickness of 25 mm.   This membrane presents an entirely different appearance to the film of *Bact. aceti*, and its behaviour with potash (it resists the action of boiling potash, whilst *Bact. aceti* is entirely disintegrated by the same solution in the cold), and the fact that it is stained a deep blue with iodine, shows that it consists largely of *cellulose.*   For this reason the name *Bact. xylinum* was given to this bacterium.   When a membrane of *Bact. xylinum* is examined microscopically, it is found to consist of bacteria, arranged more or less in lines, and lying embedded in a transparent structureless film.   When stained

* Journ. Chem. Soc., 1886, p. 442-439 ; and 1887, p. 643.

with aniline-violet, the bacteria are seen to occur in chains of rods, the rods being about 2 μ in length. In old cultivations the rods are largely replaced by micrococci about 0·5 μ in diameter. The swollen involution forms of *B. aceti* are never observed. Under some conditions bodies are seen in the cells resembling spores. The most favourable temperature for the growth of this ferment is about 28° C. (82° F.); above 36° C. (99° F.) it refuses to grow.

Bact. xylinum produces the same chemical changes in various solutions as does *Bact. aceti ;* but in addition, it produces a large quantity of cellulose in solutions of dextrose, levulose, and mannite. Cellulose is especially formed in levulose solutions, and sometimes constitutes as much as 62 per cent. calculated on the weight of the membrane dried at 100° C. (212° F.). The cellulose is converted into dextrose by the aid of acids, and here we have a reverse action to that which occurs with *Bact. aceti*, namely, the conversion of levulose into dextrose through the action of the ferment.

The acetic ferments are perhaps the least troublesome of the bacterial enemies of the brewer. Their action is perfectly definite, and they are readily recognisable in beer before they have had time to do much damage. Their presence also is generally due to aëreal contamination, and is more apt to attack individual cases than to run through an entire brew, as do some of the more insidious organisms with which we have to deal.

(f) The so-called Beer-Sarcina.

Until quite recently, the so-called Sarcina received very little attention as regards brewing operations. In 1879 Hansen mentioned a sarcina-form as occurring among the organisms of the air, which were able to develop in wort and beer, and incidentally stated that it was probable that some one or other of the diseases of beer was due to its presence.

He described the organism as forming small spherical points, usually occurring in groups of two, three, or four—in the latter case forming symmetrical four-celled tablets. Later it is mentioned by Bersch ; but it was not until Balke, in 1884,[*] examined the question that anything very definite was known about it. He found it in a large number of beers which were suffering from a peculiar disease, consisting of turbidity, combined with an unpleasant acid taste and smell. After this, Reinke often observed it in Berlin "white beer," and here it appeared to be associated with the so-called reddening of white beer.

Then in 1886, von Huth[†] performed a number of experiments in order to ascertain the source and conditions of growth of sarcina, and although his work is of little value from a scientific point of view, yet he apparently obtained some evidence of its presence in various substances and localities.

The most important statement made by this author is that when he inoculated pure cultures of yeast and pure cultures of sarcina, derived from various sources and various nutritive liquids, into sterilised wort, he was able to produce a beer corresponding in every respect with the so-called "sarcina-beer." This is an important statement, and, if it can be relied on, furnishes the one thing wanting in the researches carried out by other workers.

In 1888, Dr. P. Lindner,[‡] of Berlin, published an elaborate communication on the sarcina organisms of the fermentation industries. In this he described six species found in various localities and materials connected with fermentation. Only the first of these—*Pediococcus cerevisiæ*—concerns us here. This organism was isolated from bottled beer, which was suffering from the disease usually associated with sarcina.

---

[*] Wochenschrift für Brauerei, 1884, p. 183.
[†] Allg. Zeitschrift für Bierbrauerei und Malzfabrikation, 1885 and 1886.
[‡] Report of the Berlin Brewing School, 1888.

Pediococcus cerevisiæ was separated by culture in meat-extract gelatine; when separated, the microscopical examination showed that the organism occurred only in the form of cocci, diplococci, and tetrads—the typical sarcina forms were entirely absent (Plate VIII., fig. 1). The diameter of the single cells varies between $0\cdot9$–$1\cdot5\ \mu$; but under certain conditions and in old cultures large abnormal forms are observed. These Lindner regards as involution forms; they closely resemble those of *Bacterium aceti* and *Bact. termo.* In old cultures there are also formed flocks of cells of a brownish colour; these flocks tend to unite and fall to the bottom, where they form a dirty-white viscid and slimy sediment. In no case, were the typical packets of sarcina observed. It appears to form no resting-spores, and 8 minutes at $60°$ C. ($140°$ F.) suffices to kill the cells.

Pediococcus cerevisiæ grows more or less freely in nutritive gelatin; meat-extract gelatin, Agar-Agar, or neutral malt-extract gelatin being the more favourable media. It does not grow well in acid media, unless it has been previously growing in an acid medium, and hopped-wort gelatin did not prove a suitable material, unless the inoculation was made direct from beer or neutral malt-extract. The free presence of oxygen greatly promotes the growth of colonies, which have in all cases a greyish white appearance.

Of nutritive solutions, peptonised broth forms the most suitable medium for the development of the organism. In this, a white sediment appears after some days, and then the liquid becomes turbid. It grows in a similar way in neutral malt-extract, and from this liquid may be successfully inoculated into hopped wort. Lindner states that both in malt-extract and in hopped wort the growth of *Pediococcus cerevisiæ* appears to cause no marked alteration in the smell or taste of the liquid, and the amount of acid formed is not great. All attempts to cultivate the organism in sterilised beer were unsuccessful, and Lindner therefore rightly states that it must remain doubtful whether it has any other injurious action than producing turbidity.   This is, of course, a very weak point in Lindner's work, since it is impossible to consider the case proved against *Pediococcus cerevisiæ* as the cause of the so-called "sarcina-beers," until pure cultures of the organism have produced the disease.   It will be remembered that von Huth claimed to have done this by inoculating into sterilised wort pure cultures of yeast and of sarcina, and if this really is so, the importance of the observation can hardly be over-estimated.

Lindner found *Pediococcus cerevisiæ*, in very varied places—in both top- and bottom-fermentation beer, in yeast, in the air of breweries, in horse-dung and the air of stables, in water, and in one case it was traced to a pigeon-cot which was placed near the coolers of a brewery.   It is also found on malt and in grains.

We mentioned above that *Pediococcus* is often found associated with ropiness.   Lindner mentions that he found it growing abundantly in ropy beer, and considered it probable that it might be the cause of the ropiness.

Ropiness in beer is a disease which is enveloped in a great deal of obscurity, and even when we come to examine what has been done on this subject by reliable workers we find much that is indefinite.   Pasteur, in 'Études sur la biére,' describes a ferment which is said to produce viscous wort ; this ferment occurs in the form of strings of small cells. Schutzenberger, in his book on "Fermentation," also describes this ferment ; and, in fact, throughout the literature of the subject we universally find the cause of the phenomenon described as an organism consisting of chaplets of small cells. Pasteur, in his work on the subject, states that the sugar acted upon is transformed into a kind of gum or dextrin, mannite and carbonic acid ; and indeed the reaction is said to be so definite that it can be represented by chemical equations—100 parts of sugar being said to give about 51·09

Plate. VIII.

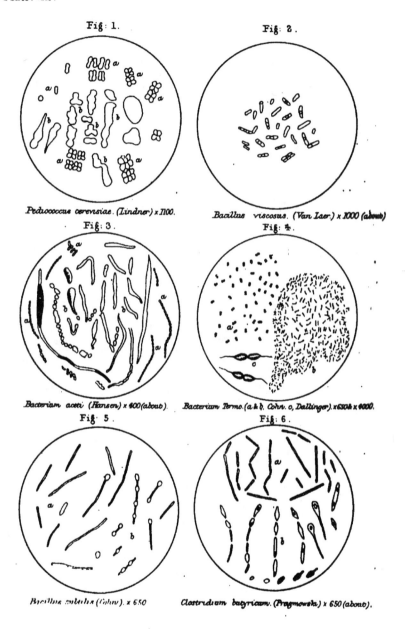

Fig: 1.

Fig: 2.

Pediococcus cerevisiae. (Lindner) x 1100.

Bacillus viscosus. (Van Laer.) x 1000 (about)

Fig: 3.

Fig: 4.

Bacterium aceti (Hansen) x 400 (about).

Bacterium Termo. (a & b. Cohn. c, Dallinger). x 650 & x 6000.

Fig: 5.

Fig: 6.

Bacillus subtilis (Cohn). x 650

Clostridium butyricum. (Pragmowski) x 650 (about).

F. & F N Spon. London & New York

The Roll h Son. Lith.

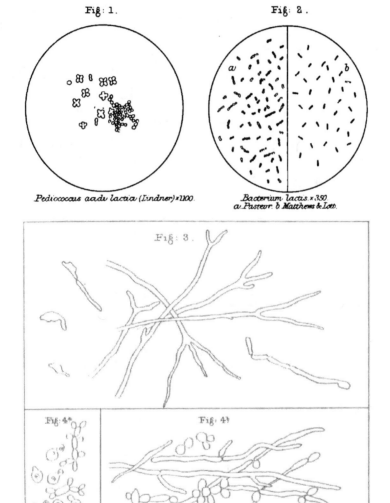

Fig: 1.

Fig: 2.

Pediococcus acidi lactici (Lindner) × 1100.

Bacterium lactis. × 350.
a. Pasteur. b. Matthews & Lott.

Fig: 3.

Fig: 4ᵃ

Fig: 4ᵇ

Fig 3. *Oidium lactis. (after Hansen) × 280 (about)*
Figs 4ᵃ & 4ᵇ *Monilia candida (after Hansen) × 500 (about)*

E. & F. N. Spon, London & New York.

parts of mannite and 45·5 of gum. Later, the gum was given the name of viscose.

We are inclined to believe, however, that this is not the action which takes place in beer, and after examining a considerable number of ropy beers, we are convinced that the organism described by Pasteur is not the usual cause of ropiness. The organism we have always found associated with this disease consists of a form closely resembling Lindner's *Pediococcus*, if, indeed, it is not the same. This is to some extent confirmed by the statement of Lindner quoted above, and indeed, we believe it is the general experience of those who have closely examined the question.

(g) Bacillus Viscosus.

H. van Laer has quite recently isolated two species of bacilli from samples of ropy Belgian beer, which he names *Bacillus viscosus* No. 1 and No. 2 (Plate VIII., fig. 2).

The former produced a viscous condition in sterile beer-wort, after cultivation for twenty-four hours at 27° C. (80° F.); and at the end of forty-eight hours the liquid is as ropy as albumin; at the same time, large quantities of carbonic acid are evolved. At the end of three days the viscosity was so great that 50 c.c. of the wort, cooled to 18° C. (64° F.), required 180 seconds in order to pass through the opening of the viscosimeter, 3 mm. in diameter. The same amount of the original wort required only nineteen seconds. With the increase of viscosity of the wort, the evolution of carbonic acid decreased; the liquid, however, remained turbid, with a chicory-like colouration, and with a smell so characteristic, that the viscous fermentation of liquids can be recognised by this alone.

The surface of this liquid is almost completely covered with yellowish viscous particles, which throw out ramifications

2 B

towards the bottom.   This peculiarity is very characteristic, and serves to distinguish the first form from *Bacillus viscosus* No. 2. In the following days, the evolution of carbonic acid ceases, and the wort becomes the colour of a mixture of coffee and milk.   The particles swimming on the surface have now completely overspread the liquid, and contain many bubbles of gas.

Bacillus viscosus No. 2 produces, under the same conditions as No. 1, a viscosity of about 70 (i.e., a wort which takes 70 seconds to pass through the viscometer), which is about that usually shown by beer in the brewery.   The evolution of carbonic acid is less vigorous, and the amount of surface growth is almost inappreciable.   The viscous action takes place much quicker in hermetically-sealed flasks, but the surface viscous growth is never produced.   This agrees with the observation made in practice, that beer in bottle becomes ropy much quicker than that in cask.   In flasks hermetically sealed, *Bacillus viscosus* No. 1 always gives rise to the above-mentioned viscous growth.   In this slimy mass the bacteria develop spores, either one only, when it is, as usual, in the centre; or when there are two, they occur at either end. Gentian-violet is the most suitable stain for showing these spores.

Both species cause viscosity in peptone-cane-sugar solutions, in milk, in Pasteur's solution, in urea-asparagine solutions, and in Mayer's solution ; the production of viscosity in the last three solutions is very noteworthy, since they contain no sugar or carbohydrate.

An examination of the conditions which favour or retard the viscous fermentation showed that an elevated temperature (42° C., 107·6° F.) completely stops the action, and a temperature of 100° C. (212° F.) for three minutes is sufficient to kill the organism.   When worts were simultaneously inoculated with yeast and the bacillus, the fermentation was injuriously influenced, directly in proportion to the number of bacilli added; when the inoculation with bacilli followed the primary fermentation, no injurious results were observed.

In a solution of sugar in distilled water no formation of viscosity takes place until a little peptone is added. The disease phenomena then occur the more rapidly and more intensely, the greater the amount of nitrogenous matter present. Inoculated wort, with the addition of 3 per cent. sterilised urea, became so extremely viscid, that the entire contents of the flask formed one single mass.

The author studied the development of the disease in solutions with an increasing amount of peptones, and found that those with the largest amount of nitrogenous substance became the most viscous.

The author concludes from these results that in practice, the beers with much assimilable nitrogenous substances will have a great tendency to undergo the viscous fermentation. He also considers that the generally accepted view of the unpeptonised albuminoids being the chief source of danger to the brewer, to be an error, because it appears to be the peptones which are especially predisposed to support the viscous fermentations.

In a series of experiments with solutions containing a constant amount of peptone, but increasing quantities of sugar, it was found that the viscous fermentation commenced the sooner the poorer the solution was in sugar. This agrees well with what is observed in practice, namely, that beers remain sound for a longer time when their attenuations are high.

The same was observed with dextrin. Acid above 0·15 per cent. is sufficient to prevent the production of viscosity. Alcohol and phosphates do not influence the fermentation when present in moderate quantity, but gypsum is stated by van Laer to distinctly favour the development of viscosity Carbonic acid and salicylic acid retard the action, and sulphurous acid used in conjunction with steam destroys the bacillus.

The author soaked pieces of wood for twenty-four hours in a nutritive solution which had become viscous, and then suspended the fragments in the vapour of water, and in sulphurous acid gas for fifteen minutes. The fragments were then introduced into sterile beer-wort, when it was found that they were completely sterile.

This experiment has an important practical bearing, since the application of this acid to brewery vessels will be sufficient to sterilise them, so far as regards the viscous ferment.

The author sums up the principal results as follow :—

A. Circumstances which favour the viscous fermentation.
  (1) The temperature of top-fermentations.
  (2) A long exposure of the worts to the air before the addition of yeast.
  (3) The employment of a contaminated yeast.
  (4) A large amount of nitrogenous substances.
  (5) Too low an attenuation.
  (6) The employment of a water containing a large amount of gypsum.
B. The circumstances which retard the viscous fermentation :—
  (1) The acidity.
  (2) Alcohol in the presence of small quantities of acid.
  (3) Salicylic acid.
  (4) Sulphurous acid gas in the presence of water vapour.

Van Laer gives some account of the products of viscous fermentation. In the presence of sugar, carbonic acid is always evolved, the liquid at the same time becoming acid. The amount of acid always increases considerably after the period when the viscosity begins to decrease. The odour produced during fermentation is very characteristic, and is the same in all media. The viscous substance mentioned above as produced by *Bacillus viscosus* I. can be collected on a filter, whilst the liquid which passes through is very viscous also, thus showing that two substances are formed, one insoluble in water, the other soluble in water ; the former is a nitrogenous substance, whilst the latter is free from nitrogen. The body soluble in water is insoluble in absolute alcohol, and is coloured yellow by iodine ; concentrated potash dissolves it in the cold, and the solution gives a yellow colouration on warming. It is not precipitated by tannic acid from its solutions.

We have every reason to believe that the *Bacillus viscosus* is not at all a common disease ferment in this country. We have never found it associated with viscosity in English beers, but on the contrary, have *always* found ropiness to be accompanied by what is apparently the *Pediococcus cerevisiae* of Lindner. Nevertheless we can fully support van Laer's conclusions that pressure appears to be particularly favourable to the development of viscosity : beers in bottle being much more prone to undergo this change, than those in cask. Acid too has a marked effect in checking the disease, and hops also appear to exert a restraining influence on the development of the organisms.

The foregoing comprise all the Schizomycetes which have any interest in connection with brewing operations. We must now glance at another group of micro-organisms which, although not directly concerned in fermentation, yet are a source of danger to the brewer at different stages of the brewing process. These organisms belong to

III. The Hyphomycetes, or Moulds.

In the foregoing pages we have mentioned the injurious results which arise from mould-growth on the malting floors. In addition to this the moulds may cause injury to the brewer by their growth on hops, on fermenting and other brewery vessels, on the walls, &c., of the brewery premises, and on the interior of the trade casks. Not only may the growth of mould be a danger in itself, but it is also a source of indirect danger, inasmuch as the thick felty colonies may harbour the cells and spores of the two classes of disease organisms we have considered above—wild yeast and bacteria—and thus serve as fruitful sources from which these organisms may be disseminated throughout the brewery. Speaking generally the moulds require for their development an abundant supply of moisture and air, and hence we usually find them flourishing in damp situations and in damp vessels.

(a) *The Mucors.*

The members of this class include several species which are of much interest to the brewer. The most frequent of the species of Mucor—*Mucor Mucedo*—forms on moist surfaces, a greyish, felt-like mass, from which a number of fine threads, (distinguishable by the naked eye) are thrown up; these threads have dark-coloured spherical heads, containing the spores from which young plants are subsequently developed.*

The spores of certain species of Mucor have the power, when cultivated in wort or saccharine solutions, of growing in the submerged state in forms closely resembling yeast-cells. In this form of growth the cells are known as *gemmæ*, and multiply by budding like the true yeasts.

When growing in the submerged form, the Mucors act as true alcoholic ferments. Hansen has lately shown that the majority of the Mucors ferment not only dextrose and invert-sugar solutions, but also bring about the fermentation of maltose. Of the species examined by him, however, only one, *Mucor racemosus*, is able to invert cane-sugar. *Mucor erectus* possesses the strongest fermentative power: it forms

---

* For a more detailed illustrated description of the Moulds, see Jörgensen's "The Micro-organisms of Fermentation." Edited from the German by G. H. Morris; and Matthews and Lott "The Microscope in the Brewery and Malt-house."

8 per cent. of alcohol in beer-wort ; it also excites an alcoholic fermentation in dextrin solutions, and converts starch into reducing sugars.

(*b*) *Penicillium glaucum.*

This species is the most widely distributed of all the Moulds. It grows freely on all moist, nutritive surfaces in the form of greenish-grey, felt-like masses, which when examined under the microscope are seen to consist of tangled and ramified mycelia. These mycelia are in the form of transparent, branched and divided threads, from which perpendicular threads are thrown up ; the latter are usually branched at the top and carry the spores in the form of a tuft.

Penicillium is very abundant on the malting floors when the grain contains any quantity of broken corns, and when the temperature and amount of moisture are allowed to rise too high. In addition to the disagreeable taste and smell communicated to the resulting malt, it is also injurious in that it attacks the starch of the grain and causes a loss in extract. *Penicillium* also readily grows in damp vessels, and especially in empty beer-casks. It does not appear to grow in a submerged form as does *Mucor*, neither has it the power of inducing alcoholic fermentation, but it secretes an invertive ferment capable of converting cane-sugar into invert-sugar.

(*c*) *Oidium lactis.*

The natural habitat of this organism is in milk, hence the name *lactis* given to it by Fresenius. It occurs, however, in many other liquids, such as yeast, wort, and beer poor in alcohol. It is capable of exciting a feeble alcholic fermentation, and we have reason to believe that it is present in beer more often than is usually supposed. What its precise function is, when growing in beer, is not known ; but it is one of those organisms the presence of which should be carefully guarded against.

The usual mode of growth of *Oïdium lactis* is shown in Plate IX., Fig. 3 ; the extremities of the forked and branched mycelial threads become separate, owing to the formation of transverse septa, and the conidia thus separated are oblong, spherical or pear-shaped, and closely resemble abnormal yeast forms.

(*d*) *Monilia candida.*

This is another species, interesting on account of the fact that, although its usual habitat is on moist surfaces, such as cow-dung or succulent fruit, where it forms a white film, yet when introduced into saccharine solutions, it develops a vigorous growth of yeast-like cells, which bring about an alcoholic fermentation. *Monilia candida* is shown in Plate IX., Figs. 4*a* and 4*b*. Fig. 4*a* shows the organism when growing submerged in beer-wort or other saccharine liquid ; the cells closely resemble *Sacch. cerevisiæ* and *ellipsoideus ;* Fig. 4*b*, shows the mould-like growth of Monilia ; the characteristic form is the chain of thread-like cells, with rings of oval cells at the joints, the latter break off and occur as individuals or in groups resembling *Sacch. conglomeratus* (Reess).

The yeast-like cells of Monilia excite a vigorous fermentation in beer-wort, and in solutions of cane-sugar, maltose, dextrose, &c. ; as much as 5 per cent. of alcohol is formed. The ferment can withstand a fairly high temperature, a vigorous growth and fermentation taking place at 40° C. (104° F.).

A remarkable feature in connection with *Monilia,* is its power of directly fermenting cane-sugar. Hansen states that it secretes no invertase but ferments cane-sugar as such.

Monilia candida is found among the air-sown ferments in wort and beer, and is on this account interesting to the brewer.

(e) *Dematium pullulans.*

Very much has been written by Pasteur and others regarding this mould. It occurs especially on fruit, where it develops a branched mycelium, from which separate yeast-like cells. De Bary and afterwards Loew have thoroughly described this organism; the yeast-like cells develop into mycelial threads, and these after a time become much thickened with numerous transverse septa, and also become dark brown or olive green. The yeast-like cells excite no alcoholic fermentation, although they grow in saccharine solutions. Hansen found *Dematium* very prevalent in the air, and we have lately observed it growing with great freedom on damp walls; it there forms dark green patches.

Lindner has recently shown[*] that *Dematium*, when grown in beer-wort, causes the liquid to become intensely viscous or ropy. We can confirm this statement of Lindner's, since the above-mentioned growth, and pure cultures obtained from it, converted beer-wort, in twenty-four hours, into an extremely ropy condition. On this account brewers should pay the greatest attention to the absence of this mould from their fermenting rooms, since its presence might result in serious trouble.

(f) *Fusarium hordei.*

This mould was first described by C. G. Matthews[†] as occurring on germinating barley; it is, after *Penicillium glaucum*, probably the most frequent mould found on germinating grain. As a rule it is found when the barley is of inferior quality, and then it appears as a crimson or pink-tinted patch on the defective corns, more particularly in the neighbourhood of the embryo. The most marked phase in the development of this mould is the crescent-shaped compound spores. When grown in beer-wort, Matthews states that this

[*] Wochenschrift für Brauerei, 1888, p. 290.
[†] Journal Royal Microscopical Society, 1883, p. 321.

mould forms cells somewhat resembling those of *Mucor racemosus* under similar conditions, and which excite a feeble alcoholic fermentation.

When present on the growing floors, this mould is not readily communicated to sound and healthy corns, and its spores, on account of their weight and adherence, do not readily spread through the growing grain.

PRACTICAL CONSIDERATIONS.

Many micro-organisms, which mainly infest worts during cooling and refrigeration, can freely develop in a well-made beer wort, and would do so even in the absence of putrescible matter, were it not that the wort is shortly afterwards sown with yeast. When the wort is properly constituted and when the yeast is vigorous, its energy of reproduction keeps down the development of these organisms, firstly, because a wort free from putrescible impurities is more suited to the development of yeast than to that of bacteria ; secondly, because the yeast-cells added outnumber the bacteria ; and thirdly, because while the yeast is added in a form capable of immediate development, the organisms introduced from the air being in the desiccated form are incapable of so readily propagating. We have said that a well-constituted beer wort is more suitable for the development of yeasts than of the fission-fungi, and this is due to several causes : firstly, its acid reaction ; secondly, its relatively high proportion of carbohydrate and low proportions of nitrogenous bodies and phosphates, and thirdly, because of the relatively low temperatures at which it is fermented.

Considered from the standpoint of bacterial contamination, the quality of the yeast may be regarded as an index to the purity of the materials and the cleanliness of the plant in a brewery. An impure yeast tells a tale which should be immediately construed into practical action. We have referred

to the connection between yeast vigour and bacterial development, saying that the more vigorous a yeast the less chance
have the organisms for propagation. This is obviously the
case; it must be remembered, however, that the increase of
contamination in a brewery must eventually lead to the
weakening of the yeast, and the successive weakening of the
yeast must in its turn weaken its power of checking the
development of the organisms.

The importance of microscopically examining yeast in the
brewery is obvious; the observations should include the configuration of the yeast itself as well as the number and
varieties of bacteria. Yeast should be well filled and round,
with a thin cell-wall and ungranulated and highly refractive
cell-contents. Angular shrivelled cells show weakness or
death of the cells. The plasma of dead cells exudes from
the cells, and may thus encourage the growth of disease
ferments, for it contains much phosphate and much nitrogenous matter and in a state ready to be immediately
assimilated by the disease ferments.

It is therefore desirable to supplement ordinary microscopic
examination of a yeast by some test which, apart from mere
appearance, enables us to form some idea as to the vitality
of the yeast-cell. To do so we inject an aqueous solution
of eosin into the preparation. The eosin stains dead plasma a
bright and distinct pink and leaves living plasma unaffected.

The troubles which wild-yeast gives rise to are, in a sense,
less controllable than those produced by bacteria, and for the
reason that they are not detected by microscopic observation.
In the present state of our knowledge it is difficult to definitely
assign certain forms of beer disease to certain forms of wild-
yeast, as has been possible by Hansen and others in respect of
many of the diseases caused in Continental beers. But there
are certain defects which are certainly associated with wild-
yeast contamination.

The tumultuous cask fermentation, or fretting of beer,

is mainly a summer trouble, and is primarily incited by an excess of wild-yeasts. These are probably of the same variety (*S. Pastorianus*) which in moderation are essential for cask condition, and which during the summer months will be likely to infect the brewings to a far greater degree than in the colder months. Fretting beers attenuate very low in cask through the excessive degradation and fermentation of the maltodextrins at the instance of the yeasts; the final beer is not only thin, but is invariably defective in flavour (" sick "), and is not infrequently liable to very early acidity. This sick flavour is probably the result of by-products of what we may term the abnormal after-fermentation; the instability is probably due to the general disturbance of the beer (see Chap. V., p. 207), and also possibly to the low amount of maltodextrins in the beer when the fret is over. The fret is in great measure governable by the constitution of the beer, for the wild-yeast will break down the very low maltodextrins (the apparent free maltose) with far greater ease than they would medium or normal maltodextrins, thus we generally find these frets to occur only in beers containing low malto-dextrins. As a matter of fact, the fret may invariably be stopped by raising the type of maltodextrins and taking other measures which are given in detail elsewhere (see p. 413).

The variety of yeast which incites frets is probably a bottom yeast; the fact that the fermentation of fretting beers in the fermenting vessel proper is generally normal, and that the fret only commences some time afterwards, is significant that the yeast is not a top yeast, but deposits on the bottom of the vessel during the primary fermentation, and comes down mechanically with the beer when this is racked into casks.

It may be sometimes overcome by adding an excess of finings to the beer in the fermenting vessel before racking into casks. By this means we imbed the bottom yeast deposit in the isinglass, and have therefore a far smaller quantity of the ferment brought into our casks. The treatment is simple, and is sometimes efficacious. We admit, however, that it is not always so, and the better way of dealing with it is to replace the lower maltodextrins in our beer, upon which the wild yeast acts so easily, by more resistant higher maltodextrins. This substitution is always a successful way of dealing with the trouble.

We must add that this and other wild-yeast troubles are not necessarily confined to the summer months. Similar troubles sometimes arise during a spell of warm weather in the colder months of the year. This is especially the case when the yeast has not been changed since the summer, but it must not be forgotten that the wood plant and the casks, especially when these are unsound, would absorb the wild-yeast forms so prevalent in summer, and thence they contaminate worts and beers at other seasons of the year.

Turbidity of beer caused by yeast or yeast turbidity, is not necessarily a result of wild-yeast, for normal yeast if cultivated in an ill-constituted wort, may refuse to work out well, and thus remain suspended in the beer. But it is probable that certain forms of wild-yeast do produce turbidity, though this seems more marked in Continental than in English breweries. These wild-yeasts are light and do not deposit.

We have within recent years had some experience of a peculiar bitter flavour to beers, especially those brewed in summer. This bitter is accompanied by turbidity, and the attenuations run very low. The cloudiness of the beer disappears on adding potash to it, and we have no hesitation in ascribing both the bitter and the turbidity to uneliminated hop resins. Normal yeasts eliminate these resins by attaching them to their surface; but in the case referred to we seem to be dealing with a yeast that refuses to perform this function. This yeast is probably identical with the *S. coagulatus* described by Matthews, and which the author finds competent to degrade and ferment the maltodextrins during the primary fermentation. This would explain the low attenuations. The author also states that contamination with this yeast leaves a clinging resin bitter in the beers; this, doubtless, explains the intense bitterness of the beers referred to.

We have found that the raising of the percentage and type of maltodextrins is of material service in dealing with this trouble. But the improvement is only gradual unless the yeast is changed; and that this is the case is confirmatory of the supposition that a wild-yeast variety is at the bottom of the trouble.

The benefits in respect to brilliancy which so frequently follow a change of yeast unaccompanied by any change in the constitution of the wort, we ascribe mainly to the new yeast consisting principally of cells which readily attach to themselves a due amount of hop resin. Conversely, when a change, similarly unaccompanied by any change in the wort, produces turbidity, we believe that this is mainly due to the fresh yeast containing those cells in abundance which refuse to eliminate the resins in this way.

In this country we use a mixture of yeasts, the bulk of which is primary. Attenuation in the fermenting vessels will proceed under normal circumstances until the whole of the maltose and fermentable sugars used as malt adjuncts have been fermented out. We are taking it for granted in the above statement that the wort contains a sufficiency of phosphates and nitrogenous yeast food, and that a sufficiency of oxygen is provided for the due reproduction of the yeasts. Under these circumstances attenuation proceeds until there is no fermentable sugar or maltose remaining, and it is the elimination of the fermentable sugars which determines the cessation of attenuation. The maltose is converted into alcohol and carbonic acid gas, small quantities of glycerin and succinic acid being produced simultaneously.

A small proportion of the carbohydrate however (about

1 per cent.), is used by the yeast-cells and converted by them into fats and cellulose.

Pasteur and Monoyer have suggested equations which claim to account for all these changes.* These equations are, however, purely hypothetical, and it has lately been questioned whether the glycerin and succinic acid are really the results of fermentation at all. It seems more probable that they are the products of the metabolism of the yeast-cells.

In addition to carbonic acid, hydrogen, nitrogen, and some volatile product of the marsh-gas series are evolved. The hydrogen arises from a decomposition of water by the yeast, and this is especially the case when the pressure is low. The nitrogen proceeds from a decomposition of the albuminoids.

We are aware that our statement, that the whole of the maltose disappears, and that it is the elimination of the maltose that determines the cessation of fermentation, is opposed to the views generally held, but for reasons which will follow, we are forced to that conclusion. We imagine that those who hold that racked beer contains free maltose, base their arguments upon the fact that the beer at that stage has a very

---

* Pasteur's careful quantitative experiments led him to express the decomposition of 100 parts of cane-sugar (corresponding to 105·26 parts of invert-sugar) as follows:—

Alcohol .. .. 51·11
Carbon dioxide .. .. .. $\begin{cases} 48\cdot89 \text{ according to Gay-Lussac's equation,} \\ 0\cdot53 \text{ excess over ,, ,, ,, (see p. 283)} \end{cases}$
Succinic acid .. .. .. 0·67
Glycerin .. .. 3·16
Matter combined with yeast 1·00

Thus, out of 100 parts of cane-sugar, 4 parts disappear and form glycerin and succinic acid. Pasteur gives the following equation to represent the decomposition of these 4 parts :—

$$49 (C_{12}H_{22}O_{11} + H_2O) = 24 (C_4H_6O_4) + 144 (C_3H_8O_3) + 60 CO_2.$$
$$\phantom{49 (C_{12}H_{22}O_{11} + H_2O) = 24 (}\text{Succinic acid.} \text{Glycerin.}$$

Monoyer has proposed a more simple equation to represent the decomposition ; it is as follows:—

$$4 (C_{12}H_{22}O_{11} + H_2O) = 2 (C_4H_6O_4) + 12 (C_3H_8O_3) + 4 CO_2 + O_2$$
$$\phantom{4 (C_{12}H_{22}O_{11} + H_2O) = 2 (}\text{Succinic acid.} \text{Glycerin.} \text{Oxygen.}$$

The excess of oxygen, shown in the equation, is supposed by Monoyer to serve for the respiration of the yeast-cells.

decided reducing action on Fehling's solution. But it is unnecessary to say that such a reducing action is abundantly explained by the maltose constituent of the maltodextrin. It is at any rate suggestive, that very shortly after we add to racked beer a fermentable sugar (as we do in priming beers) a vigorous and rapid fermentation ensues. It is clear, therefore, that there is an abundance of yeast in the beer at that stage to do fermenting work ; and that there is nothing in the beer to check or interfere with a vigorous fermentation. The cessation of the fermentation must be due, and due only to the absence of fermentable matter. A fermentation can also be readily induced in racked beer by adding diastase. This transforms the maltodextrin into maltose, which is then fermented at once.

The following experiments show that when maltose is added to a fermented beer a vigorous and rapid attenuation results. We can see no reason why, if the beer contained maltose before, that maltose should not have fermented in the same manner as that added.

The residues of three beers were taken and set to ferment as described in Chapter XII. (p. 504). When it was apparent that all fermentation had ceased, the amount of apparent maltose in each was determined in the usual way, and was found to be equal to :—

No. 1.—1·381 gram of maltose per 100 c.c.
No. 2.—1·233 ,, ,, ,,
No. 3.—1·364 ,, ,, ,,

Pure crystallised maltose was then added to each beer at the rate of 1·9414 gram of maltose per 100 c.c. Fermentation commenced most vigorously, and continued for three days ; at the end of that time fermentation had ceased, and the apparent maltose was again estimated. The values now found were : —

No. 1.—1·425 gram of maltose per 100 c.c.
No. 2.—1·275 ,, ,, ,,
No. 3.—1·392 ,, ,, ,,

It will be noticed that, in each beer, the value after the second fermentation was slightly higher than before; the difference, however, is very slight, and within the errors of experiment. These results indicate most clearly that the apparent maltose in beer cannot consist of free, fermentable maltose, for in the instances given, every condition was favourable for fermentation, and considerably more maltose than was apparently present was readily fermented when it was added to the liquid, yet the apparent maltose remained absolutely untouched by the fermenting yeast. A trace of malt-extract added to the residues after the second determination caused a brisk fermentation to again start.

When a racked beer is pitched with fresh yeast, a certain amount of sugar will disappear; generally this will amount to about 0·8 to 2 per cent. calculated on the original wort-solids; but occasionally it comes out very much higher, from 5 to 8 per cent. similarly calculated being found in some instances. This it was that so puzzled us for a very long time, since the sugar thus disappearing would appear to be free maltose unfermented during the fermentation, and we could find no adequate reason for its having thus escaped fermentation. We have now come to the conclusion that this unfermented "free maltose" is not actual, but apparent. It consists, in fact, of very low maltodextrins (compounds of very much maltose and little dextrin) which become slightly fermentable when the yeast added is very excessive in quantity, and when the conditions of fermentation are such as to promote its vigour, but which may escape fermentation when in fermenting vessel, especially where there is any yeast weakness, or where the fermentation conditions are somewhat unfavourable to the full activity of the ferment. This will explain why by adding fresh yeast to a racked beer and fermenting at high temperature (as is done in the analysis) we are able to remove substances which were not removed during the brewery fermentation.

2 C

The above point also explains to us why it is that some yeasts will, with the same wort, give a lower attenuation than others, and again, why, with the same wort and the same yeast, we can lower our attenuations by aërating, rousing, and by other similar devices for encouraging the yeast. It is clear to us that in the first case we are dealing with a yeast which is particularly competent to ferment out the lower maltodextrins in the fermenting vessel ; while in the second case, it is possible to incite a yeast sufficiently to determine an increased capacity for fermenting out these very low maltodextrins.

In talking of maltodextrins as unfermentable we have referred to the normal or medium types : under ordinary conditions these are so, or very nearly so ; and if they be fermentable at all, they are so slightly so as, from the practical standpoint, to be considered entirely unfermentable. The very low maltodextrins, however, are undeniably fermentable to some extent ; to what extent seems to depend upon the vigour, and also on the amount, of the yeast and the conditions of fermentation. Every brewer knows that some control of his attenuations is possible by altering his yeast rate ; and we believe that the increase or decrease of yeast rate is the most powerful practical factor in determining a fermentation or non-fermentation of these very low maltodextrins.

In giving the maltodextrin type on the result of analysis we naturally give the mean type, that is, the average maltose to the average dextrin proportion. But where the type does not seem abnormally low, there may be yet these lower maltodextrins present, although they are masked by other more dextrinous types which accompany them. We have generally found, however, that where the type is normal there is no evidence to suppose the existence of this very low type. But analysis of the beer will determine this, irrespective of mean type, since in analysis they will be included in what has been hitherto erroneously known as the free maltose :

that is, sugar not fermented during the brewing fermentation, but fermented on the addition of a great excess of yeast.

If, then, the yeast is sufficient in quantity, if it is supplied with sufficient air, if it is kept sufficiently agitated to prevent the cells getting coated with a film of alcohol and carbonic acid, and if there is a sufficiency of yeast-foods, the fermentation proceeds until the maltose is entirely fermented. But occasionally there is an insufficiency of maltose, and the attenuation is inadequate. In such cases, besides aëration, rousing, &c., we dress the fermentations with malt-flour or wheat-flour, or a cold-water extract of malt; the diastase in these will break down the maltodextrins, converting them into free maltose, which will then readily ferment away. Brewers sometimes have what they call boiling or "fiery" fermentation, i. e., fermentation proceeding unaccompanied by the rising of a yeasty head. Lack of aëration and excessive temperatures—conditions which incite attenuation, but not yeast reproduction—are generally the chief causes of the defect. Dressing with diastatic materials, however, frequently helps us to overcome the defect, though the rousing and aëration accompanying their addition are probably more productive of benefit than the materials themselves. The best remedies to apply, however, are thorough rousing and aëration, and expulsion of carbonic acid, together with a lowering in temperature.

Some brewers mix their yeast with unboiled wort from the mash-tun, at a somewhat elevated temperature, before pitching. The action of this is twofold. Firstly, the undestroyed diastase of the wort will be immensely invigorated by the yeast, and under its influence will rapidly degrade the maltodextrins and even the dextrin, thus giving to the yeast a sort of preparatory meal of the most assimilable carbohydrate, and presumably most assimilable nitrogenous matter. The yeast becomes invigorated, and is added to the beer in a condition to work at once.

There seems, secondly, to be a further reason why the practice should· be beneficial. Hansen has recently shown that when a mixture of normal and wild-yeasts are put to ferment in a wort, the normal yeast will work first, and the yeasty wort will contain cells of nearly entirely normal yeast during the very early stages ;.and he suggests pitching beers not with dry yeast but with yeasty wort taken shortly after the pitching of the wort. In a rough sort of way, "setting" on the yeast with wort would thus appear to be a means of getting a vigorous growth of normal yeast ; and in summer it might therefore be well to leave some of the yeasty deposit behind in the vessel used for the purpose ; for this will presumably contain the greater amount of wild-yeast. During fermentations the thorough aëration of the wort is customary and necessary. This is effected by rousing, or by transferring the wort from one vessel to another. The benefits arising from these processes are not entirely due to the invigoration of the yeast by air ; they are also in great measure attributable to the simultaneous expulsion of the excess of carbonic acid gas. This gas in excess undoubtedly restricts the vigour of the yeast, alike in assimilative capacity and in attenuative capacity ; to expel a portion of it, therefore, conduces to greater vigour.

Due regard must be had also to the temperature at which fermentations are conducted. Broadly put, high temperatures encourage attenuation, rather than the reproduction of yeast ; while low temperatures on the other hand will encourage reproduction and discourage attenuation. Attenuation is of course necessary, but reproduction is equally so. By a proper reproduction, much of the nitrogenous and phosphatic matter is removed entirely from the beers, for they are used up by the yeast and will be removed with the yeast ; such beers will be sounder than others in which the removal of these substances is only incomplete. Temperatures affect stability in other ways : the higher they are the more encouraging are they to disease-organisms, and *vice versá.* The temperature undoubtedly plays a part in the flavour of the beer: beers fermented at high heats having a characteristic vinous flavour, due, probably, to the formation of the higher alcohols and their derivations. On the whole, it may be said that sound, clean-tasting beers are best fermented at what we may term low heats, say between 58° to 68° or 70° higher; heats give characteristic flavours, which are sometimes admired, but at the expense of soundness.

The attenuation being complete, the beer lies in the fermenting vessel, or in secondary settling vessels for some days, when the yeast works out and is removed, and it is then run into the casks.

Whether or not a pure system would suit us in this country is an open question. So far experiments have not yielded decisive results, but they may yet do so. Continental brewers, brewing on the top-fermentation system, find that they can adopt the process successfully; whether or not we shall be able to do so is a question that only practical experiment can decide. If we can do so it will be a step, and a very long step, in advance; and by it one more means of control will have been given by science to brewing.

SYSTEMS OF FERMENTATION.

We may conveniently classify the systems of fermentation used in this country under the following heads:—The skimming and the cleansing systems, the stone-square system, and the Edinburgh system. There are two modifications of the skimming system, the one in which fermentation and yeast elimination are carried out entirely in one vessel; and the other, where the beer is let down at about half its original gravity into a shallower skimming vessel, from which the yeast is removed. The advantages of this latter over the more ordinary skimming system are the more thorough aëration of the beer, the more complete expulsion of carbonic acid gas, the thorough agitation of the yeast-cells, the more perfect elimination of the yeast, and consequent greater racking brightness. The surface attraction in these shallow skimming vessels is greater; consequently the deposition of insoluble matter is facilitated. This system is decidedly preferable to the one sometimes adopted of merely letting the beer down into a shallow settling or racking back a few hours previous to racking. The beer is then practically naked, and there is no final yeast-crust to prevent oxidation and the escape of gas at a stage when its expulsion is undesirable in respect to subsequent condition.

Upon the skimming system we consider either thorough rousing or a more direct method of aëration to be particularly important. The wort here exists in large bulk, and is often of considerable depth ; the necessity for agitation and expulsion of gas is therefore more than ever necessary. The range of heats allowed varies very considerably, according to the character of the beer required. The most usual range is from about 58° or 60° to 70° or 73° F. Some brewers prefer to commence lower and to confine the rise to 66° or 68° F. On the other hand others allow the final temperature to reach 75°, and even 80° F. These extremely high heats are, however, generally confined to black beers. The skimming system is one particularly adapted for the production of mild beers.

There are several modifications of the cleansing process, which, however, mainly consist of the transference of the wort, whilst in active fermentation, to smaller vessels in which the process is completed. The influence of the transference of the wort at this stage is of importance, on account of the agitation and aëration and gas expulsion, which stimulate the reproductive and therefore assimilative capacity of the yeast. The final settling of the beer in the comparatively small vessels is of course an additional advantage in regard to the attainment of racking brightness. And upon this cleansing system the influence of the surface attraction is not only at its maximum, but, unlike the case of shallow racking casks, it is exerted when the beer is entirely out of contact with the air. Another important feature of the cleansing system is the aëration of the yeast and "feed," which further promotes the clarification and purification of the beer. The most important and perfect system of cleansing is the Burton-union method. Each union is generally provided with a removable attemperator, as are also the yeast-trough and feed-back. In these vessels the progress of fermentation is admirably under control. Cleansing in loose pieces or hogsheads may be regarded as a makeshift system, and it certainly promotes dirt and waste, as each cask has to be topped up by cans. The labour involved is very great, and there is no attemperating control. The casks are rolled out to be washed, and are rolled in again, usually over a dirty cellar floor, and pick up anything in their way, upon the very part over which the yeast will afterwards flow.

There are still some brewers who cleanse direct with carriage casks. These are placed on small stillions, and are worked exactly like the larger pieces. The labour is considerable, although there is no re-racking. In some breweries the plan answers very well, and when the cleansing is thoroughly complete, fine conditioned and brilliant beers are produced. For this system to be successful the yeast has to be kept in perfect order.

Very few remarks will suffice respecting the ponto system of cleansing. These vessels are larger than Burton unions, and are set up on their heads. The yeast flows over a sort of lip at the top, and drops into a trough below. The feed back is here above the yeast trough, and of course commands the pontos. An automatic feed is arranged by means of a ball valve, which keeps the vessels full.

Speaking generally, the cleansing system produces cleaner

drinking and more slowly conditioning beers than the
skimming system.   The duration of the process, and the
range of temperature is about the same in either.   Both
skimmed and cleansed beers are usually allowed a week
before they are racked into cask.   It is, however, very often
possible to rack on the fifth day after pitching.   Although
the cleansing system has decided advantages, it entails much
more labour and is distinctly more wasteful than the skimming
system.   There are many vessels and mains to keep clean,
and every additional vessel means loss.

The stone-square system is practically confined to the
North of England.   These squares are constructed of stone
or slate, and have a capacity of about 30 to 50 barrels.   The
vessels consist of two parts : the lower one, which has double
sides and a space between for attemperating liquor ; and
the upper one, which is placed so that its bottom forms a
top or cover to the one below.   In this roof, or cover, there
is a man-hole with a raised collar about 6 inches high, and
also a smaller hole provided with a valve.   The lower vessel
is a little more than filled, so that the wort lies some 2 or
3 inches deep in the upper one, or yeast back.   Pumping is
resorted to at different intervals until attenuation is nearly
over.   The wort is pumped from the lower vessel into the
yeast back, the valve of course being closed.   After well
rousing with the yeast, the valve is opened, and the arrested
wort allowed to run back into the square.   The whole crop
of yeast is eventually left in the upper back.   This system is
productive of an excellent yeast crop and bright racking
beers, and of very brisk and well-conditioning beers.   We do
not consider that stability is so pronounced as in other
systems where the wort is transferred from one vessel to
another while in course of fermentation.

Upon the Edinburgh or rousing system the beer is run
down into a second square, leaving the yeast behind in the
primary vessel.   The fermentations are kept well roused,

and attenuation proceeds very rapidly indeed, although the range of temperatures employed is similar to the ordinary skimming process. The rise of heat, however, is, of course, more rapid, corresponding to the quicker attenuation. The fermentation is sometimes completed in 35 hours, when the beer is let down into the second square, where a protecting layer of yeast covers the surface. From this secondary vessel the vessel is racked about 36 hours after " squaring." The time of settling naturally varies, as does that of the fermentation. In some cases the whole process is completed in 60 hours, and occasionally in even less. The rapid fermentation has, as might be expected, an enervating influence upon the yeast, and considerable care is needed to keep it in healthy condition.

The characteristics of beers fermented under different systems will in part depend upon the structure of the vessels employed, and upon the means taken for aërating and agitating the wort and for expelling the carbonic acid. In those systems where agitation, aëration, and gas expulsion are complete we expect a cleaner tasting and sounder beer than where these conditions do not apply. By aëration we invigorate the yeast, inciting it to ferment out substances like the lower maltodextrins, which might remain unremoved when yeast is uninvigorated. By expelling the gas, we incite the reproduction of the yeast, and the consequent removal of nitrogenous matter and phosphates. By agitation we bring the cells into contact with fresh fermentable matter and remove them from the environment of alcohol.

So far as structure of vessels is concerned, depth as affecting pressure is the most important consideration. Pressure exerts an influence upon the by-products of fermentation, and will, therefore, influence flavour. When pressure is excessive natural gas evolution is restricted, and the reproduction of yeast may be interfered with. Again, the deeper the vessel, the more difficult is aëration by rousing and similar devices.

The yeast used for pitching purposes should be preserved in as dry and as cold a state as possible, and preferably in slate tanks, provided with false sides and bottoms, through which ice-cold water can percolate. These tanks should be kept in a room reserved for this purpose, and quite removed from all sources of contamination. Pitching yeast should be as fresh as possible; it is a mistake to keep it with the object of maturing.

CHAPTER VIII.

THE RACKING AND STORAGE OF BEER.

SETTLING AND RACKING.

WHEN the main fermentation is over, and when, in fact, the beer has attenuated to the proper degree, and when the bulk of the yeast has been worked out, it is run either direct to casks or first through a settling or racking back, and thence to casks after some hours' (three hours to fourteen hours) standing in that vessel. This vessel is not used in all breweries, but on certain systems, such as the Burton and the stone-square system, its employment is almost invariable. On the skimming system its use is frequent though not invariable. The vessel is certainly a convenience ; large bulks of beer lie here and are rapidly and easily treated with priming (*q.v.*), caramel (*q.v.*), antiseptics (*q.v.*), or cold-water malt-extract (*q.v.*). The addition of any of these things at this stage clearly involves very much less labour than their addition to a hundred or more casks. Apart, however, from economising labour, the settling back, by the surface it exposes to the wort, gives it an opportunity of depositing some of its yeasty and other suspended matter, and the separation of these matters at this stage will reduce the subsequent amount of sediment or " bottoms " in the casks. In many trades this reduction of bottoms is a thing to be aimed at. On the other hand, the vessel has its disadvantages—at any rate during the summer months ; for, whatever precaution be taken, the use of the vessel renders inevitable a certain agitation and consequent aëration of the beer. Aëration during fermentation and under due

control is required ; but aëration of beer—and the consequent
expulsion of carbonic acid—must always tend to a resuscita-
tion of yeasts, which at that stage should be somewhat torpid ;
and in summer time, when there is naturally a predisposition
towards fretfulness, the aëration of beers at this stage is not
without danger.   Then, too, in the summer time exposure to
the air will mean exposure to infection.   Considering the
advantages and disadvantages of the vessel, it may be said,
that except in the height of summer, its use is at least
desirable—the benefits outweighing the disadvantages.   Even
in summer, when the transference of wort is so conducted
as to limit agitation and aëration to a minimum, the vessel
may be safely used, if beer is not exposed there for more
than a few hours.

   After settling in this vessel for the desired time the beer
is run off into casks.   In this process the only precaution
necessary is to avoid unnecessary aëration and agitation of the
beer, and unnecessary fobbing.   By having the racking hose
or sleeve long enough to reach the bottom of the cask, we
sufficiently guard against these risks.   When the beer is
racked straight from the fermenting vessel into casks, the
same precautions must be taken.

   In the case of running ales, the casks into which beer is
racked from fermenting vessels or settling casks are the trade
casks.   In the case of stock beers, the casks into which the
beer is run are frequently larger than trade casks—butts or
puncheons ; then, as beer is wanted, it is drawn from these
larger casks.   In many concerns, however, the beers of all
classes are racked at once into trade casks.   In selecting one
of these alternative plans the following points must be considered.   When much of the trade is in very small casks, such
as 4½ gal., 6 gal. and 9 gal. casks, it is not desirable, in the
case of store beers, to rack beers direct into such small casks.
Putting aside the undesirability of storing beer in very small
bulks (which will be considered presently), it must be remem-

bered that the smaller the cask the greater the area of wood
surface do we expose to the beer. As the wood surface is
always likely to be contaminated, it is clearly unwise to preserve beers for any time in small vessels. When, on the other
hand, the beer goes out in barrels, it is desirable to rack beers
direct into these vessels. In such vessels the wood surface
exposed is not excessive, while the bulk is sufficiently great
(see p. 393). It is clear, too, that where the racking from
store casks into trade casks can be avoided, it is best to so
avoid it—for the racking at this point will necessarily mean a
certain loss of beer, and an inevitable agitation and aëration
and expulsion of carbonic acid, and the double danger of
contamination in store cask and trade cask.

Practical men are well aware that bulk has a most important influence upon the flavour, condition, and, in some degree,
stability, of beer, the superiority lying certainly with beer
stored in large bulk. That this is so is in all probability due
in great measure to the influence of the pressure connoted by
bulk rather than of bulk itself. H. T. Brown showed that the
products of fermentation varied somewhat with the pressure,
this applying more particularly to the by-products than to the
main products of the fermentative change. Thus at high
pressure less aldehyd and acetic acid were produced, under
otherwise the same conditions, than at low pressure. Now, it
is upon these by-products of fermentation and of after-
fermentation that the flavour of beer is in a great measure
dependent. There is every reason to suppose that pressure
will exert an influence upon cask fermentation similar to its
influence upon primary fermentation, and it is by no means
unlikely that it is to the influence of the greater pressure upon
the cask fermentation products that is due the superior flavour
of beers stored and conditioned in relatively large bulks.
Again, Müller showed that diastase is far more active under
great than under low pressure. Now, although we have no
diastase to deal with in our finished beer, yet the ferment

secreted by the secondary yeasts, and upon which we depend
for the gradual degradation of maltodextrins into maltose,
is undeniably a ferment very strongly akin to diastase, and it
is therefore not unreasonable to suppose that the activity of
this ferment will be appreciably assisted by pressure. If that
be so, it will not be hard to understand why beer stored in
great bulk (i. e., under great pressure) should condition better
than that stored in small bulk. Apart, too, from the more
rapid degradation of the maltodextrins, the gas produced
under high pressure will be far more likely to actually dissolve in the beer than that developed under low pressure.

So far as stability is concerned, we know that a satisfactory conditioning will of itself conduce in a very marked
degree to the checking of bacteria, and therefore to soundness ; but, as has been previously stated, any given volume of
beer stored in large vessels will be exposed to a very much
smaller wood surface than the same bulk stored in small
vessels ; and since the wood surface is usually more or less
contaminated, the smaller the area of it in contact with beer
the better for stability.

When beer is racked there is still a considerable amount of
residual normal yeast in it which can be well dispensed with.
Brewers are in the habit of leaving the bungs of their casks
out, or loose, for some time, in order to give the yeast
an opportunity for clearing itself out. The yeast makes
its appearance in the form of a foam, and it and the beer associated thus fall over the side of the casks. This waste is
made up for by topping-up the beer with fresh beer. The
topping-up process, however, is not invariably carried out ;
but in the case of running ales which it is decided shall
fine shortly after racking, it is a good and useful plan. When
the beer has purged itself of the excessive yeast, the cask is
shived down tight, vent being given when necessary by a
porous spile. The porous spile is afterwards driven in tight
when no further vent is needed. Some brewers shive down

their beer at once after racking.   This system gives enormous
condition, but it is generally only permissible when beers
have been fined in the fermenting vessel (see p. 413). By
fining at that stage most of the yeast which is usually transferred with beer into casks is retained in the fermenting-tun.
There is, however, more to be said against this system than
in its favour.

There are probably no two breweries where the system of
managing the beer in cellar is precisely the same, for every
brewer will have to arrange his cellar management according
to the degree of briskness desired by his customers, to the
length of time beer is in cellar, and according to various
trade considerations of that sort. To discuss each system
would not be possible; it will be sufficiently seen, from what
has been said that by opening or closing bungs immediately
after racking, by topping-up, by porous spiling, that we have
the management of the beer under control.

Of the cellars themselves the main points are a low and
even temperature, and absolute cleanliness. Upon the temperature of the cellar will depend, in great measure, the
correct heat at which beer should be racked. It is desirable
in winter to rack it one or two degrees below cellar temperature, so that it shall not chill (and therefore cloud) after
racking. In summer the racking temperature should about
equal cellar temperature; if, as in winter, it be appreciably
below it, there will be a risk of fret incident to this rise. The
brewery cellar heat, however, must not be chosen quite irrespective of the inferior cellar accommodation found on the
consumers' premises. Consumers' cellars are generally hot in
summer and cold in winter. Although the brewery cellar
should be even in temperature, it should not be too low,
otherwise frets in summer can certainly be anticipated. An
even temperature of about 55° F., will generally be found to
best meet practical requirements.

THE DRY HOPPING OF BEERS.

The addition of hops to the trade cask is a very usual practice in the case of all beers which are kept some time, and in which a distinct hop aroma and flavour are desired. The practice, therefore, more particularly applies to pale ales than to mild ales, or stouts and porters; but even in these other classes of beer it is sometimes carried out, though not very frequently. The hops are added at racking, or shortly afterwards, the quantity used being from $\frac{1}{4}$ to 1 lb. per barrel. For this purpose it is customary to employ only high-class and thoroughly sound hops, since impurities introduced with hops at this stage would tell to the full, not being subject to the purifying influence of boiling, as they are when added to the copper. This precaution is imperative. It is also usual to choose hops of rather a delicate and mild character; such, for instance, as Worcesters; hops from this particular district enjoy a deservedly high reputation for this purpose.

The continued length of contact between the beer and the cask-hops serves to extract the bitter principle to some extent, but more particularly the oils, to which, it will be remembered, the delicate aroma and hop-flavour are more particularly due; in the copper these oils are in great part dispelled, unless some of the hops are reserved till shortly before turning out the worts. But in the cask the full yield is obtained when the beer is stored for a moderate time. So far as the resins are concerned, it is probable that only a small amount of them is extracted by the finished beers. Yet such amounts as are extracted would appear to be of an antiseptic character, for there is no doubt that dry hopping much conduces to the preservation of the beer, and there is no other constituent extracted from the hops at this stage which could account for the marked effect so obtained.

The influence of dry hopping is principally in the way of improving the flavour and the stability of beers; but it exerciscs other functions as well. The hops have adherent to their surface, besides some kinds of bacteria, several varieties of wild-yeast, and these varieties will promote after-fermentation by assisting the ordinary cask-yeast to degrade the malto-dextrins. It is at any rate certain that a dry-hopped beer will condition more rapidly than the same beer not treated in this fashion. Besides this, the cask hops are of the greatest · help in preserving casks. This is of special importance in the private trade, and in agency trades, where casks are frequently kept some time after their contents are consumed before they are returned. So far as brilliancy is concerned, unless the cask hops added amount to over ⅛ lb. per barrel, they do not apparently delay clarification by fining. When, however, the hop rate exceeds ⅛ lb. a rather long time should be allowed before the finings are added ; but this constitutes no objection to the process, since such large rates as these would only be used for stock ales to which finings in the ordinary course would not be added until some time after the introduction of the cask hops.

THE CONDITIONING OF BEER.

Broadly put, the conditioning of beer depends upon the degradation of the maltodextrins into maltose at the instance of the cask-yeasts, the so-formed maltose being then fermented by these yeasts. The wort containing a sufficiency of malto-dextrins, and the beer a sufficiency of cask-yeasts, condition-ing will proceed with regularity and sufficient persistency. At the same time circumstances sometimes arise which interfere with condition, even in the presence of an adequate amount of maltodextrins. In some cases, when the worts are unduly aërated during fermentation, the beer is racked too " clean," that is, containing an insufficient quantity of yeast—cask-yeast in that event being proportionately reduced. Again, the yeast-cells, especially in heavily hopped beers, get frequently so

coated with hop resins as to expose but a small surface to the
beer itself; and when the casks remain at rest for some time
and the beer is therefore quiescent, the cells are in addition
surrounded by a film of the alcohol which they have ejected
during the earliest stages of cask-fermentation, and which
prevents the contact between the ferment and the fermentable matter. It is not surprising, therefore, when beers are
stubborn to condition, that vigorously rolling the casks is
found in practice to be an excellent remedy. By so doing
we disturb the alcoholic film surrounding the yeast-cells,
bringing them into direct contact with the fermentable matter;
and, so far as the hop-resin coating is concerned, we give the
exposed portions of the cells more chance of doing their
work by changing their position in the beer.

It may be safely said that conditioning will always be
accentuated by agitation, and no safer plan can be adopted
than cask rolling when beers are defective in this respect.
Brewers know that a sample cask of their beer in their own
cellars will often be dull and unsatisfactory, while other
casks of the same brewing sent into the trade will be quite
the reverse. That this is so is primarily due to the thorough
agitation the casks get during their journey. It has been
suggested that brewers should judge their beer not from a
quiescent cask, as they now do, but from the behaviour of a
sample cask that has been purposely sent on a journey and
then returned to the sample cellar.

Temperature will of course play some part in promoting
or retarding condition ; sudden chills, or the protracted storage
of beer at unduly low temperatures will of course tend to stop
the change ; while unduly high temperatures will tend to
convert it into a fret.

Again, the point at which finings are added will have an
appreciable influence upon the rapidity or otherwise of condition—the early addition of finings retarding condition by
enveloping the cells in a gelatinous mass, thus preventing

their contact with the beer. In all cases of stubborness of conditioning the addition of finings should be postponed until the condition has started in good earnest, either naturally or artificially. On the other hand the addition of finings at an early stage is a convenient means for dealing with beers that tend to become unduly brisk.

But when all is said and done, it is the carbohydrate proportions which primarily determine the satisfactory or unsatisfactory conditioning of a beer. With a lack of malto-dextrin there will always be difficulties ; for even if there be a sufficiency of the very low maltodextrins, which appear on analysis as free maltose, the condition due to these will always be transient, and after a short period of evanescent briskness, such beers as these will become flat, and will then tend to become unsound.

Beers consumed quickly may, however, be safely given a species of cask-fermentation, by adding to them some sort of fermentable sugar in solution. Such additions as these are termed priming, and it is customary to add them in the form of a syrup of about 1150 sp. gr., of which the Excise permit the employment at the rate of two quarts per barrel of beer. The sugars generally employed for this purpose, are invert-sugar, glucose, and cane-sugar ; of these, invert-sugar is the favourite, as, in addition to ready-fermentability, it imparts during the early stages, a luscious sweetness which confers an apparent extra gravity on the beer. Glucose is also used, but more particularly in pale ales, where a dry flavour is more desired than the sweet fulness of invert-sugar. The very appreciable amount of unfermentable matter contained in glucoses, however, renders them less suited for priming pur-poses than are invert-sugars. Cane-sugar is not often used as priming except in stout and porter brewing. The fermen-tation of the priming sugar throws no burden upon the secondary yeasts, since it is readily fermentable at the in-stance of the residuary primary yeasts ; it is at any rate a form of cask-fermentation which can always be relied upon, generally making its appearance within three or four days of the addition of the sugar.   In the case of running mild ales the priming is added as a rule to the settling square; thus saving time and labour, and sufficiently answering the purpose for such beers as these, especially where the trade is a quick and uniform one.   In the case of pale ales, or superior ales generally, it is best to add the priming to each individual cask just before it is sent out into the trade.   By so doing, the customer gets the beer at its best, and with moderately quick draught the after-fermentation, due to the priming, will last out the required time.   If in these cases the priming were added to the settling back, it would probably be all over by the time the beer was required in the trade.   Priming solutions should always be boiled; they should be made with the purest material, and made fresh at least once a week in winter, and every three days in summer.

Some brewers prefer rather to assist the yeast in degrading the maltodextrins into maltose, than to add extra fermentable sugar.   To do this they add to the beer a small quantity of cold-water extract of malt, which, containing active diastase, will supplement the action of the secondary yeasts in degrading the maltodextrins.   Only a small quantity needs to be used; larger quantities would produce too rapid a removal of the maltodextrins, and this excessively early disappearance would be obviously dangerous in regard to the stability of the beer. Pale ales will stand one pint of concentrated extract per one hundred barrels of beer, added at racking; running ales double that amount.[*]   The condition produced in this way is a steadier and more lasting condition than that produced by priming, and it is in most respects preferable to it.   It depends,

[*] Take two parts by weight ground pale malt, five parts by weight cold water; soak for six hours, stir periodically, then strain off and filter the solution through flannel.   One pint of the solution to be used as above stated for 100 barrels of pale ale, and one quart to the same amount of running ale.   The extract to be made fresh on each occasion.   The grains are to be discarded.

however, upon having a sufficiency of maltodextrin naturally present ; priming, on the other hand, is independent of this. When the extract cannot be conveniently added to the beer in racking vessel, it may be added to the individual casks. In that case, it is previously diluted with beer to such an extent as to permit of the quantity due to each cask being sufficiently appreciable to be conveniently measured.

THE CLARIFICATION OF BEERS.

Beer nowadays is demanded in absolutely brilliant condition, and however good it may be in other respects, it will be returned to the brewer as unsaleable if it is in the least cloudy or turbid. That such is the case is probably due, first, to the importation into this country of Lager and Pilsener beers (which are always brilliant), and to the substitution in public houses and restaurants of the old-fashioned mugs by glasses. The brewer has therefore to strain every nerve to send out beer which will be absolutely brilliant within a very short time of its delivery at the customer's cellars. It is owing to this demand for brilliancy, and also owing to a demand for fresh rather than for stored beer, that clarification is now almost entirely artificially effected. Many beers now brewed are capable of going bright spontaneously if kept sufficiently long ; but such beers would, although bright, not come up to the required standard of brilliancy, while the storage necessary for their arriving at the moderate stage of brightness, would in many trades not be permissible on account of the undesired maturity of flavour it would lead to, and on other grounds of a commercial character.

The fining material used in this country is almost exclusively isinglass, "cut" with acid of some sort. Many brewers make their own finings by cutting isinglass in returned sours, the acetic and other acids of returned beer serving to soften and dissolve the isinglass. The beer used for this purpose is, or ought to be, stored until it has spontaneously brightened. All beer turbid with decayed yeast and acid-forming bacteria would be obviously dangerous. When the old beer has become perfectly bright, it is capable of making finings possessed of great clarifying power, and for running ales such finings are permissible. They are, however, undesirable for pale and other delicate ales. Apart from the danger of infecting the beers with organisms derived from the old beer, there is always a tendency for beer fined by these old beer finings to show up the acid flavour after it has been kept for some time. The amount of acid added with these finings becomes perceptible on the palate when the beer, after continued storage, has lost most of its body.

The safer way of softening and dissolving the isinglass is to use a mixture of sulphurous and tartaric acids. Sulphurous acid used alone gives finings of somewhat poor clarifying properties ; tartaric acid used alone gives finings likely to become mouldy during manufacture or storage, although the finings themselves have good clarifying properties. But finings made from a mixture of the acids have satisfactory clarifying properties, and they keep thoroughly well.

In making finings there are two points to which the strictest attention should be paid: first, the scrupulous cleanliness of the vessels and implements employed ; second, the absence of all hurry or haste in the manufacture. The necessity for cleanliness is obvious, for isinglass in solution is a highly putrescible body. The vessels used for dissolving and cutting, and storing, should be of slate, and kept in a cool place out of contact with the dirt and dust of the brewery.

So far as the necessary slowness of the process is concerned, experience has shown that when the finings are prepared hurriedly, that is, when the only partially softened and dissolved glass is prematurely forced through sieves, the product is only apparently of uniform gelatinousness ; as a matter of fact it then consists of numerous little masses of dissolved gelatine with shreds of undissolved isinglass. Such preparations as these give very unsatisfactory results; in the beers they are apt to break up into small bits, which instead of sinking to the bottom of cask, or working out of it, are suspended throughout the bulk of the beer, and when the beer is tapped they come through the tap with it. This "bittiness," as brewers term it, is a frequent source of trouble, and the cause of many barrels of returned beer.

A full month should be given for the preparation of finings, and it is well to give it a double sieving, the first through a rough sieve at the end of say twenty-five to twenty-six days from the time of soaking; a second through the fine sieve, three or four days afterwards. So far as the quantities of material are concerned, the following recipe of Southby's has been found to give excellent results in practice.

To every seven pounds of isinglass, we require one pound of tartaric acid and one pound of sulphurous acid (of about 5 per cent. strength). The isinglass is placed in an unheaded hogshead, provided with a cover. It is covered with water, and the tartaric acid and sulphurous acid added, the tartaric acid being previously dissolved in a little warm water. The mixture is kept covered down and well stirred from time to time, more water being added as the isinglass swells, so that it may always be covered with liquid. When the isinglass is thoroughly softened (in about twenty-five to twenty-six days) it is passed through a coarse sieve into another unheaded hogshead (also provided with a cover); it is now, after two or three days, passed through a fine sieve, and then diluted to one hogshead.

The clarification of beer by finings is still a somewhat obscure process. The finings are in solution, or in a state bordering on solution, and when these are introduced into a fluid-like beer, containing particles in suspension, the finings seem to come out of solution, and coagulate; and in coagulating imbed in themselves the suspended particles in question.

The coagulum will thus rise or fall according to the specific gravity of the liquid, and according as to whether the liquid is evolving gas. In the latter case, the coagulum will rise even though the specific gravity of the liquid, may not be greater than that of the coagulum ; in this case, the ascent of the coagulum is due to the particles of gas attaching themselves to it and buoying it up until it reaches the surface.

It would appear that the reaction of the liquid to be fined has an important bearing upon the coagulation of the finings, for so long as the liquid is as acid as the finings themselves there would be no reason for them to change their dissolved or semi-dissolved condition. But when they are introduced into a fluid less acid than themselves, there will be a tendency for the finings to come out of solution. So far as beer is concerned the degree of acidity certainly does seem to play a part in determining the rapidity of clarification ; for, other things being equal, a beer brewed with a high proportion of saccharine will fine more readily than an all-malt beer ; and in a saccharine-brewed beer the acidity will be reduced proportionately to the amount of saccharine employed. Again, an acid beer is invariably difficult and even impossible to fine ; and such a beer may be frequently made to fine by neutralising the whole or a part of the acidity. It may well be that the transference of the finings existing in a decidedly acid medium into a less acid medium is the exciting cause of their precipitation or coagulation.

It was believed for some time that the precipitation of the finings was due to the tannin left in the beer derived from hops added to copper. But, as has been previously explained, there is no evidence to prove this. In fact, such evidence as there is, would negative such a hypothesis. The coagulation of finings would appear to be not at all a chemical process, but purely a physical one.

The nature and quantity of the particles in suspension, however, play a great part in determining rapidity and

efficiency of clarification. A beer which is almost fine
will not so readily become partially brilliant as one which
is moderately cloudy. The finings would appear to require
something in the nature of an obstacle upon which to exert
themselves. Again, the nature of the suspended particles is
of material influence. When they are very fine, like the
particles of hop resins previously referred to, the finings seem
powerless to separate them ; even very small yeast-cells and
bacteria offer difficulties in the way of clarification. But
ordinary sized yeast-cells seem not at all resistant to finings,
even when the beer is quite milky with them.

DEFECTIVE FINING OF BEERS.

The causes of defective fining may be summarised thus :—

      (*a*) Hop-resins.
      (*b*) Yeast.
      (*c*) Bacteria.
      (*d*) Flatness of beer.

(*a*) *Hop-resin Turbidity.*—Primarily this may arise from
some obscure and unascertained defect in the hops, but as we
have to take hops as they are given to us, it becomes necessary
to ameliorate any such defects by indirect means. The excess of resin in the beer (putting aside the question of hops)
must be attributed to the defects on the part of the yeast, in
point of not eliminating them as they should do, during the
primary and secondary fermentation. This very frequently
is the fault of the yeast, which should in such cases be
changed ; or it may be due to the yeast directly and the wort
indirectly ; that is, the yeast has become abnormal through
being cultivated in abnormal worts. In this case, too, the yeast
should be changed. In many cases, however, it is directly
due to defects in the carbohydrate proportions in the worts—
more particularly to a lack of maltodextrins. When this is
the case, the beers do not condition ; and there being nothing

so effective for the elimination of resin as a brisk cask-fermentation, the absence or the incompleteness of this process leaves a great deal of uneliminated resin suspended in the beer. In all cases of hop-resin turbidity, then, the materials and mashing system should be so adjusted as to secure an abundance of maltodextrin. But condition artificially induced is not without use, and the addition of priming sugar as a temporary measure is of service in such cases. When there is a sufficiency of maltodextrins, and when owing to some defect in the yeast it appears unable to degrade them, the addition of some diastase solution (cold-water malt-extract), or of a very little yeast (about a teaspoonful per barrel) from the very last risings will be found of use. Yeast which rises quite towards the end of fermentation contains a very large proportion of secondary varieties.

But attention must also be paid, in all cases of hop turbidity, to the hot aëration of the wort; for, as has been explained, it is the aëration of wort hot which best promotes those changes by which the hop-resin particles are rendered denser and more coherent.

It is more particularly mild beers that give rise to trouble in the direction of hop turbidity—not that such beers are inherently more predisposed to it than pale ales; in fact they are less so, since they contain less hops. But the fact that they are required bright almost immediately, outweighs the above point. Many mild beers which refuse to fine at the time they are required, would easily do so if kept for a month or so, like pale ales are kept. But the demand for fresh mild beers and the relative instability of running ales, renders their storage impracticable. The principal difficulty in the way of getting a rapid fining action in mild beers affected in this way is that the condition has had no opportunity for eliminating hop resins. Of course directly finings are added the condition is in great measure interfered with, and thus the very means upon which the elimination of the resins depends is in great measure checked. In all such cases, then, and until things are put right, the beers should be fined at as late a point as possible, and the beers should be kept in the brewery cellars for as long a time as is practicable. They should certainly not be fined until just before being sent away. Of course everything should be done while the beers are in the cellar to promote condition : by rolling, priming, and, if necessary, by diastase solution, and even by the addition of late yeast— of all these, rolling, being the most natural, should be first thoroughly tried before recourse is had to the other more artificial stimulants.

In the winter time hop-resin turbidity makes itself felt in another form : in the cloudiness that affects beers which are at all chilled, even after they have been fined satisfactorily at an earlier stage. In great measure this can be avoided by racking so low that subsequent chilling is averted, but, as before, brisk conditioning should be aimed at by rolling, priming, and the other devices mentioned, when cloudiness due to chilling is perceived. Obviously, too, the storage temperature should be somewhat increased.

Beers brewed from a change of yeast are frequently turbid. This may be due to causes subsequently to be referred to, but it is also frequently attributable to the fresh change of yeast not properly eliminating hop-resins. Similarly, a change of yeast will frequently produce bright beers, materials and system being unchanged in other respects. This is due to the increased capacity of the fresh yeast to eliminate hop resins.

Beers turbid with hop resins go absolutely brilliant on being made slightly alkaline with potash. They go bright on digestion at 120° F. for 20 minutes, and on cooling down become again cloudy. They go bright (or very nearly so) when agitated with ether. These tests will as a rule, serve to indicate hop-resin turbidity.

(b) *Yeast Turbidity.*—Yeast turbidity may arise in several

ways; very frequently it is not in any way connected with the yeast itself, being due to the excessive viscosity of the wort preventing the proper eilmination of the yeast during fermentation, and also preventing the free action of the finings when they are subsequently added. This excessive viscosity is due, as we know, to an excess of maltodextrins. Yeast turbidity, however, also occurs when the worts are not viscous. In this case the turbidity is mainly due to small immature yeast cells, which either through inherent weakness or through the weakening effects of contamination, or through an ill-nourishing wort medium, have not developed to maturity.

It would appear, too, from Hansen's researches, that there are certain varieties of yeast (*Sacch. ellipsoideus*) which, when contaminating the pitching yeast, or otherwise gaining access to the worts in sufficient quantities, invariably produce yeast turbidity. Whether or not yeast turbidity in this country is ever due to this cause is, however, an open question, and one upon which the writers of this book are rather doubtful. Yeast turbidity will occur, too, when the beers are fretting—when, in fact, the beer in cask is under-going a violent fermentation, accompanied, as it generally will be, by an excessive reproduction of yeast.

The means of overcoming yeast turbidity are relatively simple. When arising from excessive viscosity of wort, the maltodextrins must be reduced permanently by lowering curing or mashing heats, or employing a higher sugar-rate; temporarily by "dressing" fermentations with diastase solutions, or diastatic substances (such as malt, wheat, or other flour). The beer in cask can also be dressed with a little diastase solution, and thoroughly rolled. When yeast turbidity is due to small yeast-cells, the following temporary measures are useful as promoting the development of these cells into maturity: adding wort from a fresh fermentation, topping up, priming, and rolling. Fermentations in progress, too, should be well aërated at frequent intervals. As permanent measures, the yeast or wort will require to be altered, according to which is in fault.

Yeast turbidity due to frets, can only be satisfactorily combated by avoiding an excess of the very low maltodextrins which cause them ; the reasons of this will be clear on reference to Chapter VII., p. 378. As temporary measures, the following are useful in the case of frets due to very low maltodextrin (those which appear as free maltose on analysis) : keeping beers thirty-six hours longer in fermenting vessels, using more yeast, giving a greater range of fermenting heats, and fining as late as possible. The type of maltodextrin should be raised by using higher dried malt, or by mashing higher ; the sugar rate to be restricted, and no " dressing " used during fermentation. It is better, on the ground of palate, to attain the required maltodextrin by using higher-dried malt than by very high mashing heats.

(*c*) *Bacterial Turbidity.*—It is very seldom that beers become turbid through bacterial development so early in their career as the fining stage. When they do there is no temporary remedy, and the brewer should without delay set about discovering the causes of the infection.

(*d*) *Turbidity of Flat Beer.*—Flat beer very often refuses to fine for a reason quite other than the mere fact of its containing particles which are not readily removable by finings. The reason referred to is a mechanical reason, and not a chemical one. When a beer is in fair condition the finings poured into it are brought by the natural movement of the beer into contact with the great bulk of the liquid. But when the beer is quite flat, this is clearly not the case, and in these circumstances the finings are brought to operate on only a portion of the beer. When flat beers have to be fined, it is of service to inject the finings through an injector, or, failing an injector, to syringe them in with an ordinary syringe. By this means the finings are artificially distributed more or less over the whole bulk of the beer.

THE BOTTLING OF BEER.

Beer destined for bottling should be made of high class materials, and be brewed in winter. It ought to possess sufficient stability to be storable without deterioration, until it shall clarify naturally. Beer made on these lines should be permitted to go through the ordinary cask fermentation, and then flatten. When thus flattened (but still just about saturated, although not supersaturated, with gas) it should be bottled, and the bottles stored in a warmish place (about 65° to 70° F.). The warmth of the bottle store will soon resuscitate the residual yeasts, and the beer will then come into the required degree of briskness. Bottle fermentation induced in this manner produces the pleasant pungent flavour characteristic of bottled beer, and it will be sufficiently brisk. Beer bottled earlier, i. e., during the actual secondary fermentation, becomes tumultuous in bottle. It gives a tremendous foaming head, but the beer itself is flat and flavourless, and frequently turbid. Beer of this sort, of which of course there is much turned out in these days of competition, has to be fined, for bottled when it is, there has not been time for natural clarification.

Bottled beer as made in this country must necessarily always have a sediment. Fermentation in bottles must necessarily be accompanied by a reproduction of yeast, so that even if bottled fine, a sediment will inevitably be formed after storage. To obviate this sediment, or at any rate retard it, an alternative system of bottling has lately been introduced. On this system the requisite briskness of beer is produced, not from the gas produced during the fermentation in bottle, but from gas artificially produced and pumped into the beer by machinery similar to that used in the manufacture of aërated waters. Bottled beer made on these lines requires no storage, and on that account does not demand the high-class materials necessary in the case of beer brewed in the usual way. The system answers well for light, cheap, bottled ales, for which there is now a demand; but the flavour is very different from beer bottled in the usual manner. The characteristic pungent flavour is entirely wanting; this flavour is probably due to the actual solution of gas into the beer which occurs when it is slowly produced, and under the combined influence of pressure and time.

THE USE OF CARAMEL.

Local considerations frequently compel brewers to darken the shade of the beer as naturally produced. This is sometimes done by using a little coloured or patent grist in the mash-tun or copper, but more frequently by adding caramel either to copper or to the beer itself. There is no scientific change to be studied in this connection; the amount of caramel added being solely dictated by the depth of shade it is desired to attain. Caramel added to beer should be of good quality, that is, its tint should be permanent during storage and during fining, and it should never be thrown out as a fine powder, as sometimes happens with inferior products.

Caramels are sometimes added to stouts and porters, less with the object of adding to colour than of conferring fulness. Caramels made at relatively low temperatures from dextrose sugars contain a substance which is unfermentable by yeast, both primary and secondary, and which possesses a sweet flavour. This substance may possibly be an anhydride of dextrose—dextrosan—at any rate its properties, chemical and physical, would appear to agree with those of the body named.

A useful instrument for determining the amount of caramel to be added in order to produce a given shade is Lovibond's Tintometer. This appliance standardises tints of beer, by certain selected tinted pieces of glass, which are graduated and numbered according to their depths of tint, and which build up with fair accuracy to form deeper shades. These

coloured glasses are permanent in colour, and they are therefore superior, for purposes of comparison, to standard beers
or caramel solutions.

ANTISEPTICS.

Chemical substances possessing an antiseptic action have
now been used for some time at various stages of the brewing
process—sometimes in the mash, sometimes in copper, but
much more frequently in the finished beer; and it is on this
ground that we shall treat of them in this chapter. The
antiseptics usually employed are various forms of sulphite—
bisulphite of lime being the most popular. Other sulphites
are: sulphite of sodium (Beane's material), sulphite of potassium ("K.M.S."), sulphite of magnesia, &c. The only
substance other than a sulphite which is at all largely employed is salicylic acid.

Bisulphite of lime is a solution of sulphite of lime in
sulphurous acid, and on the whole it may be regarded as the
most satisfactory of these chemicals. In some breweries it
is unusable, on account of the tendency to yield sulphuretted hydrogen—familiarly known as stench. This action
is due to the reduction of the sulphurous acid by bacteria. In
most breweries, however, bisulphite is usable, and without
any apparent tendency towards decomposition. It may be
broadly said that when a brewery is able to use bisulphite,
its plant is cleaner and its materials sounder than when its
use leads to the unpleasant smell of sulphuretted hydrogen;
for it is a fair presumption that in the former breweries the
infection is low, and that the beers form an indifferent
medium for the development of bacteria.

When bisulphite is inadmissible, the other sulphites
mentioned may be usable without harm, for, in proportion
to their inferior preservative powers, they are less liable to
be decomposed.

Salicylic acid is a powerful antiseptic, and is more useful for running than for stock ales. It is liable during lengthy storage to decompose, and to form unknown substances of objectionable flavour; but during short or moderate storage this change does not occur. It is open to the objection of flattening beer. That this is so is probably due to its restricting the ferment, secreted by cask-yeasts, upon which we depend for a breaking down of maltodextrins. At any rate salicylic acid has a very marked restricting influence upon diastase, and it is a fair presumption that it will similarly react upon the ferment in question, which in so many of its properties is similar to diastase.

We consider that antiseptics have their use, and in due moderation we believe that they are harmless to the consumer and of service to the brewer. But we very strongly protest against their abuse, and we very strongly object to their employment as a means to mask defects of material, system, or plant cleanliness. Used in this illegitimate way, they must inevitably lead to disappointment and trouble, and we are bound to say that any such trouble is abundantly deserved. The brewer who doses his beer with chemicals to hide his ignorance or carelessness, or the inferiority of his materials, will never succeed in his object; for under these circumstances the resulting beer will be nauseous in two directions: nauseous from the mere flavour of the substances added, and nauseous from their decomposition products.

Why these substances should possess an antiseptic action we do not know. In the case of the sulphites, it is ascribed to their oxygen-absorption capacity; yet on the other hand, they are capable of parting with oxygen as well as absorbing it, as is evidently the case when they are reduced to sulphuretted hydrogen. In the present state of our knowledge, it is advisable only to say that they are actively antiseptic, without venturing on any explanatory hypothesis.

2 E

INTERPRETATION OF THE ANALYSIS AND EXAMINATION OF BEERS.

From the results of the analysis and examination of beer, it is possible to form an opinion as to the soundness and general quality of that beer; and where it is defective, we get some indication of how the defects have arisen, and therefore how they can be avoided in future brewings.

So far as the analytical figures are concerned, we consider that the percentage and the type of the maltodextrins are the most important points. In beers just racked and when no sugars have been used, we like to see about 12 to 16 per cent. of total amyloïns referred to original wort solids. We have found that any less amount of these constitutes a deficiency, and that beers containing any such deficiency, will be unsatisfactory in point of fulness, condition, and stability. Any excess over 18 per cent. would, as a rule, constitute an excess, resulting frequently in turbidity. Due regard, however, must be paid to the original gravity of the beer, for the above percentages referring to solid matter, beers of different gravity will actually contain very different rates of amyloïns although the percentage of them may work out the same on wort solids. We should therefore demand a higher rate in a weak, than in a strong beer, and *vice versâ*, but the limits of 12 to 16 per cent. on wort solids, will generally cover variations due to this source, and will apply to all malt beers recently racked.

In beers brewed with malt substitutes (sugars) we expect a lower maltodextrin rate; but in no circumstances do we consider it safe to have a lower percentage than 8–10 for running ales, and 10–12 per cent. for stock ales.

So far as mean type is concerned, we require a higher type for store beers of all kinds, than for running beers. So far as freshly racked beers are concerned, we regard any mean type below $\left\{ {3 \atop 1} {M \atop D} \right.$ as dangerously low, while any type higher than $\left\{\begin{smallmatrix} 1 & M \\ 3 & D \end{smallmatrix}\right.$ may be considered excessively high. In comparing the types of amyloïns in beers, and those in the worts which yielded them, we have observed the former decidedly lower than the latter. For instance, in well-brewed pale ale worts, there is a general tendency towards a mean type of $\left\{\begin{smallmatrix} 1 & M \\ 2 & D \end{smallmatrix}\right.$; the resulting beers, however, contain amyloïns of about $\left\{\begin{smallmatrix} 1 & M \\ 1 & D \end{smallmatrix}\right.$ mean type. There is clearly a degradation of type occurring during the brewery fermentation or during the analysis of beers ; and due allowance must be made for this fact. Of course, during storage, there is a very decided degradation in type, and a lowering of the maltodextrin percentage ; but we presuppose that beers are analysed immediately after racking, and the above remarks are made on that assumption. When, for any reason, beers are analysed after considerable storage, great allowances must be made on that account. We do not consider it necessary to enter upon this point however, since beers analysed for general character, &c., should be examined before they have undergone the changes which storage inevitably induces.

The very low maltodextrins which are unfermented during the brewery fermentation, but which ferment out during the laboratory fermentation (see p. 385), and which have therefore been hitherto erroneously regarded as free maltose, are evidently substances which should be avoided, or at any rate, be reduced to a minimum, in beers brewed in summer weather, or for summer consumption. For such beers will evidently tend to fret when the conditions of temperature and the nature of the yeast present are such as lead to their very rapid degradation and fermentation. We therefore consider that stock beers and summer-brewed running ales should not contain more than 2 per cent. (on original wort solids) of these bodies. On the other hand, winter-brewed running ales may contain

2 E 2

considerably more (say up to 6–7 per cent.) without harm :
and not only so, but they would be useful in contributing to
an early condition, which in the winter months is otherwise
frequently difficult to attain.

The lowering of the specific gravity during forcing is connected with the above point ; we prefer for stock ales and
summer-brewed mild ales, a not excessive lowering in gravity
(say not over 3–4 degrees during the periods suggested for
forcing these beers). For winter-brewed mild ales, on the
other hand, we should not object to a greater decrease (say
6 degrees).

The total dextrin rate is of less importance than the
maltodextrin rate, but is still worth noting, for, better than
any other, this estimation leads us to an opinion as to whether
the beer was all malt, or malt and grain brewed on the one
hand, or partly sugar brewed on the other.

The acids in a new beer are seldom beyond the normal
rates ; but the increase in acid as forcing is worth noting as
a supplementary test to the microscopic examination.

Some experience has shown us that the microscopic
examination of a forced beer sediment does not tell us all we
should know regarding its stability ; for a given number of
organisms in one beer will prove fatal, while in another which
is inherently more stable, we find that the same organisms in
apparently similar or greater number, will not be so. If,
however, we supplement the microscopic examination by
determination of acidity, and by tasting and smelling the
beer, we get some indication, not only of the degree of infection, but of the result of that infection during storage on the
practical scale. In the periods of forcing advised for the
various classes of beer, there should be no appreciable
increase in acidity, nor should the flavour be at all deteriorated, other than in respect of such deterioration which
must naturally follow from gas expulsion and warmness of
the sample.

The interpretation to be placed upon the microscopical examination of sediments, we have attempted to deal with in the analytical section (p. 508).

The comparison of original with specific gravity gives us the amount of attenuation at a glance; and in cases where the chemical determination of maltodextrins is impossible, we consider very valuable information can be learnt from the following easily acquired data :—Specific gravity of wort (original gravity), racking gravity, specific gravity at date of analysis (with due regard to period intervening between racking and examination), and specific gravity after forcing.

SECTION III.—ANALYSIS.

CHAPTER IX.

THE ANALYSIS OF WATER.

PRELIMINARY REMARKS.

IN the following directions the estimations are not given in the order of importance suggested in Chapter I.; they are now taken in the order in which they should be performed in the laboratory. In all cases those determinations should be first undertaken which the keeping of the water might tend to alter—for instance, the nitric and nitrous acids, and the ammonia, free and albuminoid; again, time is economised by at once commencing those determinations the results of which are not obtained until after the lapse of some days— for instance, the sugar test. Many tests are now suggested which, on account of their relative unimportance when separately considered, were not mentioned in Chapter I. Taken together, however, with the main and subsidiary tests there referred to, they are competent to afford some considerable information.

It is in all cases advisable to insist on samples being collected in glass stoppered vessels, previously thoroughly cleansed. "Winchester Quarts" answer the purpose admirably. Two of them will generally hold about sufficient for a full analysis. The vessels should, after efficient cleansing, be rinsed two or three times with the water to be analysed prior to being filled with it.

New supplies should be pumped for at least a week before analysis; if examined earlier the earthy and other impurities normal to a water in this condition are apt to induce an unfairly pessimistic opinion of the water.

In analysing a water for brewing purposes, we advise its being collected from the liquor tank. We have had experience of many really pure waters contaminated during the passage from the well to the tank, or in the tank itself. Under these circumstances it is better to test the water in its impure than in its pure state, for the impurity due to the uncleanliness of pipes and vessels will count as organic contamination in brewing purposes. In the event of doubt arising as to the source of impurity in such cases, the comparison of the organic analysis of the water as taken from the well or main direct, and from the tank, will give the requisite information.

The more knowledge in possession of the analyst regarding the history of the supply, its surroundings, the depth of well (if it be from a well), and the geological formation from which it is drawn, the more definite and useful will be his opinion.

THE EXAMINATION OF WATERS.

(1) Observe the colour and character of the water; i. e., note whether it is clear, turbid, or very turbid; whether it is bluish, yellowish, or colourless. Fill a cylinder of not less than 18 inches in height, place it over a white surface, and take observations by looking down through the water on to the white surface.

Pure waters (if not from newly made wells) are generally clear, and either colourless or bluish in tinge. Equally pure but hard waters (especially those containing much carbonate of calcium and magnesium) are frequently somewhat turbid, through a separation of these bodies due to expulsion of the gas holding them in solution, on pumping the well. Yellowish looking waters, especially if turbid, are suspicious;

at the same time the tinge may be due to harmless vegetable matter.

(2) The taste and smell of a water should be taken ; the smell after gently warming (say at 80° F.) in a closed vessel for ten minutes, and then opening the vessel suddenly. Pure water should be odourless ; the taste should be agreeable. At the same time mineral matters are apt to give a decided taste which must not be confused with such as are sufficiently polluted as to give any indication of this contamination to the palate.

(3) Take the reaction of the water by litmus paper. Most waters are neutral or slightly alkaline. In the majority of cases the alkalinity is due to the earthy mineral matters. It is sometimes due to alkaline carbonates characteristic of waters of Class III. In these cases the alkalinity is more marked after boiling the water in a test tube for a few minutes than in its natural state. The carbonic acid which in part neutralises the alkalinity is expelled on ebullition.

Acidity of water—especially if it be not destroyed on boiling—throws some suspicion on the water being polluted with industrial by-products, such as paper-mill refuse, &c.

(4) *Microscopic Examination.*—Matthews and Lott [*] recommend (and the recommendation is a good one) that the containing vessel should be well shaken, about half-a-pint of the sample be poured into a glass funnel large enough to receive the bulk, and closed at the lower end by an inch or two of india-rubber tubing terminating in a small tube closed at the lower end by fusion. After about six to twelve hours' settlement, the bulk of water may be siphoned off, and the little tube (containing the sediment) quickly detached. This is shaken up, a drop abstracted by a pipette, and the drop microscopically examined.

It is well to let all waters stand about the same time, and not to protract this period of standing. Stagnant waters

[*] Matthews and Lott : ' The Microscope in the Brewery,' Bemrose & Son.

propagate bacteria freely, and after some days more bacteria would be found in a water, than if examined say after twelve hours. About six to twelve hours should as a rule be allowed. The top of the funnel should be covered with a glass plate.

A magnification of about 350 diameters is desirable.

Considerable care is requisite to discriminate between minute particles of mineral and vegetable matter and bacteria. To dissolve the mineral matter (if such be present), a drop of dilute hydrochloric acid (decinormal acid) may be mixed with the sediment in the glass before examination. A drop of dilute potash (normal strength) may be similarly used to dissolve the vegetable matter. Both these reagents leave the organisms unaffected. The potash and acid must not, however, be used together, for they would in part neutralise one another. Three drops of the sediment should be examined, the first untreated, the second mixed with acid, the third with potash. A comparison of these will give decisive information.

To detect dead vegetable or other protoplasmic matter, the drop of sediment may be mixed before examination with a drop of an aqueous solution of eosin. The eosin will stain the dead matter pink, leaving the other matters unaffected.

(5) *The Sugar Test.*—Two stoppered bottles of about 200c.c. capacity are thoroughly cleansed by rinsing with nitric acid, and then washed free from acid by distilled water. Into one of the bottles is poured some of the sample, so as to nearly fill it. To this is added a teaspoonful of pure crystallised cane-sugar and the stopper replaced. The bottle is then agitated until the sugar dissolves. It is labelled "unboiled." and the date and hour of the experiment noted. It is now put on the forcing tray, and observed every few hours during the daytime for seventy-two hours, until it becomes turbid. The time when the turbidity first makes its appearance is noted.

Another portion of the sample is boiled in a flask for

thirty minutes, a plug of cotton wool being inserted at the end of this period, to exclude outside contamination while the water is cooling from the boiling point to about 80° F. When it has cooled to about this point it is quickly filtered into the second bottle; the same quantity of sugar as previously mentioned is added, and after being labelled as before (but with the word "boiled" in place of "unboiled") it is placed on the forcing tray and observations taken. The exact time when turbidity makes its appearance is noted. The observation, as before, to extend only up to seventy-two hours.

(6) *Ammonia:* (a) *free* [and *saline*]; (b) *albuminoid.*—500 c.c. of the water are boiled with 10 c.c. of a saturated solution of sodium carbonate in a retort connected with a condenser. The distillates are received in white glass cylinders, capable of holding about 100 c.c., and marked to 50 c.c.

To the first distillate of 50 c.c. add 1 c.c. of Nessler's solution. The colour obtained after standing for three minutes is equalised by adding a definite volume of standard ammonium chloride solution to 50 c.c. of distilled water, and coloured by 1 c.c. of Nessler; the Nessler to be added after thorough intermixture of the standard ammonium chloride solution and the 50 c.c. of water. The standard solution of ammonium chloride contains 0·01 milligram NH_3 per c.c.

A second distillate of 50 c.c. is matched in a similar way, and the process continued until the distillate gives no reaction with Nessler's solution.

When this is the case, the flame is removed, the retort stopper opened, and 50 c.c. of alkaline potassium permanganate solution introduced by a pipette, the stopper replaced, and ebullition recommenced.

The distillation and colourisation estimation of the ammonia thus expelled is proceeded with in the same way as before.

In matching the colour, not only the distillates, but the mixture of distilled water and ammonium chloride should remain in contact with Nessler solution for 3–4 minutes before the analyst decides as to the similarity or dissimilarity of the colours. If dissimilar, another 50 c.c. of distilled water with a greater or less amount of ammonium chloride solution, must be experimented upon. It is not possible to get accurate results by increasing the ammonium chloride, after once adding the Nessler.

For full information regarding this process, the reader should consult the treatise on water analysis by Wanklyn and Chapman (Trübner & Co.).

Example.—500 cc. of water taken:—

Free (and saline) ammonia	{	1st distillate equalled by 3 c.c. AmCl solution
		2nd ,, ,, 1 c.c. ,,
		3rd ,, — no ammonia.
Albuminoid ammonia ..	{	1st distillate equalled by 7 c.c. AmCl solution
		2nd ,, ,, 2 c.c. ,,
		3rd ,, ,, 1 c.c. ,,
		4th ,, ,, 0 c.c. ,,

Calculation.—The total free ammonia is thus equalled by 4 c.c. of AmCl solution. Each c.c. contains 0·01 mgm. ammonia; hence, in 500 c.c. water, we have 4 × ·01 = 0·04 mgm. ammonia.

A litre of the sample will thus contain 0·08 mgm. ammonia. Milligrams per litre are equivalent to parts per million. If therefore, we state our result in this way, the sample will contain 0·08 parts per million of free (and saline) ammonia. To convert this into grains per gallon, multiply by 0·07; this particular sample would therefore contain 0·0056 grain of ammonia per gallon.

The albuminoid ammonia is calculated quite similarly. Adding up the total number of c.c.'s of ammonium chloride used to equalise the distillates in the example before us, we find that we have used altogether 10 c.c.

Five hundred c.c. of our sample contain,[*] therefore, 0·10

---

[*] The expression "contains" so and so much albuminoid ammonia is incorrect, and for the reason given in Chapter I. (p. 37). It is used here, however, for convenience, and to avoid a longer phrase.

mgm. ammonia, a litre will therefore contain 0·20 mgm. Hence the sample contains 0·20 parts per million of albuminoid ammonia, equivalent to 0·014 grains per gallon.

In performing the above tests, it is clear that our utensils and reagents should be free from ammonia. The distilled water, sodium carbonate solution, and alkaline potassium permanganate solution, should all be thoroughly tested by distilling small quantities from each, and testing the distillates for ammonia by Nessler. If the faintest colouration is produced, they must be purified until quite free from ammonia.

The apparatus must be tested by boiling in it some distilled water previously found pure by experiment. Until the distillates are quite free from ammonia the apparatus cannot be used, and it should be purified by boiling in it pure water and alkaline potassium permanganate until the distillates indicate complete purity. All contact between the evolved steam and india-rubber connections should be avoided, and when out of use, the interior of the vessels (condenser, &c.) must be protected from dust, which is always more or less ammoniacal.

(7) *The Oxygen Test.*—Two quantities of 200 c.c. each of the water sample are poured into a clean stoppered bottle (capable of holding 300 c.c.), and to each is added 10 c.c. of standard solution of potassium permanganate (containing 0·1 mgm. of available oxygen per 1 c.c.). Immediately after the addition of the permanganate, add to each, 10 c.c. of standard sulphuric acid solution (5 per cent. acid). The bottles are now put aside at about 80° F. (either on the forcing tray or in a warm cupboard). At the end of fifteen minutes the first bottle is taken and five drops of a standard solution of potassium iodide are added (10 parts of potassium iodide in two parts water). To this is added about 1 c.c. of clear starch-paste; the bottle (the contents of which will be blue) is placed under a burette containing standard

solution of sodium thiosulphate (1 gram of the crystallised salt per litre of water), and the standard solution run in, with continual shaking of the bottle, until the blue colour just disappears. The number of c.c.'s of thiosulphate solution required is carefully noted.

A blank experiment is now made in order to check the thiosulphate solution, which is unstable. To do this, take 200 c.c. of distilled water and proceed just as before, without, of course, putting the distilled water and permaganate aside to stand.

At the expiration of the four hours the second bottle put aside on the tray, or in the warm cupboard, is taken out and treated in precisely the same manner as that given, a blank experiment to check the thiosulphate being performed at the same time.

Example.—Bottle No. 1, taken at the end of fifteen minutes, 24 c.c. of thiosulphate solution used ; for blank experiment 25 c.c. In this case, therefore, 200 c.c. of sample absorbed an amount of available oxygen equivalent to that absorbable by 1 c.c. of sodium thiosulphate. Now 25 c.c. of sodium thiosulphate would absorb the *whole* of the available oxygen in the permanganate used, hence the ferrous and nitrous bodies (see p. 38) have absorbed one-twenty-fifth of this amount. Now the 10 c.c. of permanganate contained $1\cdot0$ mgm. of available oxygen, hence the quantity absorbed by 200 c.c. of the water was $\frac{1}{25}$ mgm. = $0\cdot04$ mgm. A litre of water would therefore absorb $0\cdot2$ mgm., or in other words, the water would absorb (through ferrous salts and nitrites), $0\cdot2$ part per million, or $0\cdot014$ grain per gallon.

Bottle No. 2 required 20 c.c. of sodium thiosulphate, the blank as before requiring 25 c.c. The water therefore abstracts as much oxygen as would 5 c.c. of the sodium thiosulphate solution. As 25 c.c. abstract the oxygen from 10 c.c. permanganate solution, the water abstracts $\frac{5}{25} = \frac{1}{5}$ of this amount. This amount, as before explained, equals $1\cdot0$ mgm. Hence

200 c.c. require $\frac{1}{5} \times 1\cdot0 = 0\cdot2$ mgm. oxygen, or a litre, 1 mgm ; or a gallon, 0·07 grain. Hence, stated in grains per gallon, our result would be as under :—

Oxygen absorbed by organic matter, ferrous salts and nitrites   0·07   grain
Oxygen absorbed by ferrous salts and nitrites only ..   ..   .. 0·014   ,,

Hence oxygen absorbed by organic matter   ..   ..   .. 0·056   ,,

It is very frequently sufficient to determine the oxygen absorbed in three hours without making any allowance for that absorbed by ferrous salts and nitrites. The test should only be regarded in conjunction with the other tests ; and the refinement of allowing for the oxygen due to inorganic constituents is not therefore necessary, unless the total amount absorbed is sufficiently excessive to call for explanation. When the total amount absorbed is small or moderate, there is no need for estimating that abstracted by these inorganic constituents.

(8) *Nitrous and Nitric Acids.*—These acids are generally estimated together; for we have no evidence before us to support the supposition that the effect of nitrites is different from that of nitrates in brewing operations. It is possible, however, to estimate the nitrites separately, as well as together with the nitrates, from which two estimations the exact amount of nitric acid and nitrous acid is obtained. Although not of frequent necessity, the method for separately estimating the nitrites is given below.

The nitrites and nitrates are estimated by reducing them to ammonia and determining the ammonia so produced. This process—due to Gladstone and Tribe—although giving as a rule results slightly below the truth, is quite sufficiently accurate for all practical purposes.

Evaporate 250 c.c. of the sample with two or three drops of pure potash solution in an evaporating basin, until about nine-tenths of the liquid have been expelled. During evaporation, stir frequently, lest the solids separate out on the sides of the vessel; when sufficient has been expelled by evaporation, transfer the contents of the basin to a test tube, using ammonia-free distilled water for rinsing purposes.

During the evaporation, proceed as follows :—

Into a conical flask, which fits on to a condenser, place some pure zinc clippings, and pour on to them some copper sulphate. Place the flask on the forcing tray. After 20 minutes, pour off the supernatant blue fluid, carefully taking care not to disturb the surface of metallic copper precipitated on to the zinc. Wash the metal " couple " two or three times with warm distilled and ammonia-free water.

Now wash the contents of the test tube (containing the residue of the evaporated water) into the flask containing the metal with ammonia-free distilled water, and fit the flask on to the condenser. The receiver should be a flask graduated at 100 c.c., and containing a drop of hydrochloric acid.

The flask is kept at a gentle heat for $1\frac{1}{2}$ hours, at the expiration of which about 100 c.c. of ammonia-free distilled water, and six drops of potassium hydrate solution are added to the flask, through a stop-cocked funnel which is provided for this purpose. Proceed to raise the temperature of the flask-contents until distillation commences, and then distil over rather under 100 c.c. Now remove the receiver, make up exactly to 100 c.c. with ammonia free-distilled water, shake it well, and abstract 10 c.c. with a pipette. The 10 c.c. is put into a Nessler jar, diluted to 50 c.c., and Nesslerised in precisely the same manner as that described in estimating the free and albuminoid ammonia.

If the 10 c.c. abstracted gives too deep a tint discard it, and abstract a less amount from the receiver; if the tint is insufficiently distinct, operate on a larger bulk abstracted from the receiver.

Example.—250 c.c. of the sample taken; 10 c.c. of distillate gave same colour with Nessler as that obtained by taking

4 c.c. of standard ammonium chloride solution (0·01 mgm. ammonia)—see determination of ammonia, p. 426.

Hence 10 c.c. of distillate contain 0·04 mgm. NH_3. Hence 100 c.c. of distillate (the whole distillate) will contain 0·4 mgm. NH_3.

We will now calculate this back into nitric acid, preferably taking the anhydrous form of the acid (N_2O_5) for this purpose. To convert ammonia into nitric acid, we must multiply by the molecular weight of N_2O_5, and divide by twice that of NH_3. Now $\frac{108}{34}$ is practically equal to 3·18. If, therefore, we multiply 0·4 by 3·18, we get the milligrams of N_2O_5 in 100 c.c. of distillate, or in other words in 250 c.c. of sample ; 0·4 × 3·18 = 1·272.

Now if

250 c.c. of sample contain	1·272 mgm. N_2O_5	
1 litre ,, will contain	5·088 ,,	,,

or

1 gallon ,, ,, ...	5·088 × 0·07 = 0·35 grain, N_2O_5.	

[It simplifies calculation to remember that if in operating with the above quantities we abstract 10 c.c. from the distillate, and if this is matched by 10 c.c. of ammonium chloride solution, the amount of N_2O_5 is 0·89 grain per gallon in the sample; thus if we apply this in the above case, the amount of N_2O_5 in grains per gallon will be 0·89 × $\frac{4}{10}$ = 0·35, as previously found.]

If it be desired to calculate the whole of the ammonia into N_2O_3 (anhydrous nitrous acid), we have to use the factor 2·23 instead of 3·18 ; 2·23 representing $\frac{N_2O_3}{2NH_3}$.

In estimating nitrites (Griess' method) separately, we proceed as under: 100 c.c. of the sample are poured into a glass cylinder of such dimensions that the above quantity of water will rise to a height of 16 to 18 inches. To this is added 1 c.c. of sulphuric acid (1 part acid to two parts water), and immediately afterwards 1 c.c. of metaphenylenediamine solution (5 grams of metaphenylenediamine per litre, decolourised if necessary by charcoal, and acidified with sulphuric acid).

The pink colour developed when the water contains nitrites is now matched by taking a definite number of c.c.'s of a standard solution of potassium nitrite (containing 0·01 mgm. N_2O_3 per c.c.), diluting to 100 c.c., and adding sulphuric acid and metaphenylenediamine as before.

Example.—100 c.c. of water give a tint matched by taking 2 c.c. of the standard potassium nitrite solution treated as above. 100 c.c. of water will therefore contain 0·02 mgm. of N_2O_3. A litre will therefore contain 0·2 mgm. N_2O_3, and a gallon, 0·014 grain of N_2O_3.

Now let us suppose that our sample contains, according to the previous example, 0·35 of N_2O_3 and N_2O_5, both calculated as N_2O_5. We now know that 0·014 N_2O_3 will be included in that amount calculated as N_2O_5. To see how much N_2O_5, 0·014 N_2O_3 would correspond to, we must multiply it by $\frac{108}{76} = 1·42$. Therefore the 0·014 N_2O_3 would correspond to 0·074 × 1·42 = 0·019 (say 0·02) N_2O_5.

Hence the true N_2O_5 in our water would be 0·35 less 0·02, equal to 0·33; the exact result, therefore, being

N_2O_5 .. .. 0·33 grain per gallon
N_2O_3 .. .. 0·014 ,,

In giving the method for estimating nitrites we have omitted many necessary precautions to ensure accuracy, but those who wish to perform the determination will find full particulars in Frankland's treatise on 'Water Analysis' (p. 40).

(9) *Chlorine.*—250 c.c. of the sample are poured into a white evaporating basin, and five drops potassium chromate

solution (10 per cent. solution) added.   Standard silver nitrate (1 c.c. of standard solution corresponds to 0·001 gram Cl) is now run in, until the orange tint becomes just permanent. The liquid must be vigorously stirred during the progress of the test.

Example.—250 c.c. of the sample required 14·2 c.c. of standard silver nitrate.   Hence 250 c.c. contain 14·2 × 0·001 or 0·142 gram Cl; a litre will therefore contain 0·142 gram Cl and a gallon . . . 0·142 × 70 = 9·94 grains Cl. [When a water takes more than 25 c.c. of silver solution it is well to operate on 100 c.c. of the water, and so on according to circumstances.]

(10) *Total solid matter.*—A weighed platinum basin is placed over a suitable aperture in a water-bath.   500 c.c. of the water are measured and the basin filled to about two-thirds its capacity.   As the sample evaporates the basin is replenished from the 500 c.c. measure, until the whole of that bulk is evaporated ; a little distilled water is used to rinse out the last portion.   The outside of the dish is then carefully wiped and placed in the air-bath at 261° F.   This temperature is selected as sufficing to expel the water of crystallisation from the mineral compounds, while it is not sufficiently high to decompose the organic matter.   The dish is kept at this temperature for 30 minutes, and then weighed.   It is again dried at the same temperature for 15 minutes, and reweighed.   As a rule, the two weights will agree ; if not, drying at 261° must be continued until the weights are constant.

The final weight, less the weight of the platinum capsule, gives the weight of solid matters in 500 c.c. of the water.

Example :—

Weight of dish + solids	35·342 grams
Weight of dish alone	35·078 „
Weight of solids	0·264 „

If 500 c.c. of the sample contain 0·264, a litre will contain 0·528 gram, or a gallon will contain 35·96 grains (0·528 × 70).

The solids thus obtained will include the organic matter. In impure waters the weight of the solid matter will thus exceed that of the bases and acids combined, sometimes by 3 or 4 grains. In pure waters, however, the agreement is generally within 1·0 to 1·5 grain, and in such cases, this approximate identity is a sufficient proof of the accuracy of the determinations of acids and bases. When, however, the water is impure the above check is no longer of value, and in such cases it is well to adopt Heron's suggestion,* which is briefly this. The solid matter obtained as above is digested for a short time with an excess of dilute sulphuric acid, and then evaporated, dried, and ignited until the weight is constant. By this means the organic matter is destroyed, while the bases are all converted into sulphates. The bases, as separately determined, are now all *calculated* into sulphates, and their sum should be in close agreement with the solid matter after treatment with sulphuric acid.

This plan is a very excellent one for the checking of results when a water is impure; in fact it is the only check. For if we were to drive off the organic matter by strong ignition, and thus attempt to compare the mineral solids with the sum of combined acids and bases separately determined, no comparison would hold. In expelling the organic matter, we should also expel the carbonic acid from the earthy carbonates, the nitric acid, and a portion of the chlorine from the chlorides. Although the carbonic acid so expelled may be restored by digesting the ignited solids with ammonium carbonate, there is no ready means of restoring the loss due to the volatilisation of nitric acid and chlorine; hence the weight of the solids after ignition affords no guide to the weight of the combined saline matter.

* Transactions of the Laboratory Club, vol. ii., p. 22.

At the same time the solid matter obtained according to the previous instructions should always be subjected to incineration, with the object of detecting the degree of dis-colouration produced, and the nature of the fumes evolved. The interpretation to place upon these indications has already been considered; the process itself hardly requires description. The basin containing the solids, of course after the weight has been duly recorded, is placed over the flame of a Bunsen burner, in such a way that one part after another of the solids is subjected to the heat of the flame. The degree of discolouration, its transientness or persistency, and the nature of the fumes, are then recorded.

The following will serve as an example of a water, the organic impurity of which causes the solid matter dried at 261° to considerably exceed the sum of combined acids and bases as separately determined. Heron's check would there-fore be applied as follows :—

Total solid matter	36·21	grains per gallon.

Saline bodies :—

Sodium chloride..	7·20	grains per gallon.
Calcium nitrate	6·37	,,
Calcium sulphate	2·30	,,
Calcium carbonate	12·57	,,
Magnesium carbonate	4·20	,,
Silica and alumina	0·20	,,
	32·84	,,

There is therefore a difference here of between 3 and 4 grains. To test whether this is due to inaccuracy in analysis or to organic matter contained in the solids, we digest our solids with sulphuric acid, re-dry, and find the weight to be 39·81 grains.

We now proceed to calculate the bases into sulphates.

Sodium chloride ..	7·20	equivalent to sodium sulphate	.. 8·71	$\left(\dfrac{Na_2SO_4}{2NaCl} = 1·21\right)$
Calcium nitrate ..	6·37	equivalent to calcium sulphate	.. 5·29	$\left(\dfrac{CaSO_4}{CaN_2O_6} = 0·83\right)$
Calcium sulphate ..	2·30	equivalent to calcium sulphate	.. 2·30	$\left(\dfrac{CaSO_4}{CaSO_4} = 1\right)$
Calcium carbonate ..	12·57	equivalent to calcium sulphate	.. 17·09	$\left(\dfrac{CaSO_4}{CaCO_3} = 1·36\right)$
Magnesium carbonate	4·20	equivalent to magnesium sulphate	6·80	$\left(\dfrac{CaSO_4}{MgCO_3} = 1·62\right)$
Silica and alumina ..	0·20[*]		.. 0·20	

Total bases as sulphates .. .. 40·39

Total solids after treatment with sulphuric acid .. .. 39·81

The agreement between these figures is sufficient to show the accuracy of the analytical results, and to indicate that such disagreement as there was between the total solids, prepared in the ordinary way, and the sum of the saline compounds, was due to organic matter contained in these solid matters.

(11) *Determination of Sulphuric Acid.*—500 c.c. of the sample are placed in a porcelain evaporating basin, and about 12 drops of hydrochloric acid added. The evaporation is continued until about two-thirds of the water are expelled. Barium chloride is added, the whole is digested for an hour, and the precipitated barium sulphate is thrown onto a filter. The precipitate is washed, dried, ignited, and weighed in the usual manner.

In all cases where water-bulks are evaporated, care should be taken to stir the solids which separate out on the sides of the basin into the residual liquid. This can be sufficiently well done by means of an ordinary stirring rod applied frequently.

[*] The quantity of silica and alumina being small, no calculation becomes necessary.

Example :—

Weight of crucible + BaSO₄ + filter ash	..	..	18·231 grams
,, crucible	..	..	18·078 ,,

,,	BaSO₄ + filter ash	..	..	0·153 ,,
,,	filter ash	..	..	0·0015 ,,

,,	BaSO₄	..	..	0·1515 ,,

The $BaSO_4$ is preferably calculated into anhydrous sulphuric acid (SO_3), as this form is the most convenient for subsequently combining the acids and bases. To convert $BaSO_4$ into SO_3, the factor 0·34 is sufficiently accurate.

Hence 500 c.c. of the sample will contain 0·1515 × 0·34 SO_3 = 0·052 gram SO_3 ; a litre will therefore contain 0·104 gram, and a gallon, 0·104 × 70 = 7·28 grains.

(12) *Determination of Silica, Alumina, Iron, Lime, and Magnesia.*—500 c.c. of the water are poured into an evaporating basin, made distinctly acid with hydrochloric acid, and evaporated to complete dryness. The residue is allowed to cool, then 40 c.c. of distilled water, 20 c.c. of hydrochloric acid, and a few drops of nitric acid are added, and the whole digested for 30 minutes. The *silica* will remain insoluble (with any organic matter present, which, however, will be expelled during the ignition of the precipitate), the other bases will go into solution. The whole is therefore filtered, and the silica weighed as such.

The filtrate is now rendered strongly ammoniacal and boiled. The ammonia will throw down the iron and aluminium as hydrates. These are collected on a filter paper and weighed as Fe_2O_3 and Al_2O_3. To determine the quantity of iron, the precipitate is extracted with HCl and water, the insoluble porton of the filter ash removed by filtration, and the filtrate made up in a Nessler jar to 50 c.c. One c.c. of standard potassium ferrocyanide solution is added, and any colour so produced is matched by the required number of c.c. of a standard solution of ferric chloride (1 c.c. = ·001 Fe) diluted to 50 c.c., and tinted with 1 c.c. of the standard potassium ferrocyanide solution. The number of c.c. of ferric chloride solution required to give an equal tint will show the quantity of iron present, and hence by calculation, the amount of alumina. The filtrate from the iron and alumina precipitation is, if necessary, evaporated to a convenient bulk, and then heated with an excess of ammonium oxalate and some ammonia. The calcium oxalate is filtered off, dried, ignited in a platinum crucible to calcium oxide, and weighed as such.

The filtrate from the above is, if necessary, evaporated to a convenient bulk, ammonium and sodium phosphate being added. The ammonio-magnesium phosphate is thrown on to a filter, dried, ignited, and weighed as magnesium pyrophosphate.

When a water contains an excess of magnesia there is a tendency for the magnesia to come down with the lime. In such cases, the precipitate of calcium oxalate is redissolved in a very small quantity of hydrochloric acid, the solution is then rendered strongly ammoniacal, and ammonia oxalate again added. By this repetition of the precipitation, the lime is thrown down free from magnesia.

Example :—

SILICA.

Weight of crucible + SiO_2 + ash	14·323 grams
,, crucible	14·319 ,,
,, SiO_2 + ash	0·004 ,,
,, ash	0·0015 ,,
,, SiO_2	0·0025 ,,

500 c.c. therefore contain 0·0025 gram SiO_2, a litre will therefore contain 0·005, and a gallon 0·005 × 70 = 0·35 grain SiO_2.

ALUMINA AND FERRIC OXIDE.

Weight of crucible + Al_2O_3 + Fe_2O_3 + ash .. .. 19·423 grams

,,   crucible .. .. 19·417 ,,

,, Al_2O_3 + Fe_2O_3 + ash .. .. 0·006 ,,

,,   ash .. .. 0·0015 ,,

,, Al_2O_3 + Fe_2O_3 .. .. 0·0045 ,,

If 500 c.c. contain 0·0045 gram, a litre will contain 0·009, and a gallon 0·63 grain of alumina and oxide of iron. The precipitate on re-solution was matched by 1 c.c. of standard ferric chloride solution. Hence 500 c.c. of the water contain 0·001 grain Fe, or 0·0014 Fe_2O_3. One litre will therefore contain 0·0028 grain ferric oxide, and a gallon 0·19 grain.

Therefore, alumina and oxide of iron being 0·63 grain, and oxide of iron 0·19 grain, the alumina is 0·44 grain per gallon.

LIME.

Weight of crucible + lime (CaO) + ash .. .. 25·728 grams

,,   crucible .. .. 25·631 ,,

,,   CaO + ash .. .. 0·097 ,,

,,   ash .. .. 0·0015 ,,

,,   CaO .. .. 0·0955 ,,

500 c.c. containing 0·0955 gram CaO, a litre will contain 0·191 gram, and a gallon will therefore contain 13·37 grains.

MAGNESIA.

Weight of crucible + magnesium pyrophosphate

$(Mg_2P_2O_7)$ + ash .. .. 18·438 grams

,,   crucible.. .. .. 18·420 ,,

,, $Mg_2P_2O_7$ + ash .. .. 0·018 ,,

,,   ash .. .. 0·0015 ,,

,, $Mg_2P_2O_7$ .. .. 0·0165 ,,

To calculate magnesium pyrophosphate into its equivalent of magnesia (MgO) the fraction 0·36 must be used. 500 c.c. of the sample will therefore contain 0·0165 × 0·36 = 0·0059 gram MgO. A litre will therefore contain 0·0118 gram, and a gallon 0·83 grain magnesia.

(13) *Determination of Soda and Potash.*—Take 500 c.c. of the sample, evaporate with six drops of HCl in a porcelain dish, until about nine-tenths of the liquid are expelled. Turn down the flame until the liquid merely digests, and add barium hydrate until the reaction is just alkaline. Ten minutes afterwards, add ammonium carbonate, and digest at a gentle heat for one hour, and then filter.

The insoluble matter is thoroughly and repeatedly washed with hot water, after which the precipitate may be neglected. The filtrate and washings are evaporated down to dryness in a platinum basin, with ammonium chloride solution. When all is evaporated, it is gently, then strongly ignited until all ammoniacal fumes are expelled. The dish is now weighed.

The ignited residue is now extracted with hot water, the washings being passed through a filter; the filter paper so used is returned to the dish, the whole re-ignited, and then weighed.

The difference between this and the former weight is due to the sodium and potassium chlorides which were extracted by washing. The washings containing the sodium and potassium chlorides are now treated with platinum tetrachloride, and the precipitated platino-potassium chloride dried at 212° until constant, and weighed.

Having obtained the quantity of potassium chloride, and knowing the sum of the potassium and sodium chlorides, we can thus get the weight of the sodium chloride.

The above process is one which, after some experience, gives very excellent and reliable results. It does, however, require experience, especially in that portion of it consisting of the first ignition. If we ignite insufficiently, we fail to expel all the ammonium chloride, which will thus swell the apparent amount of alkaline chlorides, since it will be washed out with them, and will subsequently come down with platinum chloride and so swell the apparent amount of potassuim. If, on the other hand, we ignite too forcibly, some of the alkaline chlorides are decomposed, and our result comes out too low. Again, if the heat is applied too strongly at first, the alkalies fuse and, so to say, lock up some of the ammonium chloride, which thus avoids expulsion. The process is one, however, which can be thoroughly trusted, after the manipulator has learnt how to work it. Many chemists only get at their soda indirectly, by means of various titrated estimations of combined and free carbonic acid, &c. But these processes we regard as unreliable, and greatly inferior to the direct determination of soda in the above manner ; and when obtained in this direct way, the determination of free and combined carbonic acid can be neglected, and this is a point to the good. These matters will be better understood when we consider the combination of acids and bases.

Example :—

Weight of basin and solids before extracting with water					38·214 grams
,, ,, after ,, ,,					38·102 ,,
Loss due to NaCl and KCl (uncorrected)	..	..	..		0·112 ,,
Allowance for returned filter ash	..	..	..	..	0·0015 ,,
NaCl and KCl (corrected)	..	..	..	..	0·1135 ,,

If 500 c.c. contain 0·1135 gram KCl and NaCl, a litre will contain 0·227 gram, and a gallon will contain 15·89 grains.

The solution of potassium and sodium chlorides on treatment with $PtCl_4$ gave 0·121 gram platino-potassium chloride. The factor for converting this salt into potassium chloride is 0·305. Hence, 500 c.c. of water will contain 0·305 × 0·121 = 0·037 gram KCl ; a litre will therefore contain 0·074 gram, and a gallon 5·18 grains KCl.

Now

$$NaCl + KCl = 15\cdot89$$
$$KCl = 5\cdot18$$

$$NaCl \quad .. \quad .. \quad 10\cdot71 \text{ grains per gallon}$$

It is usual, however, to return the sodium and potassium as oxides, that is as soda (Na_2O) and potash (K_2O). To calculate these we proceed as follows :—

To convert $NaCl$ into Na_2O multiply by $0\cdot53$
,, KCl into K_2O multiply by $0\cdot63$

Hence, in this case, our water will contain

Soda (Na_2O) $10\cdot71 \times 0\cdot53 = 5\cdot67$ grains per gallon
Potash (K_2O) $5\cdot18 \times 0\cdot63 = 3\cdot26$,, ,,

(14) *Phosphoric Acid.*—It is not usual to more than quite approximately estimate the always rather small amount of this acid. It is customary to evaporate 500 c.c. of the water, acidified with nitric acid, and with the addition of ammonium molybdate, until about one-half of the water is expelled. The amount of yellow precipitate is then judged and returned as a "minute trace," "trace," or "heavy trace," according to circumstances. If, however, the amount of phosphoric acid should, for some special object, be required, the yellow precipitate is dissolved in ammonia, and the phosphoric acid precipitated with magnesia. Full particulars of this determination will be found in all the text-books on quantitative analysis. In analysing waters for brewers' purposes, the quantitative estimation is of too infrequent occurrence to require description in these pages.

(15) *Hardness.*—Having a profound mistrust in the reliability of the hardness test by soap, we prefer to omit a description of the test entirely. It could have no object, even if accurate, when the bases and acids are separately determined, as should always be the case in the analysis of water for brewers' purposes. But a considerable experience of it has

convinced us of its entire unreliableness, and having regard to this and its uselessness for brewing purposes, we prefer not to include it among the determinations described. If for any special object it may seem desirable, full descriptions of the test can be found in all text-books dealing with water analysis.

THE COMBINATION OF ACIDS AND BASES.

The scheme proposed for the combining of acids and bases is one which we consider to best fit in with such actual data as chemists have at their disposal on this point. It is, of course, not intended to be final; at the same time we regard it as in all probability correct, and that is about as much as can be claimed for any scheme of the kind. It has frequently been suggested that the respective affinities which we know to exist under ordinary conditions between the various bases and acids, may be modified by the exceeding diluteness of the solution. This may or may not be so; but even if it be so, we must not forget that the water is heated to at least 155° before mashing, and that the heating of the solution would restore these affinities to what we may term their normal. If, therefore, the water in the well does not hold the bases and acids in such combinations as are here suggested as probable, it is certainly very probable that the liquor as it it used just before mashing, will contain them combined in the manner suggested; and as the effect of the saline constituents in brewing will depend upon the form they assume in the hot-liquor back or mash-tun, rather than on that prevailing in the supply as it lies in the well or reservoir, it is clear that the modes of combination apply rather to the former than the latter state, and that they on that ground give the more practically useful information.

The scheme of combination now proposed is as follows :—

A. The *chlorine* is first combined with the *sodium.*

(1) CHLORINE in excess: combine it with *calcium ;* if

still in excess, combine it with *magnesium*; if still in excess, with any excess of *potassium* over the potassium required for saturating sulphuric acid.

(2) SODIUM in excess: combine it with any *sulphuric acid* left over after saturating potassium; if still in excess, with *nitric acid;* if still in excess with *carbonic acid.*

B. If *calcium* is in excess after combination with chlorine (or if there be no excess chlorine for combination with calcium) combine it with *nitric acid ;* if still in excess, with *sulphuric acid ;* and if still in excess, with *carbonic acid.*

C. If *magnesium* is in excess after combining with chlorine, or if there be no excess chlorine for combining with magnesium), combine it first with *nitric acid,* and then if still in excess, with *sulphuric acid,* and if still in excess, with *carbonic acid.*

D. If *potassium* is in excess after combining first with sulphuric acid, then combine with chlorine, then with *nitric acid,* and finally with carbonic acid.

E. *Sulphuric acid* is first to go to potassium, any excess to sodium, then to calcium, and finally to magnesium. Any excess of potassium, sodium, calcium, and magnesium occurring here would be treated according to the rules given under B, C, and D.

F. Any *nitric acid* in excess of sodium will be combined first with calcium, then with magnesium, and then with potassium. Any excess of calcium, magnesium, and potassium occurring here will be treated according to the rules given under B, C, and D.

G. Any sodium, potassium, lime or magnesia uncombined are calculated into the respective carbonates.

H. The iron is returned as ferric oxide, and its quantity as such returned together with silica and alumina.

EXAMPLES.[*]

Example I.—The water contained—

Silica (SiO$_2$)	0·12 grains per gallon
Iron oxide (Fe$_2$O$_3$)	nil
Alumina (Al$_2$O$_3$)	0·43 ,, ,,
Lime (CaO)	27·81 ,, ,,
Magnesia (MgO)	4·12 ,, ,,
Soda (Na$_2$O)	3·16 ,, ,,
Potash (K$_2$O)	0·53 ,, ,,
Chlorine (Cl)	4·01 ,, ,,
Nitric acid (N$_2$O$_5$)	1·82 ,, ,,
Sulphuric acid (SO$_3$)	8·31 ,, ,,

The first thing to be done is to combine the sodium with chloride. Now 3·16 grains of Na$_2$O are equivalent to 2·34 grains Na ($3·16 + \frac{46}{62} = 2·34$). Working now by the equation Na + Cl = NaCl, we find that 2·34 grains of Na
23   35·5   58·5
will combine with $2·34 + \frac{35·5}{23} = 3·61$ grains of Cl, making a total of 5·95 NaCl. This then is the first combination.

We have 4·01 Cl to start with, and we have used up 3·61 grains of Cl for the NaCl. There remains, therefore, a balance of 0·40 to be dealt with. This must now be combined with calcium. Working on the following equation, Cl$_2$ + Ca = CaCl$_2$, we find that 0·40 Cl will combine with
71   40   111
$0·40 \times \frac{40}{71} = 0·23$ Ca. The total CaCl$_2$ will therefore be 0·63. CaCl$_2$ = 0·63 is our second combination.

We turn now to the potash, and convert it into potassium

[*] These examples are purely hypothetical, but they are fairly typical of the waters generally met with, and will serve well enough as examples. The waters given as types of brewing liquids in the chapters on brewing waters (pp. 3–10) will also serve as examples, the separate bases and acids being first stated and the combinations in which they would probably exist afterwards.

sulphate. Working on the following equation, $K_2O + SO_3 = 94 + 80$ K_2SO_4, we find that the K_2O at our disposal (0·53 grains) 174

will combine with $\frac{80}{94} \times 0·53$ $SO_3 = 0·45$ grain SO_3. Hence $K_2SO_4 = 0·45 + 0·53 = 0·98$. This is our third combination.

Our total SO_3 is 8·31 ; we have used 0·45 SO_3 for our K_2SO_4. There is therefore a balance of 7·86 SO_3 to be dealt with. This we convert into calcium sulphate. Working on the equation $SO_3 + CaO = CaSO_4$ we find that 7·86 SO_3 80   56   136

will combine with $\frac{56}{80} \times 7·86$ of $CaO = 5·50$ CaO. The $CaSO_4$ therefore equals $7·86 + 5·50 = 13·36$ grains. This is our fourth combination.

We now turn to the nitric acid (of which we have 1·82 grains) and convert this into calcium nitrate. Working on the equation $N_2O_5 + CaO = Ca(NO_3)_2$ we find that 1·82 108   56   164

grains of N_2O_5 will combine with $1·82 \times \frac{56}{108}$ grains CaO $= 0·94$ CaO. This $Ca(NO_3)_2$ will therefore be 2·76 grains (0·94 + 1·82). This is the fifth determination.

We have now used up all our acids, and it becomes necessary to combine the remaining bases into carbonates. Turning first to the lime. This has partly been combined with chlorine, partly with nitric acid, and partly with sulphuric acid. That used up as chloride was used in the form of *calcium*, and to find the equivalent of lime, calcium oxide (CaO) used, we must multiply the calcium used by $\frac{56}{40}$ ($\frac{CaO}{Ca}$). Now, 0·23 grain of Ca was used up for the chlorine. This quantity multiplied as above gives us 0·32 as the equivalent of CaO.

Now we know

CaO used for calcium chloride..	0·32 grains
CaO used for calcium sulphate	5·50 ,,
CaO used for calcium nitrate	0·94 ,,
Total CaO used	6·76 ,,

The total lime, however, is 27·81; we have therefore a balance of 21·05 grains to deal with. This is converted into carbonate by multiplying by the molecular weight of calcium carbonate ($CaCO_3 = 100$) and dividing by that of calcium oxide ($CaO = 56$). $21·05 \times \dfrac{100}{56} = 37·60$. Therefore $CaCO_3 = 37·60$. This is our sixth combination.

The magnesia has hitherto remained untouched; and as there are no available acids, we have to convert it straight away into magnesium carbonate. We multiply our magnesia (4·12 grains), therefore, by the molecular weight of $MgCO_3$ (84), and divide by that of MgO (40). $4·12 \times \dfrac{84}{40} = 8·65$. Therefore $MgCO_3 = 8·65$ grains. This is our seventh and last combination.

The combinations would therefore stand—

Sodium chloride (NaCl)	5·95 grains
Calcium chloride ($CaCl_2$)	0·63 ,,
Potassium sulphate (K_2SO_4)	0·98 ,,
Calcium sulphate ($CaSO_4$)	13·36 ,,
Calcium nitrate ($Ca(NO_3)_2$)..	2·76 ,,
Calcium carbonate ($CaCO_3$)..	37·60 ,,
Magnesium carbonate ($MgCO_3$)	8·65 ,,
Silica and alumina (merely added)	0·55 ,,
Total	70·48 ,

Example II.—The water contained in one gallon :—

Silica (SiO₂)	0·18 grains
Alumina (Al₂O₃)	1·01 ,,
Iron oxide (Fe₂O₃)	0·51 ,,
Lime (CaO)	8·21 ,,
Magnesia (MgO)	2·89 ,,
Soda (Na₂O)	16·10 ,,
Potash (K₂O)	1·20 ,,
Chlorine (Cl)	5·31 ,,
Nitric acid (N₂O₅)	2·48 ,,
Sulphuric acid (SO₃)	3·62 ,,

We first take the chlorine and convert it into sodium chloride ($Cl + Na = NaCl$). The 5·31 of Cl will combine,

$$55·5 \quad 23 \quad 78·5$$

therefore, with $5·31 \times \dfrac{23}{35·5}$ Na $= 3·44$ Na. 5·31 Cl added to 3·44 Na gives us NaCl $= 8·75$.

The nitric acid (2·48 grains) is now converted into sodium sulphate. Working on the equation $N_2O_5 + Na_2O$

$$108 \quad 62$$

$= 2NaNO_3$, we find that 2·48 N_2O_5 will unite with

$$170$$

$\dfrac{62}{108} \times 2·48$ Na₂O $= 1·42$ Na₂O. 2·48 N_2O_5 added to 1·42 Na₂O gives us NaNO₃ $= 3·90$.

We now convert our potash (1·20) into potassium sulphate. Working on the equation $K_2O + SO_3 = K_2SO_4$ we find that

$$94 \quad 80 \quad 174$$

1·20 K₂O will combine with $1·20 \times \dfrac{80}{94}$ SO₃ $= 1·02$ SO₃. 1·20 K₂O added to 1·02 SO₃ gives us K₂SO₄ $= 2·22$.

Our total SO₃ is 3·62, and we have used for the potassium sulphate 1·02. There is, therefore, a balance of 2·60 to be disposed of. This is converted into sodium sulphate. Working on the equation $SO_3 + Na_2O = Na_2SO_4$, we find that 2·60

$$80 \quad 62 \quad 142$$

2 G

SO_3 will combine with $2 \cdot 60 \times \dfrac{62}{80}$ $Na_2O = 2 \cdot 01$ Na_2O. $2 \cdot 60$ SO_3 added to $2 \cdot 01$ Na_2O gives us $Na_2SO_4 = 4 \cdot 61$.

All the available acids having been used up, we must convert the rest of our bases into carbonates. Turning to the soda, we have used $3 \cdot 44$ Na for chlorine. We must now find the equivalent of this in terms of NaO. This is done by multiplying by the molecular weight of Na_2O, and dividing by that of Na. Now $3 \cdot 44 \times \dfrac{62}{46} = 4 \cdot 64$. Na_2O, therefore, used for chlorine = $4 \cdot 64$. To this we must add the Na_2O used for the other combinations. The list stands—

Na_2O used for NaCl		4·64
Na_2O ,, $NaNO_3$		1·42
Na_2O ,, Na_2SO_4		2·01
Total Na_2O used		8·07

Our total Na_2O was $16 \cdot 10$, we have therefore a balance of $8 \cdot 03$ Na_2O to deal with. This is converted into carbonate by multiplying by the molecular weight of sodium carbonate ($Na_2CO_4 = 106$), and dividing by that of soda ($Na_2O = 62$). $8 \cdot 03 \times \dfrac{106}{62} = 13 \cdot 73$. Hence $Na_2CO_3 = 13 \cdot 73$.

The lime has been hitherto untouched, so we calculate all of it ($8 \cdot 21$) into carbonate, by multiplying it by 100 and dividing by 56 (see previous example). $8 \cdot 21 \times \dfrac{100}{56} = 14 \cdot 66$. Hence $CaCO_3 = 14 \cdot 66$.

Similarly our magnesia ($2 \cdot 89$) having been untouched, we convert it into the carbonate by multiplying by 84 and dividing by 40 (see previous example). $2 \cdot 89 \times \dfrac{84}{40} = 6 \cdot 07$. Therefore $MgCO_3 = 6 \cdot 07$.

The silica, iron oxide, and alumina are added together ($0 \cdot 18 + 1 \cdot 01 + 0 \cdot 51 = 1 \cdot 70$).

Our combinations therefore stand :—

Sodium chloride (NaCl)	8·75	grains per gallon
Sodium nitrate (NaNO₃)	3·90	,, ,,
Potassium sulphate (K₂SO₄)	2·22	,, ,,
Sodium sulphate (Na₂SO₄)..	4·61	,, ,,
Sodium carbonate (Na₂CO₃)	13·73	,, ,,
Calcium carbonate (CaCO₃)	14·66	,, ,,
Magnesium carbonate (MgCO₃)	6·07	,, ,,
Silica, alumina, and oxide of iron } .. .. (Si₂O, Al₂O₃, Fe₂O₃)	1·70	,, ,,
Total	55·64	,, ,,

CHAPTER X.

THE ANALYSIS OF MALT AND WORT.

PRELIMINARY.

In analysing a malt, due regard must be given to the fact that it is not a homogeneous material, and that the chemical composition of a mouldy, undervegetated, or non-vegetated corn will widely differ from that of a normal corn, such as composes (or should compose) the bulk of the sample. It is therefore of importance, in the weighing out of the separate little lots for the various determinations, to take them as fairly as possible, and to include in them their fair proportion of what may be termed abnormal grains, when such are present. To do so, the sample (which should always be stored in a well-fitting tin, or stoppered bottle) should be thrown out on to a piece of paper, and the lots for weighing selected from different parts of the heap, which should previously be as thoroughly mixed as possible by means of a clean spatula. We have known of very divergent results arising from want of attention to these precautions. It is unnecessary to say, in addition, that as malt is very distinctly hygroscopic, it is imperative not to expose it for any time to the atmosphere during the course of the analysis.

The Diastatic Capacity of Malt.

The only process by which the diastatic capacity can be at once accurately and conveniently determined is that devised

by Lintner.[*] It is based upon the researches of Kjeldahl
(see p. 149) who found that the relative strength of a diastase
solution is estimable by the amount of maltose produced from
a uniformly prepared starch, on condition that the maltose
so produced does not exceed 50 per cent. of the total conversion products of that starch. In other words, so long as
the conditions of the test are such as to prevent the above
limit being exceeded, we can compare one solution of diastase
with another, or one malt with another, saying, for instance,
that the first is twice as strong in diastase as the second,
supposing that the same amount of maltose is produced in
the one case by half the quantity of diastase solution or malt
used in the other case.

This comparability is a point of the utmost importance ;
for if the comparison does not hold, any strict relation
between malts, as regards their diastatic power, will be
impossible, and the result will be useless and unreliable.
This point is insisted upon because other methods have
been proposed for the determination of diastatic capacity,
which fail through neglect of Kjeldahl's limits. Thus, for
instance, Salamon[†] has recently proposed a method of this
sort, where the quantity of maltose produced may exceed the
limits in question. In criticising and condemning this method,
one of us [‡] clearly showed by actual experiment that no
relation existed between the maltose formed and the diastatic
strength of the malt.

The test is performed by simply observing the volume of
a cold-water extract of a malt sample which reduces a given
volume of Fehling's solution, or rather, which produces from
starch that constant amount of maltose which is necessary
to exactly reduce the Fehling's solution. It is clear, if a given
volume of Fehling's solution is reduced by say 1 c.c. of the

---

[*] ' Wochenschrift für Brauerei, 1886 ; pp. 733 and 753,' translated, Brewing
Trade Review, 1887, p. 204. [†] Cantor Lectures, 1888.
[‡] G. H. Morris, Brewing Trade Review, 1888, p. 435.

cold-water extract of a malt sample A, while the same amount of Fehling's solution is similarly reduced by 0·5 c.c. of the diastase extract of a malt sample B, that sample B has twice the diastatic capacity of sample A. In expressing the result, we may either refer all malts to any given malt, which has, for instance, given us the best practical results ; or we may refer all malts to a fixed arbitrary standard. The latter plan is preferable, inasmuch as practical success depends on factors other than the diastatic capacity of malt, and if we select an arbitrary and fixed standard, we eliminate all such variable factors.

Lintner proposes the following as a standard : the diastatic capacity of a malt is to be regarded as 100 when 0·1 c.c. of a solution made according to the directions afterwards given, reduces 5 c.c. of Fehling's solution. This is the standard by which all results are expressed in this book, and its meaning will become clearer when the examples are worked out.

In extracting the diastase from malt, the ground malt is soaked in cold water. By this means the diastase is dissolved out, but at the same time we extract certain sugars in the malt (maltose, invert-sugar) which will also reduce Fehling, and which will thus raise the apparent diastatic capacity of malts. It is true that the amount of reducing sugars so dissolved is not very appreciable, and is very fairly constant in amount, and the error due to them is therefore of not much moment. In very highly cured malts, however, the amount of these bodies is raised, due to a caramelisation of the starch and to the reducing action of some of the products of caramelisation. On this ground, a short and simple method is appended by which a suitable correction for the sugar dissolved with the diastase can be made ; at the same time it is, as a rule, a process which can be neglected.

For Lintner's method we require soluble-starch, which must be of uniform composition, and Fehling's solution. The preparation of Fehling's solution will be found in the Appendix ; the soluble-starch is prepared as follows, and it is always well to prepare a good quantity at a time (say 1 to 2 lbs.), for its preparation, whether in large or small bulks, is a somewhat long and tedious operation.

The required amount of starch ("pure" potato starch of commerce) is mixed in a basin with a 7·5 per cent. solution of hydrochloric acid (sp. gr. 1037), sufficient acid being used to cover the starch. After standing for seven days at 60° F., or for three days at 104° F., the starch is rendered soluble. It is now washed free from acid by decantation with cold water until the wash water shows not the least acidity with litmus paper. The water is now drained off as far as possible, and the starch dried by exposure to the air. The dry starch is kept in a stoppered bottle, and, as it is required, 2 grams are abstracted, dissolved in hot water, cooled and diluted to 100 c.c.

In extracting the diastase from malt samples, the malts are treated as follows : 25 grams of ground malt * are digested with 500 c.c. cold water at the ordinary temperature (56°–70° F.) for five hours. The mixture is then filtered, and filtration continued until the solution is perfectly bright. It is this bright solution which we require, and only a little of it being wanted there is no need to filter through anything like the whole of the extract. The determination is made as follows :—

Ten test-tubes are placed in a suitable stand, and to each is added 10 c.c. of the 2 per cent. solution of soluble-starch. Then to each of the series we add a gradually increasing amount of the bright malt-extract : 0·1 c.c. to the first, 0·2 to the second, and so on, so that the tenth tube receives 1·0 c.c. They are now thoroughly shaken, and allowed to stand at the ordinary temperature for exactly one hour.† It is necessary

* In weighing out ground malt it is advisable (in order to ensure fair proportion of husk and flour) to weigh out roughly the required quantity of malt in the unground state. This is then ground, and any deficiency made up by the addition of a little ground malt, prepared for the purpose.

† Since writing the above we have found that the temperature at which the tubes containing the various quantities of diastase solution and starch are digested

to see that this period is not appreciably shortened or prolonged.

At the end of this hour's stand, 5 c.c. of Fehling's solution are added to each tube, the tubes again shaken, and the whole series placed in boiling water for ten minutes. The tubes are now observed, and the degree of reduction noted. Some will be over reduced (as shown by their yellow colour) some will be under reduced (as shown by their blue colour). We select the two consecutively numbered tubes, the one of which is slightly over reduced, and the other of which is slightly under reduced. The number of c.c. of malt-extract to have produced exact reduction will then lie between the number of c.c. added to the one, and that added to the other. Occasionally, of course, one tube will be neither blue nor yellow, showing that the number of c.c. of malt-extract added produced exact reduction. In this case, the number in question is noted; if, however, the point lies between any two tubes, say between the 0·3 and the 0·4, we call it 0·35. Sometimes, we should call it 0·32 or 0·38, according as to whether the point of exact reduction lay either nearer to the 0·3 or the 0·4. A little practice soon gives the necessary experience for judging with sufficient correctness the intermediate second decimal.

THE DETERMINATION OF MALTOSE, DEXTRIN, &c.

The method which we employ for this purpose closely follows that described by Heron, which was based upon the

---

for one hour makes a very considerable difference in the result; the higher the temperature the higher the apparent diastatic capacity. Indeed, the difference in temperature between two laboratories is frequently sufficient to cause very serious divergencies in results when the same malt is operated upon. In order to eradicate this source of error, we therefore recommend that the tubes be exposed not to the air; but that they be digested in a water-bath adjusted to a constant temperature, winter and summer, and kept at that temperature during the digestive period. Since it is easier to heat water than to cool it, we consider that a temperature of 70° F. best meets the case.

methods of O'Sullivan ;[*] and although we make some modifications in it which we consider advantageous, the principle remains the same. The method in question differs very appreciably from that given in existing text-books ; these, in ignoring certain most important points, are quite unreliable and valueless. There is no need to criticise these methods, since they are now generally admitted to be faulty; but it is perhaps, desirable to very briefly call attention to the distinguishing points of the newer methods, in order to bring out certain important factors which have been ignored in malt analysis till comparatively recently.

In the first place (as has been already pointed out in Chapter II., p. 133), malt contains certain ready-formed sugars (maltose, cane-sugar, invert-sugar). In previous methods of analysis no account was taken of these sugars ; the malt was mashed and the whole of the reducing sugar found was then calculated as maltose. But the reducing sugar consists not only of maltose due to the conversion of starch, but (1) of maltose ready formed in the malt, and (2) of invert-sugar ready formed in the malt. Therefore the maltose result as ordinarily obtained will be erroneously augmented by the maltose in the malt, and by a totally different sugar : invert-sugar. Seeing that the amount of maltose and invert-sugar in malts varies, it is clear that the result ordinarily arrived at is, on that ground, unreliable; it is further obvious that when a totally different sugar, i. e., invert-sugar, is reckoned as maltose that the error is increased, for the reducing powers of these sugars are different.

What we principally want to know in analysis is how the malt-starch breaks up under standard conditions of conversion ; that is, the actual proportions of starch transformation products ; and it is clear that the older method fails to give us the maltose, and we shall see that it is similarly unreliable in respect to the dextrin.

[*] Journal Society Chemical Industry, 1688, p. 267.

As we know, malt-starch is broken up according to circumstances into maltose, dextrin, and amyloïns (maltodextrins). We know, however, that all the transformation products of starch are capable of expression in terms of maltose and dextrin, and it is the varying proportions of these that we must first arrive at before estimating the amyloïns. The errors of the ordinary means of determining the maltose have been touched upon ; still greater errors attend the usual manner of determining the dextrin. Analysts have hitherto attempted to determine it by converting the whole of the dextrin and the maltose present in the wort (by boiling with acid) into dextrose, which is then estimated. The maltose, having been determined, its equivalent of dextrose is calculated, and this, subtracted from the total dextrose produced, gives the dextrose due to dextrin. By a calculation the dextrin is thus determined. This involves a series of errors.

(1) In the first place the figure subtracted to represent the glucose equivalent of the maltose must be wrong, inasmuch as the maltose, as ordinarily determined, is incorrect, as has been already explained.

(2) The malt contains an appreciable quantity of cane-sugar. This sugar does not reduce Fehling as such, but during the boiling with acid it becomes inverted, and the invert-sugar will be entered in the result as dextrose formed from dextrin, and will thus be calculated into dextrin, the true proportion of which will be thus augmented, as a rule, by some 4 to 5 per cent.

(3) During the boiling with acid one of two things must occur : either some of the dextrin is not converted, or else, if it be so, a portion of the first formed dextrose is converted into numerous bodies which exert no influence upon the Fehling's solution. In either case, therefore, there must be a loss, and experience shows it to be considerable.

(4) During boiling with acid the albuminous bodies suffer some form of decomposition, and become partially converted

into substances which exercise a reducing action on Fehling's solution. This forms a source of error on the plus side.

It is therefore evident that these methods for getting at the starch-transformation products are useless, and it becomes necessary to describe how these may be correctly determined.

The first thing to do is to make the necessary correction for the ready-formed sugars. This can be readily done by extracting the malt with cold water for 2 to 3 hours.* The cold water will leave the starch unaffected, and the cold extract will therefore contain none of the starch or the starch-transformation products. We may regard the cold extract as containing the ready-formed sugars, the acid, the nitrogenous matter soluble in cold water (including the diastase), the mineral matters soluble in cold water, together with mere traces of colouring matter, &c.

If, therefore, we make the cold-water extract with a definite bulk of water, we can get the total amount of ready-formed sugars by estimating firstly the total solid matter in the solution, and subtracting from that figure the sum of the soluble albuminoids, ash and acid. All these determinations are easy enough.

Having obtained our total reducing sugars it is necessary to find in what way they will effect the analytical determination of the maltose and dextrin in the hot-water mash, which will include these starch-transformation products as well as the ready-formed sugars. As we shall afterwards estimate the maltose by Fehling's solution, and the dextrin by the polarimeter it becomes necessary, therefore, to determine the effect of the ready-formed sugars on Fehling and on the polarimeter. We therefore take the reducing action of the cold extract on Fehling and also determine its opticity (or rotatory capacity

* The digestion of the malt with cold water must not be prolonged, or the diastase extracted will attack the small and ruptured starch-granules. This will result in the formation of maltose, and thus cause the amount of ready-formed sugars to appear higher than really is the case.

on polarised light). These numbers obtained, we can proceed
to mash our malt by digesting it with hot water; using, of
course, the same conditions for all samples, so that the
results of various malt analyses may be comparable. Having
done so, let us consider what our hot-water extract will contain : firstly, it will contain the starch-transformation products
which we can express in terms of maltose and dextrin;
secondly, the ready-formed sugars; thirdly, the albuminoids
soluble in hot water; fourthly, the ash soluble in hot water;
and finally, traces of colouring matter which do not affect
the analysis.

The next point to determine is the reducing action of
the hot-water extract on Fehling's solution. The reduction
will be partly due to maltose formed from the malt-starch,
partly due to the ready-formed sugars. Now if we subtract
the already determined reducing action of the cold-water
extract, it is clear that the residue is due to the maltose
formed from the starch, and from that only. Similarly, if
we determine the opticity of the hot-water mash, and subtract from it the opticity of the cold-water mash, we get the
opticity due to the dextrin and maltose from the malt starch,
and from that only. These results enable us obviously to
arrive at the exact proportion of maltose and dextrin, the
proportions of which it is so necessary to know.

We can check the accuracy of the analysis in this way :
the difference between the total solid matter of the cold-
water mash, and that of the hot-water mash will be
essentially the maltose and dextrin which are present in
one and not in the other. As we determine the total solid
matter in both extracts in any case, the agreement of the
difference between these figures, with the added percentages
of maltose and dextrin as obtained by direct analysis, will
afford a proof of the accuracy of these determinations. There
is a tendency for the difference in the total solid matter of
the hot and cold mashes to somewhat exceed the added per-

centages of maltose and dextrin as determined, because the conversion of that portion of the starch which becomes hydrolysed into maltose is attended by an increase in weight. On the other hand, the ash and albuminoids, soluble in hot, are rather greater than those soluble in cold water. On these grounds the agreement is seldom absolute; still, in the majority of cases, the difference is not, and never should be, more than 1·5 per cent.

When necessary, the albuminoids and mineral matter in the hot wort may be determined, but this is not often necessary. The gravity of the hot mash measures the brewers' extract of the malt, and figures so obtained agree closely with those obtained in practice. Beyond the determinations already referred to, we must estimate the acid in the malt, and the moisture. These tests are exceedingly simple. As previously stated, methods exist for the determination of the amyloïns, and these will be given.

Statements as to the vegetation, evenness, colour, &c., of the malt should accompany the chemical examination.

These preliminary remarks will sufficiently serve to explain the scope of the analysis and its objects. We must now turn to the details of the various processes by which they may be obtained.

1. *Preparation of the Cold-water Mash.*

Twenty-five grams of ground malt (see footnote, p. 455) are digested at the ordinary temperature for three hours, with 250 c.c. distilled water. This is filtered until bright (about 120 c.c. of bright filtrate will suffice). 100 c.c. of the filtrate contained in a suitable beaker are heated on the water-bath for fifty-five minutes at 150·8° F. (66° C.), then for five minutes the temperature is raised to 168° F. (70° C.). It is now heated to boiling, and boiled for five minutes; then cooled, transferred to a flask graduated at 100 c.c., diluted to that measure, and filtered. The bright solution must be analysed as follows :—

(1) Take the specific gravity at 60° F.

(2) Determine the reducing power on Fehling's solution (cupric oxide reducing power) gravimetrically with 10 c.c. of the solution.

(3) Determine the opticity (polarimetric reading) with 25 c.c., previously adding 2·5 c.c. of alumina solution, and filtering before taking the reading.

(4) Determine the ash by evaporating 25 c.c. to dryness in a weighed platinum dish, and then igniting.

(5) Determine the albuminoids by evaporating 10 c.c. to dryness, and treating the whole of this residue according to Kjeldahl's process.

The above is what we require to do on the cold-water mash ; remarks on the gravimetric estimation of the reduced copper (No. 2), the determination of opticity (No. 3), and of the albuminoids (No. 5), will follow the sketch of the preparation of the hot-water mash.   The determinations of specific gravity, and of ash, do not require further description.

2. *Preparation of the Hot-water Mash.*

Mash 50 grams of ground malt with 400 c.c. of water at 154·5° F. (68° C.) and place in a water bath at such a heat as will keep the mash at 150·8° F. (66° C). This is best done by getting the required volume of water in a beaker up to the right heat, inserting in the water bath, and adding the ground malt slowly, stirring the while, preferably with a thermometer. The mash is kept at the temperature named for fifty-five minutes, and then raised to 158° F. (70° C.) for five minutes. This being done, the mash is cooled, poured into a flask,

graduated at 515 c.c.,* and then filtered. 250 c.c. of the
filtrate are boiled for fifteen minutes [note whether the wort
"breaks" satisfactorily], returned to a flask, graduated at 250
c.c., cooled, and made up to 250 c.c.

Now determine :—

(1) Specific gravity of the solution at 60° F.

(2) Reducing action on Fehling's solution, using 5 c.c. of
solution for this purpose.

(3) Opticity of the solution.

3. *Determination of the Cupric-reducing Power by the Gravimetric Method* (O'Sullivan's Process).

Although this process is used in all the scientific laboratories
of this and other countries, a few words as to its working may
not be out of place, inasmuch as previous text-books on brewing
analysis have seemed to give preference to the more convenient but unreliable volumetric process. The gravimetric process gives remarkably accurate results if carried out as follows.

30 c.c. of Fehling's solution are placed in a beaker, and
diluted with 50 c.c. of well-boiled distilled water. The mixture
is then heated in a boiling water-bath. After five minutes,
the required volume or weight of the substance to be tested is
added, and the whole is then heated in the boiling water-bath
for ten minutes. The precipitate is then filtered as rapidly as
possible, and washed with boiling water. It is dried in the
ordinary way, and ignited in a porcelain crucible for twenty
minutes, and weighed. The copper will now exist as cupric
oxide (CuO). To calculate the amount of *maltose* in the known
weight of substance employed, multiply the weight of the cupric
oxide by 0·7435. To calculate the weight of *dextrose* or *invert-
sugar*, multiply the weight of cupric oxide by 0·4535 or
0·4715 respectively.

* The number 515 c.c. may seem an odd one ; but it is chosen as representing
the bulk occupied by 500 c.c. water and 50 grams of malt. This number is
recommended by Heron as the average of many experiments.

4. *Determination of the Opticity, or Polarimeter Reading.*

This is not the place for anything approaching a complete or exhaustive description of the polarimeter and its uses.   For this, the reader should consult Heron's paper already referred to.   The following directions presuppose a sort of general knowledge of the polarimeter, and only its application to the determination of the opticity of the cold- and hot-water mashes will now be touched on.

As a rule, the readings are taken in a 1-decimetre tube, for the solutions are rather too deeply tinted to permit of a 2-decimetre tube being used.   When the solutions are so dark as to interfere with an accurate reading in the 1-decimetre tube, it is preferable to dilute them rather than to decolourise with charcoal.   In such cases, 25 c.c. of the solution diluted to 50 c.c. will generally give the required reduction in tint.

It is very frequently necessary to clarify the malt solutions by adding alumina solution.   In the event of a wort having to be diluted, the alumina is added after dilution; and in any case the amount of alumina should be one-tenth of the volume of solution to which it is added.   Directly the alumina is added, the wort may be at once filtered, and the 1-decimetre tube filled with the bright solution.

The alumina is prepared as follows: "alumina cream," "moist alumina," or "aluminium hydrate," as purchased, is beaten up with distilled water to a thin cream; the desired consistency is obtainable by taking one part of "moist alumina," and ten parts water.

The tube is now filled with the clear solution, diluted or undiluted, as the case may be, and the deflection read, either in a "half-shade" or Laurent instrument, or in a "transition tint," or Soleil-Ventzke-Scheibler instrument.   The former is perhaps preferable, inasmuch as it permits of a rather greater amount of colour than the latter.

In all cases this reading is referred to the deflection for

the D line of the spectrum. It would, of course, answer equally well if all results were referred to the mean yellow line of the spectrum ; but in any case one form of expression should be adhered to, and the D line result is adopted throughout the analytical section of this work.

In using the transition tint apparatus, the reading has to be taken on a straight scale, which expresses it in terms of cane-sugar. In order to connect the reading with angular degrees for the D line, we have to multiply the reading on the straight scale by 0·346.

When a half-shade instrument is used the reading is taken on the circular scale in regular measurement, and corresponds without further calculation to angular degrees for the D line.

THE DETERMINATION OF NITROGENOUS BODIES BY KJELDAHL'S PROCESS.

The wort-solids (which need not be chemically free from moisture, but must be to all appearance quite dry) are treated in the flask in which they have been dried, with 10 c.c. of a mixture of equal parts of fuming sulphuric acid and ordinary strong sulphuric acid, which mixture is kept in stock.

The whole is digested for an hour and a half over a low flame, the flask being covered with a clock-glass. At the expiration of this period, the flame is removed, and powdered potassium permanganate is dropped in from a spatula in small quantities at a time, until the mixture turns permanently green or permanently purple. The clock-glass is, of course, replaced immediately after the addition of each lot of permanganate so as to prevent loss by spurting. The mixture is now allowed to cool, diluted with 200 c.c. of ammonia-free distilled water, and the whole poured into a distilling flask connected with a condenser. The receiver is a small flask containing 25 c.c. of $\frac{N}{20}$ sulphuric acid

($\frac{1}{50}$ normal strength*), and connected to the condenser by an adapter. The thin end of the adapter dips into the sulphuric acid, and vent is given by a second hole in the cork. The apparatus being ready, a little powdered pumice stone is added to the mixture in the distilling flask, which is then made alkaline with 40 c.c. of an ammonia-free sodium hydrate solution of specific gravity 1·300.

Heat is now applied to the distilling flask, and the distillation continued till about 150 c.c. of distillate have passed into the acid in the receiver. The flame is now removed, and the contents of the receiving flask, washed with ammonia-free distilled water, poured into a large porcelain dish. A few drops of an alcoholic solution of methyl orange are added, and the solution is exactly neutralised by $\frac{N}{20}$ ammonia, which is run into it from a graduated burette. The exact point of neutralisation is indicated by the pink methyl orange turning just permanently yellow. The number of c.c. of $\frac{N}{20}$ ammonia necessary for neutralising the solution from the receiver is then noted, and gives us the required result.

The process depends upon the fact that on treating nitrogenous bodies with sulphuric acid and potassium permanganate, the whole of the nitrogen contained in them is converted into ammonia. The ammonia is absorbed by sulphuric acid being converted into ammonium sulphate. When the soda is added and the solution boiled, the ammonia is set free and passes over into the standard $\frac{N}{20}$ sulphuric acid, a part of which acid is thereby neutralised. The greater the amount of nitrogen in the original substance the more $\frac{N}{20}$ sulphuric acid will be thus neutralised, so that the quantity of

* See Appendix.

$\frac{N}{20}$ ammonia which we use to determine the unneutralised

residue of $\frac{N}{20}$ sulphuric acid will, by giving us indirectly the amount of acid neutralised, indicate the amount of nitrogen in the substances. The necessary calculations in this and the other estimations described will be explained when we consider the examples.

The results may be returned in terms of nitrogen, or at once calculated into albuminoids by multiplying the amount of nitrogen by 6·3. This factor represents the average relation between the various types of albuminoids which have been investigated, and the nitrogen which they contain.

It is clearly necessary that the whole of the reagents used in the process should be free from ammonia or from substances capable of becoming ammonia during the course of the analysis. There is no difficulty of course in procuring dis-tilled water, soda, and potassium permanganate free from all traces of ammonia, but in the greater number of cases the mixture of fuming and strong sulphuric acid does contain a little ammonia. This ammonia, unless allowed for, would, of course, go to swell the apparent percentage of nitrogen in the substance under analysis; but the necessary correction is readily made by taking a sample, when commencing a fresh stock of the mixed acids, and estimating the ammonia in 10 c.c. The quantity so found, or its equivalent in nitrogen, is of course subtracted from the quantity of ammonia ob-tained from each Kjeldahl determination.

THE DETERMINATION OF MOISTURE.

Five grams of ground malt are heated in a weighed porce-lain dish in a water oven at 100° C. (212° F.), and weighed till constant; the loss so obtained will equal the amount of moisture in 5 grams of the sample. Due care must be taken not to heat the malt after the water is expelled. In the latter case, the malt is liable to absorb oxygen and the weight is therefore increased, and hence the apparent percentage of moisture comes out lower than it should. A malt fresh off the kiln is generally dry in two hours, an ordinary stored malt in four ; a slack malt in six or seven hours.

THE DETERMINATION OF ACID.

Although the acid in malt is generally due to acid phosphates, it is usual, in expressing results, to refer the total acidity to lactic acid.

Fifty grams of ground malt are digested at the ordinary temperature with 300 c.c. of cold distilled water for twelve hours. At the expiration of this time, 150 c.c. are filtered off, and the acidity neutralised with a standard solution of ammonia, of specific gravity, 998·6 ; this is run into the filtrate from a graduated burette, litmus paper being used as the indicator. The number of c.c. of ammonia solution so used is duly noted.

We shall now take a typical example, and calculate out the results obtained on applying the above tests. The example taken will be that of a pale and rather lightly cured malt. The determination of the amyloïns is given afterwards.

(1) *Diastatic Capacity.*

The test was conducted as above described. The tube to which 0·2 of extract had been added was slightly blue ; the tube to which 0·3 had been added was slightly yellow. The latter tube was about as much over-reduced as the former was under-reduced, hence we take 0·25 as the number of c.c. of the extract which would have produced exact reduction.

Now the diastatic capacity of a malt is 100 when 0·1 c.c. of an extract causes complete reduction of the amount of Fehling's solution employed for the test (5 c.c.). Hence the diastatic capacity of our malt is $\frac{0 \cdot 1}{0 \cdot 25}$ of the standard, or $\frac{0 \cdot 1}{0 \cdot 25} \times 100 = 40$. The diastatic capacity of the malt is therefore 40.

The correction to which we previously referred for reducing sugars, extracted with the diastase, is arrived at in this way :—

Five c.c. of Fehling's solution and 10 c.c. of starch solution are diluted with 10 c.c. of water and boiled in a boiling tube ; some of the bright extract of malt, as before used, is then run into the tube from a graduated burette, until the Fehling is exactly reduced. Complete reduction of the Fehling was obtained, in this case, when 7 c.c. of the extract had been run in. Calculated out in terms of apparent diastatic capacity, the following figures would be obtained :—

$$\frac{0 \cdot 1}{7} \times 100 = 1 \cdot 43.$$

If, therefore, we wish to apply the correction we subtract this 1·43 from the diastatic capacity as previously obtained. Thus, in the first example given, the corrected diastatic capacity would be 40 less 1·43, or 38·57.

In the majority of cases the correction equals 1·4, and it is safe to apply this generally, subtracting it from the diastatic capacity as found previous to the correction. It is hardly necessary to add that starch-transformation products play no part in the reducing action obtained in arriving at the correction, since the whole is boiled and the diastase killed before any conversion of the starch would occur.

[Lintner suggests taking a double series of tubes for the diastatic determination ; if, for instance, he found the right point between, say the 0·2 and the 0·3 tube, he would take his second series, adding to the first 0·22, to the second 0·24, to the third 0·26, &c. But experience has shown us that accuracy is not to be got in this way, since there are obvious practical difficulties in accurately measuring out hundredths of a c.c. of the extract. It is preferable to take one series only, in which each difference is one-tenth of a c.c., and to judge, by the appearance of the two consecutive tubes selected, the number of c.c. which would have been required to give exact reduction.]

(2) *Cold-water Mash Results.*

(*a*) *Reducing Power.*—10 c.c. of the cold-water mash were treated with Fehling's solution in the way already described. The weight of dried CuO was 0·162.

This figure is simply recorded for subsequent subtraction from the CuO yielded by the hot-water mash. There is no need to refer it to any of the carbohydrates which have produced the reduction.

(*b*) *Opticity.*—The opticity of the extract in a 1-decimetre tube was 2·8 divisions (Ventzke-Scheibler units).

(*c*) *Specific Gravity.*—The specific gravity of the cold-water extract was 1008·05. Subtracting 1000, and dividing the difference by 3·86, we obtain the solids in solution (per cent.)

$$\frac{8\cdot05}{3\cdot86} = 2\cdot085 \text{ grams solid matter.}$$

(*d*) *Albuminoids.*—10 c.c. of the extract were evaporated to dryness and treated with sulphuric acid in the way already described.

20 c.c. of $\frac{N}{20}$ sulphuric acid required (after the ammoniacal distillate had been introduced) 12·3 c.c. $\frac{N}{20}$ ammonia.

The sulphuric acid used in the treatment was not pure, and was found by previous experiment to contain an amount of ammonia capable of neutralising 0·7 c.c. of the standard acid, then

12·3 + 0·7 = 13 c.c., and 20 − 13 = 7 c.c.; each 1 c.c.

acid neutralised corresponds to 0·0007 gram of nitrogen ; 7 c.c. will therefore correspond to 0·0049 nitrogen, or 0·0049 × 6·3 albuminoids * = 0·03087 gram. This, therefore, is the quantity of albuminoids in 10 c.c. extract ; in 100 c.c. the amount will be 0·03087 × 10 = 0·3087 gram.

Now 100 c.c. of extract were made from 10 grams malt. To get the percentage of albuminoids on the malt all we have to do is to multiply the above figure by 10. The percentage, therefore, of albuminoids (extracted by cold water) is 3·087, or say 3·09.

(*e*) *The Ash.*—25 c.c. of the cold-water extract were evaporated to dryness in a platinum dish and incinerated. The weight of the ash was 0·028 gram.

100 c.c. of the extract would therefore contain 0·112 gram. This quantity corresponds to 10 grams malt ; 100 grams malt will therefore give 1·12 ash (soluble in cold water).

Before proceeding to the hot-water mash results, we will turn to the lactic acid, and this will enable us, with the results previously obtained, to determine the percentage of ready-formed sguars.

(*f*) *Lactic Acid.*—50 grams ground malt were taken; the infusion required 6·5 c.c. of the standard ammonia solution ; 100 grams of malt would therefore require 13 c.c. Now 13 c.c. represents 0·13 gram of *acetic* acid. To calculate from this the corresponding amount of lactic acid we multiply by the molecular weight of lactic acid (90) and divide by that of acetic acid (60) :—

$$0·13 \times \frac{90}{60} = 0·19.$$

The percentage of acid, as lactic acid, is therefore 0·19.

(*g*) *Ready-formed Sugars.*—We are now in a position to calculate the ready-formed sugars. It will be seen, on reference, that the total solid matter in the extract was

* The factor 6·3 for calculating nitrogen into its containing albuminoid is not quite correct, but it is probably sufficiently accurate to ensure no great error.

2·085 per cent. This number refers to 10 grams malt. Multiplying by 10, we get the amount (20·85) corresponding to 100 grams of malt.

If we subtract from this figure the sum of the albuminoids, ash, and lactic acid, we get the ready-formed sugars.

The sum of the albuminoids, ash, and lactic acid will be found to be 4·40 (3·09+1·12+0·19); the percentage of ready-formed sugars, therefore, will be, 16·45 (20·85 less 4·40).

(3) *Hot-water Mash Results.*

(a.) *Specific Gravity.*—This was found to be 1025·46. Subtracting 1000, and dividing the difference by 3·86, we find that the total solid matter in 100 c.c. is 6·59 grams.

We require the above figures subsequently, but we may at once turn it to account for calculating the *brewer's extract.*

What the brewer requires to know is the excess gravity in pounds over 360 (the weight of a barrel of water in pounds) of a wort made from one quarter (equal to eight bushels) of malt and with one barrel of water. This is the so-called brewer's extract per quarter; and in estimating it, we shall first assume that the quarter of malt, which is a definite measure, weighs the standard amount, viz. 42 lb. per bushel.

It is easily found by calculation that 58·33 grams of malt bear the same relation to 500 c.c. of water, as one bushel of malt does to one barrel of water. The excess gravity of the 58·33 grams per 500 c.c. multiplied by eight, will therefore give us the excess gravity of one quarter of malt to one barrel of water.

Now the gravity of the hot-water mash was 1025·46; its excess gravity therefore 25·46. This was obtained by taking 50 grams malt, and 500 water; if we multiply the 25·46 by 58·33, and divide by 50, we get the excess gravity of a mash corresponding to one bushel to the barrel.

Now $25 \cdot 46 \times \dfrac{58 \cdot 33}{50} = 29 \cdot 7$.

If now we multiply this figure by 8, we get the excess gravity from a quarter of malt to the barrel : $29 \cdot 7 \times 8 = 237 \cdot 6$. So far, we have been considering excess over 1000; the brewer, however, wants excess over 360. If, therefore, we multiply $237 \cdot 6$ by 360, and divide by 1000, we get the brewer's extract in excess over 360, or, as it is termed, in pounds per barrel.

$$237 \cdot 6 \times \frac{360}{1000} = 85 \cdot 536, \text{ say } 85 \cdot 5.$$

The brewer's extract per quarter (at 42 lb. per bushel) would therefore be $85 \cdot 5$ lb. per barrel.

Suppose the malt is found by any suitable appliance, such as Dring and Fage's chondrometer, to weigh, say 40 lb. per bushel and not 42 lbs. as previously assumed; all we have to do is to multiply the extract as above found by 40 and divide by 42.

In this case, therefore, $85 \cdot 5 \times \dfrac{40}{42} = 81 \cdot 4$ lb.

In the above calculation, the assumption is made that specific gravity is exactly proportional to concentration. Strictly speaking, this is not the case; still it is an assumption that gives results sufficiently near for all practical purposes.

To save the labour of the foregoing calculations, we may use a factor which is obtained by multiplying together the various fractions necessary in the calculation. This factor is $3 \cdot 36$. For instance, the gravity of the hot-water mash being $1025 \cdot 46$, we simply multiply the $25 \cdot 46$ by $3 \cdot 36$, and at once get the $85 \cdot 5$ as the brewer's extract in pounds per barrel, taking malt at 42 lb. per bushel.

(b) *Reducing Power of Hot-water Mash.*—5 c.c. taken and treated with Fehling as described; weight of dried CuO, $0 \cdot 292$; 100 c.c., therefore, would have given $0 \cdot 292 \times 20 = 5 \cdot 84$ grams CuO.

The CuO from 10 c.c of the cold-water mash was $0 \cdot 162$; 100 c.c. of that solution would therefore have given us $1 \cdot 62$.

If now we subtract $1 \cdot 62$ (as representing reduction due to sugars extracted by cold water) from $5 \cdot 84$, we get the CuO due to maltose produced from the starch : 5·84 less 1·62 equals 4·22 grams CuO.

It has been found that each gram CuO corresponds to 0·743 maltose ; if, therefore, we multiply 4·22 by 0·743, we get the amount of maltose due to the starch transformation in 100 c.c. of the wort : 4·22 × 0·743 = 3·135 grams maltose.

This figure gives the maltose (due to starch transformation) in 100 c.c. of wort, and this corresponds to 10 grams of malt ; on 100 grams of malt the maltose will therefore be 31·35 grams.

(*c*) *Opticity.*—The reading of the hot-water mash in a 1-decimetre tube was 22·6 divisions (Ventzke-Scheibler units). This figure will give us, firstly, the opticity of the wort, and subsequently the percentage of dextrin. First, as regards the opticity of wort : we first convert the above reading into angular degrees by multiplying by 0·346 (this calculation would of course be unnecessary where the reading is taken in a half-shade instrument giving angular degrees direct) :

$$0·346 \times 22·6 = 7·82.$$

7·28 degrees is the deviation (for D line) of a solution containing 6·59 per cent. solid matter (see specific gravity). To calculate the opticity or specific rotatory power, we have to put the deviation on 100 per cent. of solid matter ; we therefore multiply 7·82 by 100 and divide by 6·59.

$$\text{Thus opticity (a) D} = \frac{7·82 \times 100}{6·59} = 118·6°.$$

The dextrin is obtained as follows from the reading. Firstly we must subtract from our reading just obtained that yielded by the cold-water mash (2·8) : 22·6 less 2·8 equals 19·8. This reading will be the deviation due to maltose and dextrin produced from the transformation of starch ; we must now find how much of it is due to the maltose.

The maltose in 100 c.c. was 3·135. If we multiply this by·

3·905* we get the deviation due to the maltose: 3·135 × 3·905 = 12·24.

The total deviation was 19·8 ; subtracting 12·24 we get 7·56 as the deviation due to dextrin. We must now divide this by 5·625,† which will give us the dextrin in 100 c.c. of solution.

$$\frac{7·56}{5·625} = 1·34.$$

If now we multiply the above by 10 we get the percentage of dextrin in the malt, 1·34 × 10 = 13·4 per cent.

(4) *Moisture.*

7·124 grams malt taken ; the loss on heating until the weight constant was 0·88.

$$\text{Percentage of water} = \frac{0·88 \times 100}{7·124} = 1·23.$$

If we add up the whole of the constituents as above obtained, and subtract the sum from 100, we get the "grains," but we must remember that the maltose existed in the malt as starch ; however, by subtracting $\frac{1}{20}$th from the percentage of maltose we get the amount of starch it corresponded to. The grains, obtained by difference, will not be absolutely accurate, because the ash and albuminoids extracted by cold water would be somewhat lower than the amounts of these groups of bodies extracted by hot water. But the exact percentage of grains is not a very important matter, and in the ordinary way they may be calculated with sufficient accuracy in the way suggested. When, however, complete accuracy is desired, the ash and albuminoids in the hot-water mash would have to be determined, and these figures substituted for those obtained by determining them in the cold-water mash.

* 3·905 is the deviation in a 1-decimetre tube (Ventzke-Scheibler units) of a solution containing 1 gram pure maltose per 100 c.c.

† 5·625 is the deviation in a 1-decimetre tube of a solution containing one gram pure dextrin per 100 c.c.

The manner in which the wort "breaks" on boiling, the physical characteristics of the malt, and other points of practical interest should also be noted ; and the report should include remarks on these matters as well as on the analysis. Where the colour of the wort is a point requiring special attention, it may be conveniently and accurately measured by Lovibond's Tintometer.

This instrument matches the tint of the wort by suitably coloured glasses, each numbered and differing from the others in depth of colour.   As a rule, however, the colour of a wort may be sufficiently described by such expressions as "very pale," " pale," " fairly pale," " dark," &c.

Taking the malt, the results of which have been calculated, the record so far would stand as follows :—

Diastatic capacity	38·57
Opticity of wort (a) D	118·6
Colour of wort (hot mash)	very pale
Behaviour of wort on boiling..	"broke" well
Brewer's extract at 42 lb. per bushel	85·5 lb. per barrel
Brewer's extract at actual natural weight (40 lb. per bushel)	81·4 lb. per barrel

Analysis expressed on 100 parts of malt :—

Moisture	1·23
Acid as lactic acid	0·19
Albuminoids (soluble in cold water)	3·09
Ash (soluble in cold water)	1·12
Grains..	34·69
Maltose	31·35
Dextrin	13·44
Ready formed sugars (cane - sugar, invert - sugar, maltose)	16·49
	101·56*

As was previously mentioned, and for the reasons then stated, the sum of the maltose and dextrin should approxi-

---

* The total must exceed 100, since the maltose is heavier than the starch from which it was formed. A due allowance can be made for this by subtracting $\frac{1}{10}$th of the maltose.

mately equal the difference between the totals of the hot-water,
and of the cold-water mash. In this case, this difference
amounted to 45·05 calculated on 100 parts malt (6·59 less
2·085 multiplied by 10). The sum of the maltose and
dextrin is 44·79. These figures of course approximate
sufficiently closely.

THE IDENTIFICATION AND ESTIMATION OF THE MALTODEXTRINS.

So far, we have ignored the maltodextrins in the wort,
having referred these bodies partly to maltose and partly to
dextrin. But this involves a very serious inaccuracy, and
completely reliable results can only be obtained by estimating
the maltose and the dextrin locked up in the form of maltodextrin, and subtracting them from the previously found total
maltose and dextrin. The principles of the method to be
employed are these : Firstly, we ferment out the free maltose
and other fermentable sugars ; the reducing sugar remaining
after fermentation will then be due, firstly to maltose locked
up as maltodextrin, and secondly, to certain other substances,
partly caramelisation products, and partly salts of organic
acids which are unfermentable and unaffected by diastase ;
a correction is made for these substances. We then degrade
the wort with a diastase solution (cold-water malt-extract) by
which we convert the whole of the maltodextrins into
maltose ; the increase in the amount of maltose so produced
will then give us on calculation the amount of dextrin originally locked up as maltodextrin. This, added to the maltose
previously found as maltodextrin, gives us the total amount,
and the ratio of the two will give us the type of maltodextrin.

(*a*) *Determination of Maltose in Maltodextrins.*—50 c.c. of the
hot-water mash (prepared as before described) are "pitched"
with 0·25 gram yeast, and the fermentation allowed to proceed at 80° F. for 72 hours. The mixture is then made up

to 200 c.c. and filtered bright; 25 c.c. of the filtrate are now taken for the determination of the reducing power, as described above. In order to arrive at a suitable correction for the reducing bodies previously mentioned, we set another 50 c.c. of the hot-water mash to ferment with the same amount of yeast, and add to it 0·25 c.c. of a normal cold-water malt-extract. The conditions of fermentation are precisely the same as before described. At the expiration of the fermentation the mixture is made up to 200 c.c., filtered bright, and 25 c.c. of filtrate taken for the determination of the reducing action.

The cold-water malt-extract will convert the whole of the maltodextrin into maltose, which will ferment away during the fermentation. The reducing action in this case will therefore be entirely due to the reducing substances previously referred to, and if we subtract the reduction from that obtained in the case of the first fermentation, the difference will be due to the combined maltose only.

Example.—50 c.c. of the hot-water mash were fermented as above, and 25 c.c. of the fermented fluid gave 0·032 gram CuO. 50 c.c. of the same mash were similarly fermented with the cold-water extract, and 25 cc. of the fermented liquid gave 0·010 gram CuO. The reduction due to combined maltose is therefore 0·032 − 0·010 = 0·022 gram CuO. This calculated into the corresponding amount of maltose (by multiplying by 0·743) gives the figures 0·016346 maltose in 25 c.c., or in 100 c.c., 0·0732. This is the amount in 100 c.c. of the diluted solution (50 c.c. diluted to 200 c.c.), therefore in 100 c.c. of the original hot-water mash, the amount of maltose in the maltodextrins would be 0·0732 × 4 = 0·29 gram. On the malt, therefore, the percentage of combined maltose is 2·90.

(*b*) *Determination of Dextrin in Maltodextrins.*—25 c.c. of the hot-water mash are taken, and 2·5 c.c. of normal cold-water extract of malt are added. The whole is digested for one hour at 130° F., then cooled, diluted to 100 c.c., and 10 c.c. taken for the determination of the reducing action. The reducing action of 1 c.c. of the cold-water extract is taken, after first diluting to about 10 c.c., and then digesting for one hour at 130° F.

Example.—On reference to p. 473 it will be found that the total reducing action of the hot-water mash was 0·292 gram CuO per 5 c.c.; or on 100 c.c., 5·84 grams CuO.

In the above determination of dextrin in maltodextrin, 10 c.c. of the hot-water mash after degradation and dilution gave 0·17 gram CuO. From this we at once subtract the CuO due to the malt-extract used. By a preparatory determination 1 c.c. is found to yield 0·08 gram CuO, and the 10 c.c. of hot-water mash used above will contain one-fourth of this, or 0·02. After correction for the cold extract, 10 c.c. will therefore give 0·17 − 0·02 = 0·15; 100 c.c. of the diluted extract will therefore give 1·5 gram; this corresponds to 25 c.c. of original wort; 100 c.c. of original wort will therefore give 6·0 grams CuO.

Now the same amount of original wort previous to degradation gave 5·84; the increase due to degradation of combined dextrin is therefore 0·16. This is calculated into dextrin by multiplying by the factor 0·706: 0·16 × 0·706 = 0·11 gram. To calculate this on 100 parts malt we multiply by 10. The malt therefore yields 1·10 per cent. of dextrin as maltodextrin.

We have previously found that the combined maltose on 100 parts malt is 2·90; the combined dextrin is 1·10. The total maltodextrin is therefore 4·00 per cent.

From the previous analysis we know that the total maltose is 31·35. Of this, 2·90 is combined; the balance (28·45) will be free. The total dextrin was 13·44; 1·10 of this is combined; the balance (12·34) will be free or stable dextrin.

The full results will therefore be (on 100 parts of malt):—

Maltose (free)	28·45 per cent.
Dextrin (free)	12·34 ,,
Maltodextrins {2·90 maltose} {1·10 dextrin}	4·00 ,,

These figures are generally calculated, and stated, on 100 parts of solid matter in the wort, so as to exclude the variable factors of moisture and grains. In this case the wort-solids were 65·64 per cent. ; we get our required result by multiplying the previous figures by 100, and dividing by 65·64. The free maltose, free dextrin, and maltodextrins will, therefore, in this case be—

$$
\begin{array}{llr}
\text{Maltose (free)} & \dots \dots \dots \dots \dots \dots \dots \dots \dots & 43\cdot19 \\
\text{Dextrin (free)} & \dots \dots \dots \dots \dots \dots \dots \dots \dots & 18\cdot80 \\
\text{Maltodextrins} \left\{ \begin{array}{l} 4\cdot42 \text{ maltose} \\ 1\cdot68 \text{ dextrin} \end{array} \right\} & \dots \dots \dots \dots \dots & 6\cdot10
\end{array}
$$

It will be noted that the maltodextrins produced from the sample of malt under the mashing conditions adopted, approximate closely to the $\left\{ \begin{array}{l} 3 \text{ M} \\ 1 \text{ D} \end{array} \right.$ type.

Preparation of Normal Malt-extract.—400 grams of ground pale malt are mixed with a litre of distilled water to which 5 c.c. of chloroform have been added. The malt is allowed to soak for 48 hours. The malt is then strained off, and the solution filtered bright.

The chloroform preserves the solution from decomposition for about a fortnight ; when no chloroform is used it has to be made fresh each time, and this necessitates determining afresh the necessary correction for the reducing power due to the carbohydrates dissolved by the cold water. Preserved with chloroform the same corrections apply until the solution becomes turbid. Directly turbidity makes its appearance, another solution must be prepared and the correction again determined.

The mode of determining the correction for the reducing sugars in the solution has been given.

WORT ANALYSIS.

Wort analysis is necessarily similar to malt analysis, but is the simpler in so far as the wort has been prepared by the brewer ready for examination ; in the case of a malt, the resulting wort has to be made in the laboratory specially for the purpose.

In the ordinary way, all we need determine in a wort is its opticity and the percentage of maltodextrins, the latter are usually calculated on the wort-solids ; in the majority of cases the opticity may be neglected.

(1) *Opticity of the Wort.*

To obtain this result, we take the specific gravity of the wort at 60° F. From this we get the solid matter in the wort by subtracting 1000 from the gravity, and dividing the difference by the factor 3·86.

The optical activity of the wort is then read in a 1-decimetre tube, after clarifying by moist alumina (see p. 464). When very dark, the wort may be suitably diluted, and the necessary correction made for the dilution. It is impossible, however, to polarise stout and porter worts. These are so dark as to necessitate an extent of dilution which would vitiate the result, while any attempt at removing the colour by charcoal is useless, for in the process of decolorisation a very large proportion of the carbohydrate matter is removed. In these worts, however, we can always determine the amount of maltodextrin.

Example.—Specific gravity, 1062·7.

$$\frac{62\cdot7}{3\cdot86} = 16\cdot24 = \text{grams of wort-solids per 100 c.c.}$$

Opticity.—50 c.c. of the wort, diluted to 100 c.c., gave a deviation in the 1-decimetre tube of 27·5 (Ventzke-Scheibler

2 I

units)* ; 27·5 divisions are converted into angular degrees by multiplying by 0·346.

$$0·346 \times 27·5 = 9·515°.$$

This figure must be doubled to allow for the dilution $9·515 \times 2 = 19·03°$.

To get the opticity we divide this number by the solids (16·24), and multiply by 100.

$$19·03 \times \frac{100}{16·24} = 117·1°.$$

The opticity $[a]_D$ is therefore, 117·1°.

(2) *Determination of Maltodextrins in Wort.*

(*a*) *Estimation of Combined Dextrin.*—2 c.c. of wort are taken for the total reducing action. Amount of CuO yielded, 0·30 gram.

20 c.c. of the wort are now digested with 2·5 c.c. of normal cold-water extract for one hour at 130° F. The mixture is then cooled, diluted to 100 c.c., and 10 c.c. of the diluted liquid taken for the determination of the reducing power. This amount of degraded and diluted wort gave 0·34 gram CuO. From this we subtract the correction for reducing action of cold extract (1 c.c. = 0·08 gram CuO).

From 0·34 we therefore subtract 0·02 (see p. 478), giving 0·32. The CuO on the same quantity of original wort (2 c.c.) was 0·30. The increase is therefore 0·02, due to degradation of combined dextrin ; on 100 c.c. the increase will be 0·02 $\times$ 50 = 1·00 gram CuO ; equivalent to combined dextrin, 1·00 $\times$ 0·706 = 0·706 gram.

(*b*) *Determination of Combined Maltose.*—50 c.c. of the wort are " pitched " with 0·25 gram yeast ; another 50 c.c. treated in the same way with yeast, and 0·25 c.c. malt-extract

* These calculations are given in detail, and explained, under malt analysis, see p. 474.

added. Both worts are fermented in the manner previously described. After fermentation, both are diluted to 200 c.c., filtered, and 25 c.c. of each taken for the determination of reducing action.

Reducing action of wort without diastase on above quantity .. 0·13 gram CuO
    ,,   ,,   with diastase.. .. .. 0·07   ,,
Reduction due to combined maltose .. .. 0·06   ,,

To calculate this on 100 c.c. of original wort, we multiply 0·06 by 16; CuO from 100 c.c. of original wort will therefore be 0·96 gram CuO. The combined maltose will therefore be (on 100 c.c. wort) $0·96 \times 0·743 = 0·71$ gram.

We now know the combined maltose and the combined dextrin in 100 c.c. of wort; to calculate these on the wort-solids we multiply our results by 100, and divide by the solid matter per 100 c.c. (16·24 grams).

Combined maltose on 100 parts of wort-solids will therefore be $0·71 \times \dfrac{100}{16·24}$ and the combined dextrin $0·706 \times \dfrac{100}{16·24}$.

The results therefore stand (on 100 parts of wort-solids) :—

Maltose in maltodextrin	4·37
Dextrin in maltodextrin	4·34
Total maltodextrin	8·71
Type	{ 1 M { 1 D

If it is desired (though this is seldom necessary) to determine the precise proportions of free maltose and dextrin, as well as those of the maltodextrin in a wort, it is necessary to proceed precisely as in the analysis of a malt. The wort will replace the hot-water mash, and the opticity and reducing power must be corrected by those of a cold-water mash, made as in malt analysis, and analysed as there described. When, as is often the case, the wort has been prepared from more than one malt, it will be necessary to take each malt used, make cold-water mashes of each, and then strike a proper mean for the corrections on the wort. It would also be necessary to make hot-water mashes of the malts, and compare their mean gravity with that of the wort under analysis ; in the event of the wort, for instance, being twice as heavy as the hot-water mashes, the mean correction applied should be doubled.

The term mean correction is intended in this way : suppose three malts were used in the proportion of 2, 6, and 10 quarters respectively, we should then multiply the corrections of the first malt by 2, of the second by 6, and the third by 10. If we added these products together, and divided by the total (18), we should get an average.

As previously mentioned, it is exceedingly seldom that a full analysis of wort is needed. The estimation of the malto-dextrins only gives us, as a rule, all the information required.

CHAPTER XI.

THE ANALYSIS OF SUGARS AND HOPS.

INVERT-SUGAR.

IN samples of invert-sugar we require to determine the actual invert-sugar, the uninverted cane-sugar, the mineral matter, the albuminoids, the acidity (if any), the so-called "inert matter," and the moisture.

(1) *Invert Sugar.*

Take about 0·1 gram of the sample, dissolve in water, and pour into a beaker containing 30 c.c. of Fehling's solution and about 60 c.c. of water previously heated. The cuprous oxide precipitated is oxidised and weighed in the usual way. 1 gram of cupric oxide corresponds to 0·4715 gram of invert-sugar.

Example.—Took 0·165 gram of sample. The cuprous oxide, after oxidation, weighed (as cupric oxide) 0·245 gram.

$$0·245 \times 0·4715 = 0·1156 = \text{invert-sugar in amount of sample taken.}$$
$$\frac{0·1156 \times 100}{0·165} = 70·06 = \text{percentage of invert-sugar in sample.}$$

(2) *Uninverted Cane-sugar.*

When cane-sugar is digested at 150° F. for twenty minutes with a little hydrochloric acid, it is completely inverted; and the amount of it is estimable by determining the opticity before and after such inversion.

26·048 grams of cane-sugar dissolved in 100 c.c. of water will read in a 2-decimetre tube at 0° C., + 100°; the same after complete inversion and also at 0° will read, — 42·5°. The total change in reading incident to the complete inversion of the sugar is therefore 142·5° at 0° C. But this change in rotation is one degree less for every 2° C. of temperature over 0°. We must therefore make a suitable correction for temperature. Taking the amount of sugar above stated (26·048 grams in 100 c.c.) the percentage of cane-sugar in it is found by comparing the proportion of the change obtained to the maximum change of pure cane-sugar, after making suitable allowance for the temperature.

Example.—26·048 grams of the sample were dissolved in 100 c.c., and the reading determined in a 1-decimetre tube. The temperature was 61° F. (16° C.) ; the resulting reading was — 1·4°. 50 c.c. of the same solution were digested with 5 c.c. HCl at 150° F. for 20 minutes ; and the reading determined in a 1-decimetre tube at the same temperature as above, viz. 61° F. (16° C.). The reading was — 1·7°. To this — 1·7°, we have to add 10 per cent. of the reading, in order to correct for the dilution of the solution by the hydro-chloric acid, thus 1·7 + 0·17 = — 1·87°.

Difference in reading therefore in 1-decimetre tube 1·87 less 1·4 = 0·37 ; or in a 2-decimetre tube it would be 0·94.

We now make the correction for temperature on the number representing the change of pure cane-sugar. The temperature at which the readings were taken was 16° C. We must therefore divide 16 by 2, and subtract the result from 142·5. Thus, 142·5 less 8 = 134·5°.

Since both the standard reading for pure cane-sugar and the reading of the sample refer to solutions of the same concentration (26·048 grams in 100 c.c.), it will follow that the percentage of cane-sugar in the sample will be

$$\frac{0·94}{134·5} \times 100 = 0·69.^{*}$$

---

* The inversion of the cane-sugar can also be effected by digesting it with a small quantity of yeast for 1 hour at 133° F. (See p. 490).

(3) *Mineral Matter.*

About 5 grams of the sample are carefully incinerated and the ash weighed.

Example.—5·11 grams of sample were taken, and 0·073 gram of ash obtained. The percentage of mineral matter therefore equals $\dfrac{0.073}{5.11} \times 100 = 1.43$.

(4) *Albuminoids.*

About 2 grams of the sample are taken and treated with 10 c.c. of mixed acids, and the Kjeldahl process proceeded with in precisely the same manner as in the determination of albuminoids in malt (p. 465).

Example.— 2·845 grams taken. 20 c.c. $\dfrac{N}{20}$ H_2SO_4 in receiver required, after distillation, 8·6 c.c. $\dfrac{N}{20}$ NH_3 for neutralisation.

20 less 8·6 = 11·4
11·4 × ·0007 = ·00798 gram = N in sample taken
00798 × 6·3 = ·05027 gram = albuminoids in sample
or expressed as a percentage. $\dfrac{.05027 \times 100}{2.845} = 1.77$.

(5) *Moisture.*

The direct estimation of the water is not feasible, since in any attempt to eliminate it under ordinary pressure, there is an inevitable decomposition of the sugar. To perform the operation *in vacuo*, would involve trouble and inconvenience. The estimation is, however, accurately made by determining the specific gravity of a solution of known strength, and then calculating therefrom the solid matter in solution, from which by a further calculation we get the water.

About 20 grams of the sample are weighed out, dissolved in warm water, the solution cooled, and made up to 100 c.c. at 60° F. The specific gravity is then carefully taken in a specific gravity bottle.

Example:—Taking 20 grams, the specific gravity of the solution made up as above was 1061·76.

$$\text{Solid matter in solution } \frac{61·76}{3·86} = 16·00 \text{ grams.}$$

Solid matter from 100 grams sample, $16·00 \times \dfrac{100}{20} = 80·00$.

Water percentage, 100 less 80 = 20.

(6) *Inert Carbohydrates.*[*]

The so-called inert carbohydrates are obtained by adding together the above constituents, as determined by direct analysis, and then subtracting the sum from the solid matter. In this case, invert-sugar, cane-sugar, ash, and albuminoids add up to 73·95. The solid matter is 80. The inert matters are therefore 6·05 per cent.

(7) *Brewer's Extract*

is a function of the water percentage, to which it will be roughly indirectly proportional. It can be calculated from the gravity of the 20 per cent. solution, thus :—

Gravity of 20 per cent. solution	..	..	..		1061·76
Excess gravity over 1000..	..	..	..	..	61·76
„ „ „ 360[†]	..	..	..	..	$61·76 \times \dfrac{360}{1000}$

$$= 22·22 \text{ lb. per barrel.}$$

To get from this the extract due to 2 cwt. dissolved per barrel, we must multiply the above extract by 62·22 and divide by 20.[‡] Thus $\dfrac{62·22}{20} \times 22·22 = 69·12$.

---

[*] The method given above for the estimation of this group of bodies is probably erroneous. On the point of going to press, one of us has found that these bodies possess a distinct reducing action, and their *x* would appear to closely approximate to that of levulose. If this is confirmed by further experiments it will follow that the invert-sugar percentage as ordinarily estimated is too high, and that of inert carbohydrates too low.

[†] Excess over 360 gives lb. per barrel, since one barrel contains 36 gallons, each gallon weighing 10 lb.

[‡] 62·22 grams per 100 c.c. approximately represents 2 cwt. per barrel.

The above, therefore, is the brewer's extract per 2 cwt. in pounds per barrel.

In determining the brewer's extract, it is necessary to determine the gravity of a solution of the sample which shall be fairly near that of the ordinary wort. That this is necessary follows from the fact that excess gravity is only *approximately* proportional to concentration. Thus, if, for instance, we took a 10 per cent. solution of a sample, and on the other hand a 33·3 per cent. solution, and if in both cases we referred gravities as determined to pounds per 2 cwt. by calculation, then the gravity as determined on the weaker solution would apparently be greater than that on the stronger.

To be quite accurate, it would be advisable to make a special solution for extract determination, preferably 15·555 grams per 100 c.c. In the case of ordinary sugars, the gravity of such a solution would just about tally with that of the average wort, and by subtracting 1000, then multiplying by 0·36, and then by 4, we should at once get the brewer's extract per 2 cwt.

As a rule, however, it is unnecessary to make a special solution for this purpose, the result obtained from the 20 per cent. solution, prepared for the determination of water, giving results quite sufficiently near for practical purposes.

The sample of invert-sugar referred to, would stand thus :—

Invert-sugar	70·06
Cane-sugar	0·69
Inert carbohydrates	6·05
Ash	1·43
Albuminoids	1·77
Acidity*	none
Water	20·00
	100·00

Brewer's extract	69·12 lb. per 2 cwt.

* Ordinary sugars contain either no acid or only a mere trace of it. When it is sufficient to be estimable, 10 grams of sample are taken, dissolved in water, and neutralised with $\frac{N}{100}$ ammonia.

Remarks would follow here as to the colour of the sugar, consistency (fluid or semi-solid), brilliancy (or otherwise) of solution; &c., &c.

CANE-SUGAR.

The analysis of cane-sugar for brewers' purposes is conducted on the same lines as that laid down for invert-sugar—the constituents being the same, although, of course the cane-sugar will be very much more, while the reducing sugar (generally put down to glucose) is far less than in invert-sugar. On these grounds the quantities taken for analysis are rather different.

(1) *Glucose.*—Take about 1 gram of the sample, estimate the reducing power on Fehling's solution, and calculate as glucose, by multiplying the CuO by 0·4535 (for details, see determination of maltose in malt-worts, p. 463, and invert-sugar in invert samples, p. 485).

(2) *Cane-sugar.*—Exactly 26·042 grams are taken; the method employed is precisely the same as that described for the estimation of cane-sugar in inverts.

Instead of using the acid method described under invert-sugar, for determining the cane-sugar, we may make use of the invertive action of yeast, or of prepared invertase (p. 158). In the former case the neutral cane-sugar solution is heated to 55° C., (131° F.), and pressed yeast, to the extent of about one-tenth of the weight of cane-sugar, added. The whole is well stirred and the mixture allowed to stand for four hours at 55°; at the end of that time it is cooled, a little alumina cream added, and the bright solution filtered off. The opticity is determined at 16° C. (61° F.) before and after inversion, and the amount of cane-sugar present deduced from the fall of angle in precisely the same way as described above. The amount of cane-sugar thus found is calculated as a percentage on the amount taken in the usual way.

Ash, albuminoids, water, brewer's extract, as in inverts.

In determining the ash constituents, with the object of discovering whether the sugar has been prepared from the sugar-cane or the beet, stress must be laid upon the proportion of phosphoric acid, soda and potash.

(3) *Phosphoric Acid.*—A fair amount of ash having been prepared, about 0·25 gram of it is taken, and digested in a beaker for about half an hour with a mixture of one part of nitric acid and four parts of water. The mixture is then filtered, and the insoluble matter discarded. To the filtrate is added about 20 c.c. of a solution of ammonium molyblate, and digestion continued for three hours at a gentle heat. The precipitate is then washed by decantation with water containing a few drops of nitric acid ; the washings being passed through a filter. The precipitate in the original beaker is dissolved in ammonia, and that on the filter is returned to the beaker, where it is similarly dissolved. 10 c.c. of " magnesia mixture " are then added, and 50 c.c. of dilute ammonia (one part ammonia to three parts water) ; after standing for twelve hours the precipitate is filtered, washed with dilute ammonia, dried, and weighed as magnesium pyrophosphate. The weight of this multiplied by 0·64, gives the weight of P_2O_5 (phosphoric anhydride) in the amount of ash taken.

(4) *Soda and Potash.*—About 0·3 gram of the ash is taken, digested with water, an excess of barium hydrate added, and after ten minutes' digestion the whole is filtered. The filtrate containing the soda and potash is now heated with ammonium carbonate, and subsequently dealt with in precisely the same manner as has been described for the determination of the alkalies in waters (see p. 441).

GLUCOSE.

In glucoses we have to determine the dextrose (glucose), maltose, dextrin, gallisin, ash, albuminoids, water, and brewer's extract.

Since the dextrose, maltose, and gallisin have all a reducing action on Fehling's solution, it is not possible to determine these constituents separately as in the case of invert constituents; but having obtained analytical data as to cupric-oxide reducing power before and after fermentation, and opticity before and after fermentation, we are able from these to calculate the respective amounts of carbohydrates that are present.

(1) *Reducing Power.*—Take from 0·1 to 0·2 gram, dissolve, and determine cupric oxide reducing power in the usual way (p. 463).

(2) *Opticity.*—Take 10 or 20 grams of the sample (10 if a dark sample, and 20 if a pale one), dissolve and determine opticity in the usual manner (p. 464).

(3) *Reducing Power and Opticity after Fermentation.*—Take about 10 grams, dissolve in about 50 to 100 c.c. of water, add 1 gram yeast, and 0·25 c.c. of normal cold-water malt-extract.* To this is also added 10 c.c. of yeast water to afford nourishment to the yeast. The yeast water is prepared by stirring up some yeast (1 part) with water (10 parts), boiling and filtering, the filtrate only being used.

The fermenting mixture is kept at 80° F. for three days, then diluted to 200 c.c., a little alumina added, and the whole filtered. The reducing power is determined in 10 c.c. of the filtrate, and its opticity is also determined in a 1-decimetre tube.

---

* In no case have we found any difference in reducing power after fermentation between samples fermented with or without malt-extract; but to be on the safe side it is well to add it.

Example :—

Reducing Power before Fermentation.—Took 0·1201 gram of sample; CuO obtained, 0·1747 gram. This is calculated into percentage of dextrose on sample, thus—

$$\frac{0·1747 \times 0·4535 \times 100}{0·1201} = 65·97 \text{ per cent. apparent dextrose.}$$

Opticity before Fermentation.—Took 10 grams, dissolved in water and made up to 100 cc. Reading, 17·4 divs. in 1-decimetre tube.

This is calculated into $[a]_D$, thus—

$$\frac{17·4 \times 0·346 \times 100}{10} = 60°·2 = [a]_D.$$

Reducing Power after Fermentation.—10 grams taken; after fermentation, 10 c.c. of filtrate obtained as previously directed, gave 0·036 gram CuO. This is first calculated into dextrose as a percentage on sample, thus—

$$\frac{0·036 \times 0·4535 \times 200 \times 10}{10} = 3·3 \text{ per cent.}$$

Opticity after Fermentation.—Reading in 1-decimetre tube, 2·5 divs.

This is calculated into $[a]_D$, on sample, thus

$$\frac{2·5 \times 0·346 \times 200}{10} = 17·3° = [a]_D.$$

The loss in reducing power and the loss in opticity after fermentation will be due to the removal of the fermentable matter, the dextrose and maltose. We have thus—

Reducing power before fermentation	65·97 per cent.
„ „ after „	3·3 „
	62·67 „
Opticity before fermentation	60·2°
„ after „	17·3°
	42·9°

62·67 is therefore the reducing power in terms of dextrose (κ) of the mixture of dextrose and maltose in our sample; and 42·9 is the opticity ($[a]_D$) of this mixture. We can now determine the relative proportion of dextrose and maltose by the following simultaneous equations, where D and M represent respectively the dextrose and maltose in 100 parts of the sample*—

(1) 100 D + 61 M = 62·67 × 100
(2) 52·8 D + 135·4 M = 42·9 × 100.

From these equations we get—

$$D = 56·88$$
$$M = 9·50$$

The percentage of dextrose in the sample is therefore 56·88, and the percentage of maltose 9·50.

The gallisin is found from the reducing power of the solution after fermentation. The gallisin calculated in terms of dextrose was 3·3 (see p. 493.) Since the κ of gallisin is about 45, we can calculate the true percentage of gallisin, by multiplying the above figure by $\dfrac{100}{45}$; thus, $\dfrac{3·3 \times 100}{45} = 7·33$.

The dextrin is calculated as follows: the opticity due to the percentage of each of the other carbohydrates found is calculated, and the whole added together; the difference between this number and the opticity of the original sample will be due to dextrin. Thus—

Dextrose	56·88 ..	56·88 × 0·528 =	29·97
Maltose..	9·50 ..	9·50 × 1·36 =	12·93
Gallisin	7·33 ..	7·33 × 0·84† =	6·16
			49·06

Opticity of original sample	60·2°
Opticity due to carbohydrates determined	49·06°
	11·14°

---

* In these equations the figures given are the values of $[a]_D$ and κ for the respective sugars.

† These numbers and that used below in calculating the dextrin, express the opticity given by the respective sugars when 1 gram of each is dissolved in 100 c.c. and the reading taken in the 1-decimetre tube.

The opticity due to dextrin is therefore $11 \cdot 14°$, and—

$$\frac{11 \cdot 14}{1 \cdot 95} = 5 \cdot 71 = \text{percentage of dextrin.}$$

The determinations of *ash, albuminoids, water* and *inert carbohydrates* are carried out in precisely the same manner as described under these heads in invert-sugar analysis (p. 487). In this case, they came out as follow: ash, $1 \cdot 87$; albuminoids, $1 \cdot 62$; water, $15 \cdot 02$; inert carbohydrates, $2 \cdot 07$ per cent. Our analysis would therefore stand as follows :—

Dextrose	56·88
Maltose	9·50
Dextrin	5·71
Gallisin	7·33
Ash	1·87
Albuminoids	1·62
Water	15·02
Inert carbohydrates	2·07
	100·00

The brewer's extract is calculated in precisely the same manner as described in the case of invert-sugar.

HOPS.

There are only two determinations which are at all practicable in the case of hops. These are (1) the detection and determination of sulphur; and (2) the determination of tannic acid. From what has been said in a previous chapter (p. 204), it will be readily understood that any attempt to determine the other constituents of hops would be impracticable.

1. *Detection of Sulphur in Hops.*

The ordinary means of testing hops for sulphur consists in making an infusion of them in distilled water and then submitting the infusion to the action of hydrogen gas. For this purpose the infusion is poured into a flask, 2 or 3 grams of zinc added, and finally some water and about 5 c.c. of hydrochloric acid. The zinc and hydrochloric acid react upon one another with the formation of hydrogen, and if sulphur is present the hydrogen attacks it, forming sulphuretted hydrogen. The flask in which the experiment is conducted is provided with two tubes, one, passing only just below the cork, for the exit of the gas, the other passes to below the surface of the liquid and acts as a safety valve. If sulphuretted hydrogen is evolved, a strip of filter paper, moistened with a solution of lead acetate, held at the mouth of the tube by which the gas passes off will be discoloured.

Before pronouncing an opinion as to whether the hops are sulphured, any discolouration of the lead-acetate paper must be shown not to be due to the materials used. For this purpose a blank experiment is made in the same way as above, but without the hop infusion.

It must be admitted that hops which are not sulphured occasionally yield sulphuretted hydrogen, when acted upon for some time by hydrogen gas; hops which are sulphured yield sulphuretted hydrogen shortly after the evolution of gas commences.

Another, and a better method, is to distil an infusion, which is made as above. The distillate flows into a solution of iodine, in order to oxidise any sulphurous acid to sulphuric acid, and then tested for sulphuric acid with hydrochloric acid and barium chloride. If a precipitate is formed, sulphuric acid is present, and will be due to sulphurous acid in the distillate, and consequently to sulphur in the hops.

The latter method can, if required, be employed to quantitatively determine the sulpurous acid evolved. In that case a weighed quantity of hops is taken, the infusion made up to a definite volume, and an aliquot part of it taken for distillation. The sulphurous acid in the distillate is, after oxidation, determined as barium sulphate in the way described on p. 437.

2. *Estimation of Tannic Acid.*

Many methods have been proposed for the determination of tannic acid. They may be roughly divided into two groups, namely, those which depend upon the absorption of the tannin by hide finings, and those in which the tannin is oxidised, and the amount of oxygen used up then estimated. These different methods give widely discordant results, and there is, perhaps, no tannin-containing substance which gives more divergent results with the two methods than does hops. Thus, the same sample of hops, when examined by the first method, gave 4·15 per cent. of tannic acid, whilst by the second method only 1·8 per cent. was obtained.

A method, falling within the second class, has lately been proposed by Kokosinski. We have examined it, and consider that for comparative determinations it is to be recommended.

It is based on the property of tannic acid of absorbing iodine with great energy in the presence of alkaline carbonates, and is carried out as follows: *—10 grams of hops are boiled with water, and the solution made up to 500 c.c.; as hops often contain sulphurous acid, which reacts with iodine, a few drops of hydrogen peroxide are added to the hops before boiling; this oxidises the sulphurous acid to sulphuric acid, whilst the tannin remains untouched. The boiled extract of hops is carefully filtered.

Three flasks, of about 100 c.c. capacity, are now taken; to the first is added 10 c.c. of distilled water, to the second 10 c.c. of gallo-tannic acid, and to the third 10 c.c. of the hop infusion. To each flask 4 c.c. of normal sodium carbonate solution are added, and then 20 c.c. of the $\frac{1}{80}$th normal iodine solution. The iodine must be in excess, and, if neces-

* The reagents required for the determination are : normal sodium carbonate solution, normal sulphuric acid solution, $\frac{1}{10}$th normal iodine solution, $\frac{1}{80}$ths normal sodium thiosulphate solution (9·92 grams in 1 litre), a solution of pure gallo-tannic acid containing 0·05 gram in 100 c. c. ; and a freshly prepared starch solution.

2 K

sary, a further quantity must be added.   The iodine is allowed
to act for five minutes in each case, and the experiment must
be so arranged that this limit is not exceeded.   4 c.c. of
normal sulphuric acid are then added to each flask, and
immediately afterwards 10 c.c. of $\frac{4}{50}$ths normal sodium thiosulphate.   A few drops of starch solution are then added to
each flask, and the excess of thiosulphate is titrated back
with iodine solution.

The amount of iodine solution which has been used up in
the first flask gives the quantity of iodine absorbed by sodium
carbonate, by light, and by starch, and the amount lost by the
different values of the solutions of iodine and thiosulphate.

The second flask gives, in addition to that lost in the first,
the amount of iodine absorbed by the 5 milligrams of gallo-
tannic acid.   By subtracting the iodine value of the first flask
from this, and dividing the residue by 5, we get the amount
of iodine absorbed by 1 milligram of gallo-tannic acid.   From
these data, and the amount of iodine solution used up in the
third flask, the amount of tannic acid in the extract can be
calculated.   It is only necessary to subtract from the number
of c.c. of iodine solution found in the third flask, the number
of c.c. found in the first, in order to obtain the number of
c.c. of iodine solution which have combined with the hop-
tannin ; by dividing this difference by 5, the number of
milligrams of tannin contained in 10 c.c. hop-extract is
obtained.   Then, if

E = the number of c.c. of iodine solution in the first flask,
t = ,,   ,,   ,,   second flask,

and

h = ,,   ,,   ,,   third flask,

the percentage of tannic acid in the sample will be obtained
by the following equation :—

$$\frac{5(h - E)}{2(t - E)} = \text{percentage of tannic acid.}$$

Example.—10 grams of hops were boiled with water, after the addition of a few drops of hydrogen peroxide, and the extract made up to 500 c.c., and the determination proceeded with as described above. In the flask with 10 c.c. of water, 0·1 c.c. of iodine solution was used up; in the flask with 10 c.c. of gallo-tannic acid solution, 11·8 c.c. of iodine solution; and in that with 10 c.c. of hop-extract, 10·6 c.c. of iodine solution. The percentage of tannic acid in the hops is then

$$\frac{5 \times 10·5}{2 \times 11·7} = 2·24 \text{ per cent.}$$

CHAPTER XII.

THE ANALYSIS OF BEER.

THE analysis of a beer resolves itself into two parts: the chemical composition, and the microscopic appearance after forcing. Chemical analysis can also be applied with advantage to the beer after forcing; for the present, however, we will only deal with beer prior to forcing.

The points it is desired to know are :—

The original gravity (i. e., the gravity of wort previous to fermentation), the specific gravity, the percentages of malto-dextrins (and their type), of stable dextrin, and of low type fermentable maltodextrins (hitherto called free maltose), of fermentable sugars which have disappeared during the brewery fermentation, of ash, of albuminoids, and of acid, free and volatile.

The Original Gravity.—The gravity of the original wort was due to the solid matter in it; part of this has been fermented away, and part of it remains in the beer. The alcohol produced during the fermentation will be a measure of the amount of solid extract fermented away, and if, by previous experiment, we have determined the amount of solid extract corresponding to the alcohol formed, and if to this we add the extract actually remaining in the beer, we shall then know the total extract contained in the original wort, from which its gravity can be calculated.

Such previous determinations of corresponding alcohol proportions and losses of extract are provided in the Excise tables, which have been prepared on the results of a large number of experiments. These tables give us what we want in terms of, on the one hand, degrees by which water is lightened by the alcohol in the beer; and, on the other, degrees conferred by the corresponding amount of malt-extract.

To obtain the requisite data we must first find the gravity of the beer as due to unfermented solid-matter in it. This can be done by distilling off the alcohol from the beer, making up the residue to the original bulk, and determining the gravity. This gravity will be clearly only due to the unfermented extract in beers.

If during the distillation we collect the spirit, and make it up to the original bulk of the beer with water, and then determine the gravity, the difference between 1000 and that gravity will be proportionate to the alcohol produced, and roughly so to the solid extract lost during fermentation. The tables are so arranged that we can refer the above difference between 1000 and the weight of the alcoholic solution to the corresponding degrees conferred by the solid which yielded it, previous to its removal by fermentation.

The manipulation is as follows: the beer is well shaken to expel carbonic acid and 100 c.c. are measured at 60° F. This quantity is then transferred to a distilling flask, and the measuring vessel well washed and the washings added to the beer; the total volume should, however, not exceed 150 c.c.

The beer is now distilled, the 100 c.c. measure being used as the receiving flask. When about three-fourths is over the distillation is stopped. The distillate is rapidly cooled to 60°, made up with water at 60° to 100 c.c. and its specific gravity determined at 60° F.

The residue in the distilling flask is also cooled to 60°, made up with water at 60° to 100 c.c. (in the same measuring flask), and its specific gravity determined at 60° F.

Example :—

Specific gravity of distillate	990·11
„ „ residue	1031·05

Now 1000 less 990·1 = 9·9 (spirit indication).

> 9·9 by table .. .. ∴ .. .. 43·70
> Specific gravity of residue .. .. 1031·05

Therefore,

> Original gravity .. .. 1074·75

The following is the Excise table which is used for the purpose.

SPIRIT INDICATION, WITH CORRESPONDING DEGREES OF GRAVITY LOST IN MALT WORTS BY THE "DISTILLATION PROCESS."

Degrees of Spirit Indication.	·0	·1	·2	·3	·4	·5	·6	·7	·8	·9
0	..	·3	·6	·9	1·2	1·5	1·8	2·1	2·4	2·7
1	3·0	3·3	3·7	4·1	4·4	4·8	5·1	5·5	5·9	6·2
2	6·6	7·0	7·4	7·8	8·2	8·6	9·0	9·4	9·8	10·2
3	10·7	11·1	11·5	12·0	12·4	12·9	13·3	13·8	14·2	14·7
4	15·1	15·5	16·0	16·4	16·8	17·3	17·7	1·2	18·6	19·1
5	19·5	19·9	20·4	20·9	21·3	21·8	22·2	22·7	23·1	23·6
6	24·1	24·6	25·0	25·5	26·0	26·4	26·9	27·4	27·8	28·3
7	28·8	29·2	29·7	30·2	30·7	31·2	31·7	32·2	32·7	33·2
8	33·7	34·3	34·8	35·4	35·9	36·5	37·0	37·5	38·0	38·6
9	39·1	39·7	40·2	40·7	41·2	41·7	42·2	42·7	43·2	43·7
10	44·2	44·7	45·1	45·6	46·0	46·5	47·0	47·5	48·0	48·5
11	49·0	49·6	50·1	50·6	51·2	51·7	52·2	52·7	53·3	53·8
12	54·3	54·9	55·4	55·9	56·4	56·9	57·4	57·9	58·4	58·9
13	59·4	60·0	60·5	61·1	61·6	62·2	62·7	63·3	63·3	64·3
14	64·8	65·4	65·9	66·5	67·1	67·6	68·2	68·7	69·3	69·9
15	70·5	71·1	71·7	72·3	72·9	73·5	74·1	74·7	75·3	75·9
16	76·5									

When a beer has become very acid with acetic acid, it is necessary to make an allowance in respect of it; firstly, because some of the alcohol formed has been lost by being converted into acetic acid, and secondly, because the acetic acid will distil over with the alcohol, and raise the specific gravity of the distillate, and consequently reduce the apparent spirit indication and also the original gravity.

In such cases a second distillation is performed with sufficient alkali added before distillation to neutralise the acid, and this will prevent the distillation of the acetic acid. To make a correction for the loss of alcohol by conversion into acid we determine the acetic acid as follows : 100 c.c. of beer are taken and titrated with ammonia of 998·6 specific gravity, red litmus paper being used as an indicator. Each c.c. of ammonia used represents o·1 of acetic acid. This will give us the *total acid* of the beer calculated as acetic acid. We then take 100 c.c. of beer, evaporate it to dryness on a water bath, redissolve the residue in water, and titrate as before. This gives us the *fixed acid* calculated as acetic acid. The difference between the two gives us the volatile acid, or real acetic acid. The above percentage of acid is then referred to a table for corresponding loss of spirit indication, and this spirit indication is then added to that obtained by distillation.

Acetic Acid.	·00	·01	·02	·03	·04	·05	·06	·07	·08	·09
·0	..	·02	·04	·06	·07	·08	·09	·11	·12	·13
·1	·14	·15	·17	·18	·19	·21	·22	·23	·24	·26
·2	·27	·28	·29	·31	·32	·33	·34	·35	·37	·38
·3	·39	·40	·42	·43	·44	·46	·47	·48	·49	·51
·4	·52	·53	·55	·56	·57	·59	·60	·61	·62	·64
·5	·65	·66	·67	·69	·70	·71	·72	·73	·75	·76
·6	·77	·78	·80	·81	·82	·84	·85	·86	·87	·89
·7	·90	·91	·93	·94	·95	·97	·98	·99	1·00	1·02
·8	1·03	1·04	1·05	1·07	1·08	1·09	1·10	1·11	1·13	1·14
·9	1·15	1·16	1·18	1·19	1·21	1·22	1·23	1·25	1·26	1·28
1·0	1·29	1·31	1·33	1·35	1·36	1·37	1·38	1·40	1·41	1·42

Example :—

100 c.c. of beer required for neutralisation, 15 c.c. of standard ammonia solution
   ,,   ,,   after boiling .. .. .. 10 c.c.   ,,   ,,   ,,

Total acid as acetic acid .. .. 0·15
Fixed acid .. .. 0·10
————
Acetic acid .. .. 0·05
————

0·05 by table = 0·08 ; 0·08 would then be added to the spirit indication as determined by distillation and the total referred to the first table.

Specific Gravity of Beer.—This is determined on the beer after expulsion of gas, by agitation at 60° F.

Determination of Maltodextrin ; Dextrin ; Fermented Sugars ; and Fermentable low-type Maltodextrins (apparent free Maltose) unfermented. For this we require the following estimations :

Total Reducing Power.—Take 5 c.c. (2 to 3 c.c. for very strong beers, and 10 c.c. for very weak beers) and determine reducing power on Fehling's solution in the usual way (see p. 463).

Opticity.—50 c.c. of the beer are taken, a little alumina added, the whole diluted to 100 c.c., filtered, and the opticity determined in the usual way in a 1-decimetre tube.

Degradation by Malt-Extract.—50 c.c. of beer are evaporated to about half its bulk, and transferred to a 100 c.c. measure ; 5 c.c. of cold-water malt-extract (normal strength) added, and the whole digested at 125° F. for one hour. It is then cooled, diluted to 100 c.c., and the reducing power determined in 10 c.c.

The reducing power of the cold-water malt-extract is estimated in 1 c.c., and the figure thus obtained is applied as a correction.

Fermentation.—50 c.c. of the beer are evaporated to about 25 c.c., diluted to about 100 c.c. with cold water, and about 0·25 gram yeast added. Ferment for 72 hours at 80° ; then add some alumina, dilute to 200 c.c., filter, and estimate the reducing power in 25 c.c. of filtrate.

To correct the above for the non-degradable, unfermentable reducing substances present in the beer, we ferment a duplicate of the above in precisely the same manner, the only difference being the addition, prior to fermentation, of 0·25 c.c. of cold-water malt-extract. The reducing action is, as before, taken in 25 c.c., after dilution to 200 c.c., and filtration.

The *ash* is determined in 100 c.c. of the beer in the usual way.

The *albuminoids* are determined in 10 c.c. of the beer evaporated to dryness ; the residue is treated as in ordinary Kjeldahl operations.

The fixed and volatile acids have been already treated of.

Example :—

Original gravity (as stated p. 502)	1074·75
Specific gravity	1021·38

Reducing Power.—2 c.c. taken, CuO obtained, 0·083 gram. The above is calculated into total maltose on 100 c.c. of beer—

$$0·083 \times 0·743 \times 50 = 3·08 \text{ maltose grms.}$$

Opticity.—50 c.c. of the beer diluted to 100 c.c. ; reading in 1-decimetre tube, 12·4 divs. To get the dextrin, we must double the reading to correct for previous dilution.

$$12·4 \times 2 = 24·8.$$

Reading due to maltose will be $3·08 \times 3·905 = 12·01$ divs.
„   dextrin „ $24·8 - 12·01 = 12·79$ „
Per cent. of dextrin in beer $\cdot \dfrac{12·79}{5·625} = 2·27$.

Degradation.—50 c.c. of beer taken, and 5 c.c. of cold extract used ; the whole after degradation diluted to 100 c.c., CuO from 10 c.c. = 0·2985 gram. The CuO per 1 c.c. of cold-water extract was found to be 0·08 gram.

* For explanation of these factors, see ' Malt Analysis,' p. 475.

Before correcting, we will calculate the total maltose after degradation in 100 c.c. of the degraded mixture, thus

$$0.2985 \times 0.743 \times \frac{100}{10} = 2.216 \text{ grams maltose.}$$

We correct this for the CuO given by the malt-extract as follows:—

$$0.08 \times 5 \times 0.743 = 0.297 \text{ gram maltose.}$$

Then 2.216 less $0.297 = 1.919 = $ maltose after degradation in 100 c.c. of the degraded mixture; or,

$$\text{in 100 c.c. beer, } 1.919 \times \frac{100}{50} = 3.838 \text{ grams maltose.}$$

The above figure represents total maltose, plus the maltose formed by degradation of the combined (or amyloïn) dextrin.

Combined or amyloïn dextrin is therefore, 3.838 less total maltose, 3.08, after deducting the increase in molecular weight due to conversion into maltose, thus

$$3.838 - 3.08 = 0.758.$$

And $0.758 \times 0.95 = 0.72$ gram combined dextrin in 100 c.c. of beer.

Fermentation.—(a) Without malt-extract: 50 c.c. taken; after fermentation and dilution to 200 c.c., 25 c.c. taken for reducing action gave 0.174 gram CuO.

(b) With malt-extract: the same quantities gave 0.12 gram CuO.

CuO due to combined maltose $0.174 - 0.12 = 0.054$.

$$0.054 \times 0.743 \times \frac{200}{25} \times \frac{100}{50} = 0.64 \text{ gram combined maltose (amyloïn}$$
maltose) in 100 c.c. of beer.

We can now calculate the apparent free maltose, i. e., the low-type maltodextrins unfermented during brewery fermentation, but fermented during the laboratory fermentation. To do this we subtract from the total maltose, the maltose unfermented during fermentation without malt-extract; the latter

figure will be $0.174 \times 0.743 \times \dfrac{200}{25} \times \dfrac{100}{50} = 2.07$; then

Total maltose	3.08 grams
Maltose after fermentation	2.07 ,,
Fermentable maltodextrin, calculated as maltose ..	1.01 ,,

The free dextrin will be the difference between the total dextrin and combined dextrin, i.e., $2 \cdot 27$ less $0 \cdot 72 = 1 \cdot 55$.

The whole of the previous results, which now exist as grams in 100 c.c. of beer, must now be calculated as parts per 100 parts of original solid-matter. The original gravity having been found to be $1074 \cdot 75$, the solid matter per 100 c.c. is $\frac{74 \cdot 75}{3 \cdot 86} = 19 \cdot 36$.

On 100 parts wort-solids.

$$\text{Fermentable maltodextrins calculated as} \atop \text{maltose (so called free maltose)} \Big\} \quad 1 \cdot 01 \times \frac{100}{19 \cdot 36} = 5 \cdot 22$$

Malto-dextrin $\begin{cases} \text{Combined maltose} \quad .. \quad .. \quad .. \quad .. \quad 0 \cdot 64 \times \frac{100}{19 \cdot 36} = 3 \cdot 30 \\ \quad \text{,,} \quad \text{dextrin} \quad .. \quad .. \quad .. \quad .. \quad 0 \cdot 72 \times \frac{100}{19 \cdot 36} = 3 \cdot 72 \end{cases}$

Free dextrin $.. \quad .. \quad .. \quad .. \quad .. \quad .. \quad 1 \cdot 55 \times \frac{100}{19 \cdot 36} = 8 \cdot 01$

To get the matters fermented during fermentation, we take the spirit indication, which in this case was $9 \cdot 9$ (see p. 502), and refer, as before, to the Excise table; $43 \cdot 7$ is the corresponding figure, and represents the degrees lost during fermentation owing to loss of fermented extract. If we divide this by $3 \cdot 86$, we get solid extract lost, in grams per 100 c.c. of wort.

$$\frac{43 \cdot 7}{3 \cdot 86} = 11 \cdot 32 \text{ grams.}$$

To calculate on wort-solids we must, as before, multiply by $\frac{100}{19 \cdot 36}$

$$11 \cdot 32 \times \frac{100}{19 \cdot 36} = 58 \cdot 47 \text{ per cent.*}$$

The ash and albuminoids and fixed acid in this beer, calculated on 100 parts of the wort-solids, were respectively,

Ash $\quad .. \quad .. \quad .. \quad .. \quad .. \quad .. \quad .. \quad .. \quad .. \quad 1 \cdot 34$ per cent.
Albuminoids $\quad .. \quad .. \quad .. \quad .. \quad .. \quad .. \quad .. \quad 3 \ 36$,,
Acid as lactic acid $\quad .. \quad .. \quad .. \quad .. \quad .. \quad .. \quad 0 \cdot 52$,,

* There is a slight error here, due to loss of mineral matter and nitrogenous matter during fermentation. Such losses cannot be ascertained, but they are so inconsiderable as to be negligible.

The following will therefore represent the analysis calculated on 100 parts of original wort-solids.

	On 100 parts original wort-solids.	
Maltose and other fermentable sugars fermented during brewing fermentation	58·47	
Low type maltodextrins unfermented during brewery fermentation, but fermentable with active yeast (so called free maltose)	5·22	
Maltose in maltodextrins	3·30	Total maltodextrin 7·02
Dextrin in maltodextrins	3·72	
Free dextrin	8·01	
Mineral matter	1·34	
Nitrogenous matter as albuminoids..	3·36	
Acid as lactic acid	0·52	
Undetermined matter (hop-extract, glycerine, unfermentable portions of malt adjuncts, and malt worts, &c., &c.) by difference	16·06	
	100·00	

Original gravity	1074·75
Specific gravity	1021·38
Free acid in 100 c.c. of beer (acetic acid)	0·05
Fixed acid as lactic acid	0·15

THE FORCING OF BEER.

It is very essential that all beers should be submitted to the operation known as "forcing." The mode of examination was devised and introduced by Horace T. Brown, in 1875, as a result of some years' previous work on the application of the microscope to the examination of beers. Briefly stated, "forcing" consists in placing a sample of the beer under conditions which cause it to mature with much greater rapidity than in cask. The conditions embrace an elevated temperature, and an exclusion of air from the liquid.

The apparatus required for carrying out the process is very simple. First, and most important, is the "forcing tray." This consists of an oblong, flat copper vessel or tank, containing water. This may vary much in size, depending, of course, upon the space in which it is desired to place it ; but it should be at least twice as long as it is broad, and from 2 to 3 inches deep. In one corner of the tray there should be a tubular opening, with a short projecting piece of copper tube, in order to allow of the tray being filled with water. Inside the tray it is usual to fix a plate of thin copper midway between the top and bottom, and extending to within a few inches of the sides. The object of this is to promote the circulation of the water in the tray, and thus prevent any one portion of the upper surface becoming warmer than the rest, as would be the case if the water heated by the burner were allowed to impinge directly upon the upper surface of the tank. The forcing tray may be supported either on brackets fixed to the wall, or on a suitable iron-stand, and having been filled with water, preferably distilled, it is maintained at a constant temperature of about 85° F. The heat is supplied by means of a gas burner consisting of a row of small jets. This may be very conveniently made by taking a T-shaped piece of three-eighths brass tubing, and piercing a series of small holes, at equal distances, on the upper side of the T, which should be from 9 to 12 inches long, according to the size of the tray. The burner is fixed from 3 to 6 inches from the bottom of the tray, and attached to a Page's or other regulator, which is in turn connected with the gas supply. The regulator is introduced for the purpose of maintaining a constant temperature on the tray, independently of any varia- tion in the temperature of the tray itself.

The samples of beer are placed upon the tray in specially designed flasks. The original form of flask has a capacity of from 80 to 100 c.c., and is provided with a short tubulure or side-tube, projecting from the neck ; when in use, a small piece of quill tubing with an elbow bend is connected to this by means of indiarubber tubing, and the mouth of the flask is closed with an indiarubber bung. More recent forms of the flasks have been made with the side-tube and elbow-tube all in one piece, and with ground glass stoppers. We prefer, however, the old form, as the flasks with bent-over side-tubes are very apt to break in cleaning, and they do not allow of such ready cleaning as the simple straight side tube, whilst the flasks with glass stoppers require the most careful selection in order to prevent leakage of air, which is fatal to the method.

The forcing flasks, and the bottles in which the samples are to be collected, require to be most carefully cleaned and sterilised before being used. When the former have been previously used, it is sometimes necessary to use a little acid to remove the sediment; but in all cases, both flasks and bottles require to be cleaned with a little caustic soda, then thoroughly rinsed with clean water, and finally boiled with fresh distilled water. The latter operation is best effected by filling the flasks to the neck with the distilled water, and then placing them up to the neck in water in a shallow vessel with a false bottom, and boiling the water in this vessel for at least half an hour. After cooling, the flasks and bottles are emptied and at once placed in an inverted position in a draining rack, and used immediately. The indiarubber bungs and side tubes should also be boiled in water immediately before use.

The bottles used for taking samples should be cleaned as above, and should be carefully rinsed with the beer before filling. The point at which the samples should be taken varies somewhat with the different methods pursued in breweries. Where a sample cask of each brew is drawn from the racking vessel and stored for future reference, it is convenient that the sample should be taken from this cask; where this is not the case, the sample should be taken from the racking vessel immediately before the beer is racked into trade casks.

The sample should be transferred to the forcing flask and placed on the tray within twenty-four hours of being taken. Any sediment which may have settled in the bottle should be distributed through the liquid, and the flask should be rinsed out three times with successive small portions of the beer under examination. Having filled the flask to within a quarter of an inch of the side-tube, the mouth of the flask is carefully and tightly closed with a clean indiarubber bung, and the side-tube is connected to the elbow-tube by a piece of tight-fitting indiarubber tubing, both ends of which are bound with thin copper wire to the respective glass tubes. The flasks are then labelled and placed on the tray, the open ends of the elbow-tubes being placed in small beakers containing mercury; the tubes should dip at least half-an-inch into the mercury in the beaker, and six of the flasks may be arranged round each beaker. The mercury in the beakers allows the carbonic acid evolved to escape, and prevents any entrance of air.

The sample is kept on the tray according to the class of beer; we recommend four weeks for an export ale, three weeks for a stock ale, and seven to ten days for a running ale.

While on the tray, it is observed from time to time, to see whether it is brightening, whether it is conditioning, and so forth.

At the expiration of the forcing period, it is taken off, the supernatant beer being carefully poured into a beaker for further investigation, and the sediment into a small glass for microscopic examination.

The advantages to be derived from the use of the forcing tray may be stated to consist in the determination of the following:

(1) The rate of conditioning;

(2) The power of clarification;

(3) The amount and nature of bacteria latent in the beer;

(4) The condition of the primary yeast; and

(5) The amount and nature of the secondary yeast.

The results of the microscopic examination are the most important, but there are other tests which should not be ignored; these we will take first; they are performed on the main bulk of beer in the beaker.

Flavour.—The beer is tasted, and due regard paid to any deterioration in respect of (*a*) thinness, (*b*) rank bitterness, (*c*) flatness, (*d*) acidity, (*e*) putridness.

The acidity may be estimated with advantage in certain cases, and the increase during forcing determined.

Specific Gravity.—If the specific gravity be taken after forcing, and compared with that of the beer before that process, we get some idea of the conditioning of the beer; some idea whether the conditioning is normal, or whether during storage the beer will be flat on the one hand, or be inclined to fret on the other. The loss of sugar, corresponding to the decrease in gravity, can also be calculated.

Thus in the beer, the analysis of which is given above, the specific gravity was 1021·38. Suppose after forcing, the specific gravity had decreased to 1011·38. The loss in gravity would be 10°, due partly to the formation of alcohol, partly to the degradation of maltodextrins into maltose, and the subsequent loss of that maltose by fermentation. We can roughly get at the loss of gravity due to maltose, by multiplying the gross loss by 0·8.

$$10 \times 0\cdot8 = 8 \text{ degrees.}$$

To get the corresponding loss of solid maltose in grams per 100 c.c. of beer we divide by 3·86. To refer this figure to a percentage on the wort-solids we divide by the solids in the original wort (in this case 19.36) and multiply by 100. Loss of maltose during forcing, therefore

$$\frac{8}{3\cdot86} \times \frac{100}{19\cdot36} = 10\cdot7 \text{ per cent.}$$

This is a purely hypothetical case, and in the great majority of instances the loss of maltose would be far less.

We also lay stress upon the manner in which the beer brightens, and whether the sediment is compact and solid, or loose and light. The former kind of sediment is always preferable.

Plate.X.

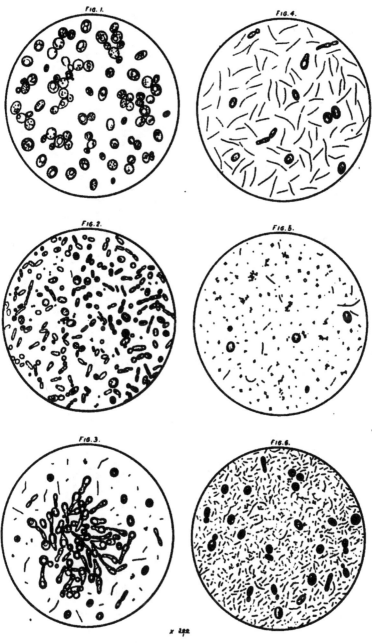

FIG. 1.   FIG. 4.

FIG. 2.   FIG. 5.

FIG. 3.   FIG. 6.

x 292

"Forced Beer" sediments
(after Matthews & Lott)

E & F N Spon, London & New York

When the beer remains turbid, we recommend the treatment of it, after forcing, with potash solution : if the turbidity is due to amorphous particles of hop-resins or albuminoids, it will go bright on addition of potash ; if to suspended yeast or bacteria, the potash will not brighten it.

In the microscopical examination, we have to lay due stress upon the state of the yeast and the number and form of bacteria. As regards the first, whether it is actively growing; in the case of the latter, whether or no they are predominant, and which species predominate. To properly interpret the microscopic appearance of forced beer sediments is more than can be taught by book work ; for the interpretation to be just and reliable, actual experience is necessary; but through the kindness of Messrs. Matthews and Lott we are able to reproduce from their valuable work, 'The Microscope in the Brewing Room,' a plate (Plate X.) of beer sediments, which we will endeavour to very briefly interpret. For this very important branch of the subject, the reader is referred to Messrs. Matthews and Lott's work, which deals very exhaustively with it.

EXPLANATION OF PLATE X.

FIG. 1.—This sediment would indicate a most satisfactory state of conditions, and a much more satisfactory one than is usually found.

FIG. 2.—Here we seem to be dealing with various sorts of wild-yeasts ; bacteria are also present, but not to any marked extent. Such a sediment as this would be found in a summer brewing, and might suggest a tendency to frettiness.

FIG. 3.—The most noticeable feature here is a marked growth of *S. coagulatus*. There are also more bacteria than should be found in a good sediment. The predominance of *coagulatus* would probably lead to a thin-tasting beer of inferior stability.

FIG. 4.—The predominant organism here is *Bacillus amylobacter*, and such a growth should direct attention to the plant. Dirty plant is generally the cause of a deposit of this kind. The beer may possibly not go acid, but it will acquire a sickly flavour, and show "smokiness," or "silkiness," rather than turbidity.

Fɪɢ. 5.—There is a considerable growth of bacteria here; a sediment of this kind would indicate instability. We notice many Sarcinæ, which under some conditions of nutriment would produce ropiness, under others lactic acid.

Fɪɢ. 6.—This is a typically bad deposit, and there is but little chance for the stability of a beer which gives such a sediment as this on forcing.

Very occasionally during forcing the mercury will be sucked back into the forcing flask; when this is so, it arises from defective conditioning, and is generally found in beers that have attenuated too low. It is always an unsatisfactory sign. Similarly, beers which have attenuated too low may sometimes develop film formations on their surface. This, too, is a bad sign.

APPENDIX.

STANDARD SOLUTIONS AND REAGENTS.

1. Distilled Water Free from Ammonia.

If, when 1 c.c. of Nessler's solution is added to 50 c.c. of distilled water in a glass jar, placed on a white surface, no trace of a yellow colour is developed in four or five minutes, the water is sufficiently pure for use. As, however, this is rarely the case, the water must be purified. For this purpose, distil from a large retort, ordinary distilled water, previously rendered distinctly alkaline with sodium carbonate. Discard the first 50 c.c. Test the remainder from time to time with Nessler's solution as above described, and when free from ammonia collect for use in a clean stoppered bottle.

2. Nessler's Solution.

Dissolve 17 grams of mercuric chloride in about 300 c.c. of distilled water. Dissolve 35 grams of potassium iodide in 100 c.c. of water. Add the former solution to the latter with constant stirring, until a slight permanent red precipitate is produced. Next dissolve 120 grams of potassium hydrate in about 200 c.c. of water, allow the solution to cool, add it to the above solution, and make up with water to 1 litre. Then add mercuric chloride solution until a permanent precipitate again forms; allow to stand till settled, and decant off the clear solution for use.

3. Sodium Carbonate.

For determination of free and saline ammonia, dissolve 100 grams of recently ignited sodium carbonate in water, and make up to 1 litre.

4. Alkaline Potassium Permanganate Solution.

Dissolve 100 grams of potassium hydrate and 8 grams of pure potassium permanganate in 1100 c.c. of water, and boil the solution rapidly until concentrated to about 1000 c.c. Cool, and preserve in a well-stoppered bottle.

5. Standard Solution of Ammonium Chloride.

Dissolve 3·15 grams of pure dry ammonium chloride in water, and dilute to 1 litre. This forms a stock solution. For use, dilute the above with water (free from ammonia) to 100 times its bulk; say, 10 c.c. to 1 litre. 1 c.c. of the latter solution contains 0·00001 gram of ammonia.

6. Standard Solution of Silver Nitrate.

Dissolve 4·79 grams of pure recrystallised silver nitrate in water, and dilute to 1 litre.

7. Potassium Chromate.

A saturated solution of the pure salt is employed.

8. Standard Solution of Potassium Permanganate.

Dissolve 0·395 gram of pure potassium permanganate in water, and dilute the solution to 1 litre. 1 c.c. = 0·001 gram available oxygen.

9. Potassium Iodide Solution.

Ten grams of the pure salt are dissolved in 100 c.c. of distilled water.

10. Dilute Sulphuric Acid (for Determination of Oxygen Absorption).

One part by volume of pure, strong sulphuric acid is mixed with twenty parts by volume of distilled water.

11. Sodium Hyposulphite Solution.

One gram of crystallised sodium hyposulphite (thiosulphate) is dissolved in 1 litre of water.

12. ZINC FOIL.

This should be cut into small pieces and preserved in a well-stoppered bottle.

13. COPPER SULPHATE SOLUTION.

Dissolve 30 grams of pure crystallised copper sulphate in 1 litre of water.

14. SOLUTIONS REQUIRED FOR DETERMINATION OF HARDNESS.

(*a*) *Standard Solution of Calcium Chloride.*—Dissolve 0·2 gram of pure calcium carbonate in hydrochloric acid. Evaporate the solution to dryness in a platinum dish on the water-bath, add a little distilled water, and again evaporate to dryness. Repeat the evaporation several times, in order to expel the excess of hydrochloric acid. Finally, dissolve the residue in water, and dilute to 1 litre.

(*b*) *Standard Soap Solution.*—Grind together in a mortar, 150 parts of "lead plaster" and 40 parts of potassium carbonate. Then add a little methylated spirit, and continue triturating until a creamy mixture is obtained. Allow to stand for a few hours, transfer to a filter, and wash several times with methylated spirit. The strong solution of soap thus obtained must be diluted with a mixture of one volume of distilled water and two volumes of methylated spirit (considering the soap solution as spirit) until exactly 14·25 c.c. are required to form a permanent lather with 50 c.c. of the calcium chloride solution (*a*), the experiment being carried out in the same way as in determining the hardness of a water.

15. FEHLING'S SOLUTION.

(*a*) *Solution of Copper Sulphate.*—34·64 grams of pure recrystallised copper sulphate are dissolved in water; about 0·5 c.c. of strong sulphuric acid is added, and the solution is diluted to 500 c.c.

(*b*) *Alkaline Solution of Rochelle Salt.*—175 grams of powdered Rochelle salt, and 90 grams of pure potassium hydrate are dissolved in water, and the solution is made up to 500 c.c.

A mixture of equal volumes of these solutions (*a* and *b*), constitutes Fehling's solution. The solution should preferably be stocked separately, and mixed as required.

16. $\frac{N}{20}$ Sulphuric Acid.

1·75 c.c. of pure concentrated sulphuric acid are diluted with water to 1 litre. The exact strength of this solution is then determined as follows: 0·1 to 0·2 gram of pure, recently ignited, sodium carbonate (accurately weighed) is dissolved in water, one or two drops of a solution of methyl orange are added, and the diluted acid prepared as above is allowed to flow in from a burette, until the yellow colour of the methyl orange is *just* changed to pink. The number of cubic centimetres required to do this must be carefully noted. A strictly accurate $\frac{N}{20}$ acid should exactly neutralise sodium carbonate in the proportion of 100 c.c. to 0·265 gram. Suppose that 0·1 gram sodium carbonate required 31 c.c. of the dilute acid, then

$$0\cdot1 : 0\cdot265 :: 41 : x$$
$$x = 82\cdot1 \text{ c.c.}$$

The above proportion shows that 0·265 gram of sodium carbonate would require 82·1 c.c. of dilute acid for neutralisation, instead of 100 c.c. as in the case of $\frac{N}{20}$ acid. All that remains to be done, therefore, in order to reduce the acid to the $\frac{N}{20}$ strength, is to dilute every 82·1 c.c. of it with water, to 100 c.c. Or, 821 c.c. may be diluted to 1000 c.c. The latter operation is performed in a graduated glass cylinder.

1 c.c. of $\frac{N}{20}$ $H_2SO_4 = 0\cdot0007$ gram of nitrogen.

1 c.c. ,, ,, $= 0\cdot00085$,, ammonia.

17. $\frac{N}{20}$ Ammonia.

Strong ammonia is diluted with water, until it is found by experiment that 50 c.c. of the diluted solution are exactly neutralised by 50 c.c. of the $\frac{N}{20}$ sulphuric acid prepared as above. 3 or 4 c.c. of strong ammonia made up to 1000 c.c. will be about the right degree of dilution to commence with.

18. Ammonia Employed for the Estimation of Acidity in Malt, Beer, etc.

Strong ammonia is diluted with water, until it has a specific gravity of 998·6 at 60° F.

> 1 c.c. of this ammonia = 0·01 gram acetic acid.
> ,,   ,,   ,,   = 0·015 ,, lactic acid.

19. Soda Solution (used in the Kjeldahl Process).

Sodium hydrate (stick soda) is dissolved in water, until the solution (when cold) has a specific gravity of 1300.

20. Sulphuric Acid (used in the Kjeldahl Process).

Mix equal volumes of pure, concentrated, sulphuric acid and fuming, or *Nordhausen*, sulphuric acid. Preserve the mixture in a well-stoppered bottle.

21. Methyl Orange.

Dissolve 1 gram of methyl orange in 100 c.c. of alcohol.

22. Potassium Permanganate (for Kjeldahl Process).

The crystallised salt is coarsely powdered.

23. Thymol.

Prepare a saturated solution of thymol in alcohol.

24. Alumina Cream.

Moist alumina is thoroughly ground with water, and then further diluted until it is of the consistency of a thin cream.

25. Ammonium Molybdate Solution.

10 grams of powdered ammonium molybdate are dissolved in 40 c.c. of dilute ammonia (one volume of strong ammonia to two

volumes of water). The solution is poured into a mixture of 120 c.c. of strong nitric acid and 40 c.c. of water. The whole is heated to about 100° F. for some hours, and the clear liquid drawn off.

26. MAGNESIA MIXTURE.

Made by dissolving 11 grams of crystallised magnesium chloride, and 14 grams of ammonium chloride in 130 c.c. of water, and diluting the mixture to 200 c.c. with a strong solution of ammonia.

INDEX.

——◆◆——

A.

B.

LONDON: PRINTED BY WILLIAM CLOWES AND SONS, LIMITED,
STAMFORD STREET AND CHARING CROSS.

1892.

BOOKS RELATING

TO

APPLIED SCIENCE

PUBLISHED BY

E. & F. N. SPON,

LONDON: 125, STRAND.

NEW YORK: 12, CORTLANDT STREET.

———•———

The Engineers' Sketch-Book of Mechanical Movements, Devices, Appliances, Contrivances, Details employed in the Design and Construction of Machinery for every purpose. Collected from numerous Sources and from Actual Work. Classified and Arranged for Reference. *Nearly 2000 Illustrations.* By T. B. BARBER, Engineer. Second Edition, 8vo, cloth, 7s. 6d.

A Pocket-Book for Chemists, Chemical Manufacturers, Metallurgists, Dyers, Distillers, Brewers, Sugar Refiners, Photographers, Students, etc., etc. By THOMAS BAYLEY, Assoc. R.C. Sc. Ireland, Analytical and Consulting Chemist and Assayer. Fifth edition, 481 pp., royal 32mo, roan, gilt edges, 5s.

SYNOPSIS OF CONTENTS:

Atomic Weights and Factors—Useful Data—Chemical Calculations—Rules for Indirect Analysis—Weights and Measures—Thermometers and Barometers—Chemical Physics—Boiling Points, etc.—Solubility of Substances—Methods of Obtaining Specific Gravity—Conversion of Hydrometers—Strength of Solutions by Specific Gravity—Analysis—Gas Analysis—Water Analysis—Qualitative Analysis and Reactions—Volumetric Analysis—Manipulation—Mineralogy — Assaying — Alcohol — Beer — Sugar — Miscellaneous Technological matter relating to Potash, Soda, Sulphuric Acid, Chlorine, Tar Products, Petroleum, Milk, Tallow, Photography, Prices, Wages, Appendix, etc., etc.

The Mechanician: A Treatise on the Construction and Manipulation of Tools, for the use and instruction of Young Engineers and Scientific Amateurs, comprising the Arts of Blacksmithing and Forging; the Construction and Manufacture of Hand Tools, and the various Methods of Using and Grinding them; description of Hand and Machine Processes; Turning and Screw Cutting. By CAMERON KNIGHT, Engineer. *Containing 1147 illustrations,* and 397 pages of letter-press. Fourth edition, 4to, cloth, 18s.

B

Just Published, in Demy 8vo, cloth, containing 975 pages and 250 Illustrations, price 7s. 6d.

SPONS' HOUSEHOLD MANUAL:

A Treasury of Domestic Receipts and Guide for Home Management.

PRINCIPAL CONTENTS.

Hints for selecting a good House, pointing out the essential requirements for a good house as to the Site, Soil, Trees, Aspect, Construction, and General Arrangement; with instructions for Reducing Echoes, Waterproofing Damp Walls, Curing Damp Cellars.

Sanitation.—What should constitute a good Sanitary Arrangement; Examples (with Illustrations) of Well- and Ill-drained Houses; How to Test Drains; Ventilating Pipes, etc.

Water Supply.—Care of Cisterns; Sources of Supply; Pipes; Pumps; Purification and Filtration of Water.

Ventilation and Warming.—Methods of Ventilating without causing cold draughts, by various means; Principles of Warming; Health Questions; Combustion; Open Grates; Open Stoves; Fuel Economisers; Varieties of Grates; Close-Fire Stoves; Hot-air Furnaces; Gas Heating; Oil Stoves; Steam Heating; Chemical Heaters; Management of Flues; and Cure of Smoky Chimneys.

Lighting.—The best methods of Lighting; Candles, Oil Lamps, Gas, Incandescent Gas, Electric Light; How to test Gas Pipes; Management of Gas.

Furniture and Decoration.—Hints on the Selection of Furniture; on the most approved methods of Modern Decoration; on the best methods of arranging Bells and Calls; How to Construct an Electric Bell.

Thieves and Fire.—Precautions against Thieves and Fire; Methods of Detection; Domestic Fire Escapes; Fireproofing Clothes, etc.

The Larder.—Keeping Food fresh for a limited time; Storing Food without change, such as Fruits, Vegetables, Eggs, Honey, etc.

Curing Foods for lengthened Preservation, as Smoking, Salting, Canning, Potting, Pickling, Bottling Fruits, etc.; Jams, Jellies, Marmalade, etc.

The Dairy.—The Building and Fitting of Dairies in the most approved modern style; Butter-making; Cheesemaking and Curing.

The Cellar.—Building and Fitting; Cleaning Casks and Bottles; Corks and Corking; Aërated Drinks; Syrups for Drinks; Beers; Bitters; Cordials and Liqueurs; Wines; Miscellaneous Drinks.

The Pantry.—Bread-making; Ovens and Pyrometers; Yeast; German Yeast; Biscuits; Cakes; Fancy Breads; Buns.

The Kitchen.—On Fitting Kitchens; a description of the best Cooking Ranges, close and open; the Management and Care of Hot Plates, Baking Ovens, Dampers, Flues, and Chimneys; Cooking by Gas; Cooking by Oil; the Arts of Roasting, Grilling, Boiling, Stewing, Braising, Frying.

Receipts for Dishes—Soups, Fish, Meat, Game, Poultry, Vegetables, Salads, Puddings, Pastry, Confectionery, Ices, etc., etc.; Foreign Dishes.

The Housewife's Room.—Testing Air, Water, and Foods; Cleaning and Renovating; Destroying Vermin.

Housekeeping, Marketing.

The Dining-Room.—Dietetics; Laying and Waiting at Table; Carving; Dinners, Breakfasts, Luncheons, Teas, Suppers, etc.

The Drawing-Room.—Etiquette; Dancing; Amateur Theatricals; Tricks and Illusions; Games (indoor).

The Bedroom and Dressing-Room; Sleep; the Toilet; Dress; Buying Clothes; Outfits; Fancy Dress.

The Nursery.—The Room; Clothing; Washing; Exercise; Sleep; Feeding; Teething; Illness; Home Training.

The Sick-Room.—The Room; the Nurse; the Bed; Sick Room Accessories; Feeding Patients; Invalid Dishes and Drinks; Administering Physic; Domestic Remedies; Accidents and Emergencies; Bandaging; Burns; Carrying Injured Persons; Wounds; Drowning; Fits; Frost-bites; Poisons and Antidotes; Sunstroke; Common Complaints; Disinfection, etc.

The Bath-Room.—Bathing in General; Management of Hot-Water System.

The Laundry.—Small Domestic Washing Machines, and methods of getting up linen; Fitting up and Working a Steam Laundry.

The School-Room.—The Room and its Fittings; Teaching, etc.

The Playground.—Air and Exercise; Training; Outdoor Games and Sports.

The Workroom.—Darning, Patching, and Mending Garments.

The Library.—Care of Books.

The Garden—Calendar of Operations for Lawn, Flower Garden, and Kitchen Garden.

The Farmyard—Management of the Horse, Cow, Pig, Poultry, Bees, etc., etc.

Small Motors.—A description of the various small Engines useful for domestic purposes, from 1 man to 1 horse power, worked by various methods, such as Electric Engines, Gas Engines, Petroleum Engines, Steam Engines, Condensing Engines, Water Power, Wind Power, and the various methods of working and managing them.

Household Law.—The Law relating to Landlords and Tenants, Lodgers, Servants, Parochial Authorities, Juries, Insurance, Nuisance, etc.

On Designing Belt Gearing. By E. J. COWLING
WELCH, Mem. Inst. Mech. Engineers, Author of 'Designing Valve Gearing.' Fcap. 8vo, sewed, 6d.

A Handbook of Formulæ, Tables, and Memoranda,
for Architectural Surveyors and others engaged in Building. By J. T. HURST, C.E. Fourteenth edition, royal 32mo, roan, 5s.

" It is no disparagement to the many excellent publications we refer to, to say that in our opinion this little pocket-book of Hurst's is the very best of them all, without any exception. It would be useless to attempt a recapitulation of the contents, for it appears to contain almost *everything* that anyone connected with building could require, and, best of all, made up in a compact form for carrying in the pocket, measuring only 5 in. by 3 in., and about ½ in. thick, in a limp cover. We congratulate the author on the success of his laborious and practically compiled little book, which has received unqualified and deserved praise from every professional person to whom we have shown it."—*The Dublin Builder.*

Tabulated Weights of Angle, Tee, Bulb, Round,
Square, and Flat Iron and Steel, and other information for the use of Naval Architects and Shipbuilders. By C. H. JORDAN, M.I.N.A. Fourth edition, 32mo, cloth, 2s. 6d.

A Complete Set of Contract Documents for a Country
Lodge, comprising Drawings, Specifications, Dimensions (for quantities), Abstracts, Bill of Quantities, Form of Tender and Contract, with Notes by J. LEANING, printed in facsimile of the original documents, on single sheets fcap., in paper case, 10s.

A Practical Treatise on Heat, as applied to the
Useful Arts; for the Use of Engineers, Architects, &c. By THOMAS BOX. *With* 14 *plates.* Sixth edition, crown 8vo, cloth, 12s. 6d.

A Descriptive Treatise on Mathematical Drawing
Instruments : their construction, uses, qualities, selection, preservation, and suggestions for improvements, with hints upon Drawing and Colouring. By W. F. STANLEY, M.R.I. Sixth edition, *with numerous illustrations,* crown 8vo, cloth, 5s.

B 2

Quantity Surveying. By J. LEANING. With 42 illustrations. Second edition, revised, crown 8vo, cloth, 9s.

CONTENTS :

A complete Explanation of the London Practice.	Schedule of Prices.
General Instructions.	Form of Schedule of Prices.
Order of Taking Off.	Analysis of Schedule of Prices.
Modes of Measurement of the various Trades.	Adjustment of Accounts.
Use and Waste.	Form of a Bill of Variations.
Ventilation and Warming.	Remarks on Specifications.
Credits, with various Examples of Treatment.	Prices and Valuation of Work, with Examples and Remarks upon each Trade.
Abbreviations.	The Law as it affects Quantity Surveyors, with Law Reports.
Squaring the Dimensions.	
Abstracting, with Examples in illustration of each Trade.	Taking Off after the Old Method.
Billing.	Northern Practice.
Examples of Preambles to each Trade.	The General Statement of the Methods recommended by the Manchester Society of Architects for taking Quantities.
Form for a Bill of Quantities.	
Do. Bill of Credits.	Examples of Collections.
Do. Bill for Alternative Estimate.	Examples of " Taking Off" in each Trade.
Restorations and Repairs, and Form of Bill.	Remarks on the Past and Present Methods of Estimating.
Variations before Acceptance of Tender.	
Errors in a Builder's Estimate.	

Spons' Architects' and Builders' Price Book, with useful Memoranda. Edited by W. YOUNG, Architect. Crown 8vo, cloth, red edges, 3s. 6d. *Published annually.* Nineteenth edition. *Now ready.*

Long-Span Railway Bridges, comprising Investigations of the Comparative Theoretical and Practical Advantages of the various adopted or proposed Type Systems of Construction, with numerous Formulæ and Tables giving the weight of Iron or Steel required in Bridges from 300 feet to the limiting Spans; to which are added similar Investigations and Tables relating to Short-span Railway Bridges. Second and revised edition. By B. BAKER, Assoc. Inst. C.E. *Plates,* crown 8vo, cloth, 5s.

Elementary Theory and Calculation of Iron Bridges and Roofs. By AUGUST RITTER, Ph.D., Professor at the Polytechnic School at Aix-la-Chapelle. Translated from the third German edition, by H. R. SANKEY, Capt. R.E. With 500 *illustrations,* 8vo, cloth, 15s.

The Elementary Principles of Carpentry. By THOMAS TREDGOLD. Revised from the original edition, and partly re-written, by JOHN THOMAS HURST. Contained in 517 pages of letter-press, and *illustrated with 48 plates and 150 wood engravings.* Sixth edition, reprinted from the third, crown 8vo, cloth, 12s. 6d.

Section I. On the Equality and Distribution of Forces — Section II. Resistance of Timber — Section III. Construction of Floors — Section IV. Construction of Roofs — Section V. Construction of Domes and Cupolas — Section VI. Construction of Partitions — Section VII. Scaffolds, Staging, and Gantries — Section VIII. Construction of Centres for Bridges — Section IX. Coffer-dams, Shoring, and Strutting — Section X. Wooden Bridges and Viaducts — Section XI. Joints, Straps, and other Fastenings — Section XII. Timber.

The Builder's Clerk: a Guide to the Management of a Builder's Business. By THOMAS BALES. Fcap. 8vo, cloth, 1s. 6d.

Practical Gold-Mining: a Comprehensive Treatise on the Origin and Occurrence of Gold-bearing Gravels, Rocks and Ores, and the methods by which the Gold is extracted. By C. G. WARNFORD LOCK, co-Author of 'Gold: its Occurrence and Extraction.' *With 8 plates and 275 engravings in the text*, royal 8vo, cloth, 2*l*. 2*s*.

Hot Water Supply: A Practical Treatise upon the Fitting of Circulating Apparatus in connection with Kitchen Range and other Boilers, to supply Hot Water for Domestic and General Purposes. With a Chapter upon Estimating. *Fully illustrated*, crown 8vo, cloth, 3*s*.

Hot Water Apparatus: An Elementary Guide for the Fitting and Fixing of Boilers and Apparatus for the Circulation of Hot Water for Heating and for Domestic Supply, and containing a Chapter upon Boilers and Fittings for Steam Cooking. 32 *illustrations*, fcap. 8vo, cloth, 1*s*. 6*d*.

The Use and Misuse, and the Proper and Improper Fixing of a Cooking Range. *Illustrated*, fcap. 8vo, sewed, 6*d*.

Iron Roofs: Examples of Design, Description. *Illustrated with 64 Working Drawings of Executed Roofs.* By ARTHUR T. WALMISLEY, Assoc. Mem. Inst. C.E. Second edition, revised, imp. 4to, half-morocco, 3*l*. 3*s*.

A History of Electric Telegraphy, to the Year 1837. Chiefly compiled from Original Sources, and hitherto Unpublished Documents, by J. J. FAHIE, Mem. Soc. of Tel. Engineers, and of the International Society of Electricians, Paris. Crown 8vo, cloth, 9*s*.

Spons' Information for Colonial Engineers. Edited by J. T. HURST. Demy 8vo, sewed.

No. 1, Ceylon. By ABRAHAM DEANE, C.E. 2*s*. 6*d*.

CONTENTS:

Introductory Remarks—Natural Productions—Architecture and Engineering—Topography, Trade, and Natural History—Principal Stations—Weights and Measures, etc., etc. :

No. 2. Southern Africa, including the Cape Colony, Natal, and the Dutch Republics. By HENRY HALL, F.R.G.S., F.R.C.I. With Map. 3*s*. 6*d*. CONTENTS:

General Description of South Africa—Physical Geography with reference to Engineering Operations—Notes on Labour and Material in Cape Colony—Geological Notes on Rock Formation in South Africa—Engineering Instruments for Use in South Africa—Principal Public Works in Cape Colony: Railways, Mountain Roads and Passes, Harbour Works, Bridges, Gas Works, Irrigation and Water Supply, Lighthouses, Drainage and Sanitary Engineering, Public Buildings, Mines—Table of Woods in South Africa—Animals used for Draught Purposes—Statistical Notes—Table of Distances—Rates of Carriage, etc.

No. 3. India. By F. C. DANVERS, Assoc. Inst. C.E. With Map. 4*s*. 6*d*.

CONTENTS:

Physical Geography of India—Building Materials—Roads—Railways—Bridges—Irrigation—River Works—Harbours—Lighthouse Buildings—Native Labour—The Principal Trees of India—Money—Weights and Measures—Glossary of Indian Terms, etc.

Our Factories, Workshops, and Warehouses: their
Sanitary and Fire-Resisting Arrangements. By B. II. THWAITE, Assoc.
Mem. Inst. C.E. *With* 183 *wood engravings,* crown 8vo, cloth, 9*s.*

A Practical Treatise on Coal Mining. By GEORGE
G. ANDRÉ, F.G.S., Assoc. Inst. C.E., Member of the Society of Engineers.
With 82 *lithographic plates.* 2 vols., royal 4to, cloth, 3*l.* 12*s.*

A Practical Treatise on Casting and Founding,
including descriptions of the modern machinery employed in the art. By
N. E. SPRETSON, Engineer. Fifth edition, with 82 *plates* drawn to
scale, 412 pp., demy 8vo, cloth, 18*s.*

A Handbook of Electrical Testing. By H. R. KEMPE,
M.S.T.E. Fourth edition, revised and enlarged, crown 8vo, cloth, 16*s.*

The Clerk of Works: a Vade-Mecum for all engaged
in the Superintendence of Building Operations. By G. G. HOSKINS,
F.R.I.B.A. Third edition, fcap. 8vo, cloth, 1*s.* 6*d.*

American Foundry Practice: Treating of Loam,
Dry Sand, and Green Sand Moulding, and containing a Practical Treatise
upon the Management of Cupolas, and the Melting of Iron. By T. D.
WEST, Practical Iron Moulder and Foundry Foreman. Second edition,
with numerous illustrations, crown 8vo, cloth, 10*s.* 6*d.*

The Maintenance of Macadamised Roads. By T.
CODRINGTON, M.I.C.E, F.G.S., General Superintendent of County Roads
for South Wales. Second edition, 8vo, cloth, 7*s.* 6*d.*

Hydraulic Steam and Hand Power Lifting and
Pressing Machinery. By FREDERICK COLYER, M. Inst. C.E., M. Inst. M.E.
With 73 *plates,* 8vo, cloth, 18*s.*

Pumps and Pumping Machinery. By F. COLYER,
M.I.C.E., M.I.M.E. *With* 23 *folding plates,* 8vo, cloth, 12*s.* 6*d.*

Pumps and Pumping Machinery. By F. COLYER.
Second Part. *With* 11 *large plates,* 8vo, cloth, 12*s.* 6*d.*

A Treatise on the Origin, Progress, Prevention, and
Cure of Dry Rot in Timber; with Remarks on the Means of Preserving
Wood from Destruction by Sea-Worms, Beetles, Ants, etc. By THOMAS
ALLEN BRITTON, late Surveyor to the Metropolitan Board of Works,
etc., etc. *With* 10 *plates,* crown 8vo, cloth, 7*s.* 6*d.*

The Artillery of the Future and the New Powders.
By J. A. LONGRIDGE, Mem. Inst. C.E. 8vo, cloth, 5*s.*

Gas Works: their Arrangement, Construction, Plant, and Machinery. By F. COLYER, M. Inst. C.E. *With 31 folding plates,* 8vo, cloth, 12s. 6d.

The Municipal and Sanitary Engineer's Handbook. By H. PERCY BOULNOIS, Mem. Inst. C.E., Borough Engineer, Portsmouth. *With numerous illustrations.* Second edition, demy 8vo, cloth, 15s.

CONTENTS:

The Appointment and Duties of the Town Surveyor—Traffic—Macadamised Roadways—Steam Rolling—Road Metal and Breaking—Pitched Pavements—Asphalte—Wood Pavements—Footpaths—Kerbs and Gutters—Street Naming and Numbering—Street Lighting—Sewerage—Ventilation of Sewers—Disposal of Sewage—House Drainage—Disinfection—Gas and Water Companies, etc., Breaking up Streets—Improvement of Private Streets—Borrowing Powers—Artizans' and Labourers' Dwellings—Public Conveniences—Scavenging, including Street Cleansing—Watering and the Removing of Snow—Planting Street Trees—Deposit of Plans—Dangerous Buildings—Hoardings—Obstructions—Improving Street Lines—Cellar Openings—Public Pleasure Grounds—Cemeteries—Mortuaries—Cattle and Ordinary Markets—Public Slaughter-houses, etc.—Giving numerous Forms of Notices, Specifications, and General Information upon these and other subjects of great importance to Municipal Engineers and others engaged in Sanitary Work.

Metrical Tables. By Sir G. L. MOLESWORTH, M.I.C.E. 32mo, cloth, 1s. 6d.

CONTENTS.

General—Linear Measures—Square Measures—Cubic Measures—Measures of Capacity—Weights—Combinations—Thermometers.

Elements of Construction for Electro-Magnets. By Count TH. DU MONCEL, Mem. de l'Institut de France. Translated from the French by C. J. WHARTON. Crown 8vo, cloth, 4s. 6d.

A Treatise on the Use of Belting for the Transmission of Power. By J. H. COOPER. Second edition, *illustrated,* 8vo, cloth, 15s.

A Pocket-Book of Useful Formulæ and Memoranda for Civil and Mechanical Engineers. By Sir GUILFORD L. MOLESWORTH, Mem. Inst. C.E. *With numerous illustrations,* 744 pp. Twenty-second edition, 32mo, roan, 6s.

SYNOPSIS OF CONTENTS:

Surveying, Levelling, etc.—Strength and Weight of Materials—Earthwork, Brickwork, Masonry, Arches, etc.—Struts, Columns, Beams, and Trusses—Flooring, Roofing, and Roof Trusses—Girders, Bridges, etc.—Railways and Roads—Hydraulic Formulæ—Canals, Sewers, Waterworks, Docks—Irrigation and Breakwaters—Gas, Ventilation, and Warming—Heat, Light, Colour, and Sound—Gravity: Centres, Forces, and Powers—Millwork, Teeth of Wheels, Shafting, etc.—Workshop Recipes—Sundry Machinery—Animal Power—Steam and the Steam Engine—Water-power, Water-wheels, Turbines, etc.—Wind and Windmills—Steam Navigation, Ship Building, Tonnage, etc.—Gunnery, Projectiles, etc.—Weights, Measures, and Money—Trigonometry, Conic Sections, and Curves—Telegraphy—Mensuration—Tables of Areas and Circumference, and Arcs of Circles—Logarithms, Square and Cube Roots, Powers—Reciprocals, etc.—Useful Numbers—Differential and Integral Calculus—Algebraic Signs—Telegraphic Construction and Formulæ.

Hints on Architectural Draughtsmanship. By G. W.
TUXFORD HALLATT. Fcap. 8vo, cloth, 1s. 6d.

Spons' Tables and Memoranda for Engineers;
selected and arranged by J. T. HURST, C.E., Author of 'Architectural
Surveyors' Handbook,' 'Hurst's Tredgold's Carpentry,' etc. Eleventh
edition, 64mo, roan, gilt edges, 1s.; or in cloth case, 1s. 6d.

This work is printed in a pearl type, and is so small, measuring only 2½ in. by 1¾ in. by
⅜ in. thick, that it may be easily carried in the waistcoat pocket.

"It is certainly an extremely rare thing for a reviewer to be called upon to notice a volume
measuring but 2½ in. by 1¾ in., yet these dimensions faithfully represent the size of the handy
little book before us. The volume—which contains 118 printed pages, besides a few blank
pages for memoranda—is, in fact, a true pocket-book, adapted for being carried in the waistcoat pocket, and containing a far greater amount and variety of information than most people
would imagine could be compressed into so small a space. . . . . The little volume has been
compiled with considerable care and judgment, and we can cordially recommend it to our
readers as a useful little pocket companion."—*Engineering.*

A Practical Treatise on Natural and Artificial
Concrete, its Varieties and Constructive Adaptations. By HENRY REID,
Author of the 'Science and Art of the Manufacture of Portland Cement.'
New Edition, *with* 59 *woodcuts and* 5 *plates,* 8vo, cloth, 15s.

Notes on Concrete and Works in Concrete; especially
written to assist those engaged upon Public Works. By JOHN NEWMAN,
Assoc. Mem. Inst. C.E., crown 8vo, cloth, 4s. 6d.

Electricity as a Motive Power. By Count TH. DU
MONCEL, Membre de l'Institut de France, and FRANK GERALDY, Ingé-
nieur des Ponts et Chaussées. Translated and Edited, with Additions, by
C. J. WHARTON, Assoc. Soc. Tel. Eng. and Elec. *With* 113 *engravings*
and diagrams, crown 8vo, cloth, 7s. 6d.

Treatise on Valve-Gears, with special consideration
of the Link-Motions of Locomotive Engines. By Dr. GUSTAV ZEUNER,
Professor of Applied Mechanics at the Confederated Polytechnikum of
Zurich. Translated from the Fourth German Edition, by Professor J. F.
KLEIN, Lehigh University, Bethlehem, Pa. *Illustrated,* 8vo, cloth, 12s. 6d.

The French-Polisher's Manual. By a French-
Polisher; containing Timber Staining, Washing, Matching, Improving,
Painting, Imitations, Directions for Staining, Sizing, Embodying,
Smoothing, Spirit Varnishing, French-Polishing, Directions for Re-
polishing. Third edition, royal 32mo, sewed, 6d.

Hops, their Cultivation, Commerce, and Uses in
various Countries. By P. L. SIMMONDS. Crown 8vo, cloth, 4s. 6d.

The Principles of Graphic Statics. By GEORGE
SYDENHAM CLARKE, Major Royal Engineers. *With* 112 *illustrations.*
Second edition, 4to, cloth, 12s. 6d.

Dynamo Tenders' Hand-Book. By F. B. BADT, late
1st Lieut. Royal Prussian Artillery. *With 70 illustrations.* Third edition,
18mo, cloth, 4*s.* 6*d.*

*Practical Geometry, Perspective, and Engineering
Drawing;* a Course of Descriptive Geometry adapted to the Requirements of the Engineering Draughtsman, including the determination of
cast shadows and Isometric Projection, each chapter being followed by
numerous examples; to which are added rules for Shading, Shade-lining,
etc., together with practical instructions as to the Lining, Colouring,
Printing, and general treatment of Engineering Drawings, with a chapter
on drawing instruments. By GEORGE S. CLARKE, Capt. R.E. Second
edition, *with 21 plates.* 2 vols., cloth, 10*s.* 6*d.*

The Elements of Graphic Statics. By Professor
KARL VON OTT, translated from the German by G. S. CLARKE, Capt.
R.E., Instructor in Mechanical Drawing, Royal Indian Engineering
College. *With 93 illustrations,* crown 8vo, cloth, 5*s.*

*A Practical Treatise on the Manufacture and Distri-
bution of Coal Gas.* By WILLIAM RICHARDS. Demy 4to, with *numerous
wood engravings and 29 plates,* cloth, 28*s.*

SYNOPSIS OF CONTENTS:

Introduction—History of Gas Lighting—Chemistry of Gas Manufacture, by Lewis
Thompson, Esq., M.R.C.S.—Coal, with Analyses, by J. Paterson, Lewis Thompson, and
G. R. Hislop, Esqrs.—Retorts, Iron and Clay—Retort Setting—Hydraulic Main—Con-
densers—Exhausters—Washers and Scrubbers—Purifiers—Purification—History of Gas
Holder—Tanks, Brick and Stone, Composite, Concrete, Cast-Iron, Compound Annular
Wrought-Iron—Specifications—Gas Holders—Station Meter—Governor—Distribution—
Mains—Gas Mathematics, or Formulæ for the Distribution of Gas, by Lewis Thompson, Esq.—
Services—Consumers' Meters—Regulators—Burners—Fittings—Photometer—Carburization
of Gas—Air Gas and Water Gas—Composition of Coal Gas, by Lewis Thompson, Esq.—
Analyses of Gas—Influence of Atmospheric Pressure and Temperature on Gas—Residual
Products—Appendix—Description of Retort Settings, Buildings, etc., etc.

*The New Formula for Mean Velocity of Discharge
of Rivers and Canals.* By W. R. KUTTER. Translated from articles in
the 'Cultur-Ingénieur,' by LOWIS D'A. JACKSON, Assoc. Inst. C.E.
8vo, cloth, 12*s.* 6*d.*

*The Practical Millwright and Engineer's Ready
Reckoner;* or Tables for finding the diameter and power of cog-wheels,
diameter, weight, and power of shafts, diameter and strength of bolts, etc.
By THOMAS DIXON. Fourth edition, 12mo, cloth, 3*s.*

Tin: Describing the Chief Methods of Mining,
Dressing and Smelting it abroad; with Notes upon Arsenic, Bismuth and
Wolfram. By ARTHUR G. CHARLETON, Mem. American Inst. of
Mining Engineers. *With plates,* 8vo, cloth, 12*s.* 6*d.*

B 3

Perspective, Explained and Illustrated. By G. S. CLARKE, Capt. R.E. *With illustrations*, 8vo, cloth, 3s. 6d.

Practical Hydraulics; a Series of Rules and Tables for the use of Engineers, etc., etc. By THOMAS BOX. Ninth edition, *numerous plates*, post 8vo, cloth, 5s.

The Essential Elements of Practical Mechanics; *based on the Principle of Work*, designed for Engineering Students. By OLIVER BYRNE, formerly Professor of Mathematics, College for Civil Engineers. Third edition, *with 148 wood engravings*, post 8vo, cloth, 7s. 6d.

CONTENTS :

Chap. 1, How Work is Measured by a Unit, both with and without reference to a Unit of Time—Chap. 2. The Work of Living Agents, the Influence of Friction, and introduces one of the most beautiful Laws of Motion—Chap. 3. The principles expounded in the first and second chapters are applied to the Motion of Bodies—Chap. 4. The Transmission of Work by simple Machines—Chap. 5. Useful Propositions and Rules.

Breweries and Maltings : their Arrangement, Construction, Machinery, and Plant. By G. SCAMELL, F.R.I.B.A. Second edition, revised, enlarged, and partly rewritten. By F. COLYER, M.I.C.E., M.I.M.E. *With 20 plates*, 8vo, cloth, 12s. 6d.

A Practical Treatise on the Construction of Horizontal and Vertical Waterwheels, specially designed for the use of operative mechanics. By WILLIAM CULLEN, Millwright and Engineer. *With 11 plates*. Second edition, revised and enlarged, small 4to, cloth, 12s. 6d.

A Practical Treatise on Mill-gearing, Wheels, Shafts, Riggers, etc.; for the use of Engineers. By THOMAS BOX. Third edition, *with 11 plates*. Crown 8vo, cloth, 7s. 6d.

Mining Machinery: a Descriptive Treatise on the Machinery, Tools, and other Appliances used in Mining. By G. ANDRÉ, F.G.S., Assoc. Inst. C.E., Mem. of the Society of Engineers. Royal 4to, uniform with the Author's Treatise on Coal Mining, containing 182 *plates*, accurately drawn to scale, with descriptive text, in 2 vols., cloth, 3l. 12s.

CONTENTS :

Machinery for Prospecting, Excavating, Hauling, and Hoisting—Ventilation—Pumping—Treatment of Mineral Products, including Gold and Silver, Copper, Tin, and Lead, Iron, Coal, Sulphur, China Clay, Brick Earth, etc.

Tables for Setting out Curves for Railways, Canals, Roads, etc., varying from a radius of five chains to three miles. By A. KENNEDY and R. W. HACKWOOD. *Illustrated* 32mo, cloth, 2s. 6d.

Practical Electrical Notes and Definitions for the
use of Engineering Students and Practical Men. By W. PERREN
MAYCOCK, Assoc. M. Inst. E.E., Instructor in Electrical Engineering at
the Pitlake Institute, Croydon, together with the Rules and Regulations
to be observed in Electrical Installation Work. Second edition. Royal
32mo, roan, gilt edges, 4s. 6d., or cloth, red edges, 3s.

The Draughtsman's Handbook of Plan and Map
Drawing; including instructions for the preparation of Engineering,
Architectural, and Mechanical Drawings. *With numerous illustrations*
in the text, and 33 plates (15 printed in colours). By G. G. ANDRÉ,
F.G.S., Assoc. Inst. C.E. 4to, cloth, 9s.

CONTENTS:

The Drawing Office and its Furnishings—Geometrical Problems—Lines, Dots, and their
Combinations—Colours, Shading, Lettering, Bordering, and North Points—Scales—Plotting
—Civil Engineers' and Surveyors' Plans—Map Drawing—Mechanical and Architectural
Drawing—Copying and Reducing Trigonometrical Formulæ, etc., etc.

The Boiler-maker's and Iron Ship-builder's Companion,
comprising a series of original and carefully calculated tables, of the
utmost utility to persons interested in the iron trades. By JAMES FODEN,
author of ' Mechanical Tables,' etc. Second edition revised, *with illustra-*
tions, crown 8vo, cloth, 5s.

Rock Blasting: a Practical Treatise on the means
employed in Blasting Rocks for Industrial Purposes. By G. G. ANDRÉ,
F.G.S., Assoc. Inst. C.E. *With 56 illustrations and 12 plates,* 8vo, cloth,
10s. 6d.

Experimental Science: Elementary, Practical, and
Experimental Physics. By GEO. M. HOPKINS. *Illustrated by 672*
engravings. In one large vol., 8vo, cloth, 15s.

A Treatise on Ropemaking as practised in public and
private Rope-yards, with a Description of the Manufacture, Rules, Tables
of Weights, etc., adapted to the Trade, Shipping, Mining, Railways,
Builders, etc. By R. CHAPMAN, formerly foreman to Messrs. Huddart
and Co., Limehouse, and late Master Ropemaker to H.M. Dockyard,
Deptford. Second edition, 12mo, cloth, 3s.

Laxton's Builders' and Contractors' Tables; for the
use of Engineers, Architects, Surveyors, Builders, Land Agents, and
others. Bricklayer, containing 22 tables, with nearly 30,000 calculations.
4to, cloth, 5s.

Laxton's Builders' and Contractors' Tables. Ex-
cavator, Earth, Land, Water, and Gas, containing 53 tables, with nearly
24,000 calculations. 4to, cloth, 5s.

Egyptian Irrigation. By W. WILLCOCKS, M.I.C.E.,
Indian Public Works Department, Inspector of Irrigation, Egypt. With
Introduction by Lieut.-Col. J. C. Ross, R.E., Inspector-General of
Irrigation. *With numerous lithographs and wood engravings*, royal 8vo,
cloth, 1*l*. 16*s*.

Screw Cutting Tables for Engineers and Machinists,
giving the values of the different trains of Wheels required to produce
Screws of any pitch, calculated by Lord Lindsay, M.P., F.R.S., F.R.A.S.,
etc. Cloth, oblong, 2*s*.

Screw Cutting Tables, for the use of Mechanical
Engineers, showing the proper arrangement of Wheels for cutting the
Threads of Screws of any required pitch, with a Table for making the
Universal Gas-pipe Threads and Taps. By W. A. MARTIN, Engineer.
Second edition, oblong, cloth, 1*s*., or sewed, 6*d*.

A Treatise on a Practical Method of Designing Slide-
Valve Gears by Simple Geometrical Construction, based upon the principles
enunciated in Euclid's Elements, and comprising the various forms of
Plain Slide-Valve and Expansion Gearing ; together with Stephenson's,
Gooch's, and Allan's Link-Motions, as applied either to reversing or to
variable expansion combinations. By EDWARD J. COWLING WELCH,
Memb. Inst. Mechanical Engineers. Crown 8vo, cloth, 6*s*.

Cleaning and Scouring : a Manual for Dyers, Laun-
dresses, and for Domestic Use. By S. CHRISTOPHER. 18mo, sewed, 6*d*.

A Glossary of Terms used in Coal Mining. By
WILLIAM STUKELEY GRESLEY, Assoc. Mem. Inst. C.E., F.G.S., Member
of the North of England Institute of Mining Engineers. *Illustrated with
numerous woodcuts and diagrams*, crown 8vo, cloth, 5*s*.

A Pocket-Book for Boiler Makers and Steam Users,
comprising a variety of useful information for Employer and Workman,
Government Inspectors, Board of Trade Surveyors, Engineers in charge
of Works and Slips, Foremen of Manufactories, and the general Steam-
using Public. By MAURICE JOHN SEXTON. Second edition, royal
32mo, roan, gilt edges, 5*s*.

Electrolysis : a Practical Treatise on Nickeling,
Coppering, Gilding, Silvering, the Refining of Metals, and the treatment
of Ores by means of Electricity. By HIPPOLYTE FONTAINE, translated
from the French by J. A. BERLY, C.E., Assoc. S.T.E. *With engravings*,
8vo, cloth, 9*s*.

Barlow's Tables of Squares, Cubes, Square Roots,
Cube Roots, Reciprocals of all Integer Numbers up to 10,000. Post 8vo,
cloth, 6s.

A Practical Treatise on the Steam Engine, containing Plans and Arrangements of Details for Fixed Steam Engines,
with Essays on the Principles involved in Design and Construction. By
ARTHUR RIGG, Engineer, Member of the Society of Engineers and of
the Royal Institution of Great Britain. Demy 4to, *copiously illustrated
with woodcuts and 96 plates,* in one Volume, half-bound morocco, 2l. 2s.;
or cheaper edition, cloth, 25s.

This work is not, in any sense, an elementary treatise, or history of the steam engine, but
is intended to describe examples of Fixed Steam Engines without entering into the wide
domain of locomotive or marine practice. To this end illustrations will be given of the most
recent arrangements of Horizontal, Vertical, Beam, Pumping, Winding, Portable, Semi-
portable, Corliss, Allen, Compound, and other similar Engines, by the most eminent Firms in
Great Britain and America. The laws relating to the action and precautions to be observed
in the construction of the various details, such as Cylinders, Pistons, Piston-rods, Connecting-
rods, Cross-heads, Motion-blocks, Eccentrics, Simple, Expansion, Balanced, and Equilibrium
Slide-valves, and Valve-gearing will be minutely dealt with. In this connection will be found
articles upon the Velocity of Reciprocating Parts and the Mode of Applying the Indicator,
Heat and Expansion of Steam Governors, and the like. It is the writer's desire to draw
illustrations from every possible source, and give only those rules that present practice deems
correct.

A Practical Treatise on the Science of Land and
Engineering Surveying, Levelling, Estimating Quantities, etc., with a
general description of the several Instruments required for Surveying,
Levelling, Plotting, etc. By H. S. MERRETT. Fourth edition, revised
by G. W. USILL, Assoc. Mem. Inst. C.E. 41 *plates, with illustrations
and tables,* royal 8vo, cloth, 12s. 6d.

PRINCIPAL CONTENTS :

Part 1. Introduction and the Principles of Geometry. Part 2. Land Surveying; com-
prising General Observation—The Chain—Offsets Surveying by the Chain only—Surveying
Hilly Ground—To Survey an Estate or Parish by the Chain only—Surveying with the
Theodolite—Mining and Town Surveying—Railroad Surveying—Mapping—Division and
Laying out of Land—Observations on Enclosures—Plane Trigonometry. Part 3. Levelling—
Simple and Compound Levelling—The Level Book—Parliamentary Plan and Section—
Levelling with a Theodolite—Gradients—Wooden Curves—To Lay out a Railway Curve—
Setting out Widths. Part 4. Calculating Quantities generally for Estimates—Cuttings and
Embankments—Tunnels—Brickwork—Ironwork—Timber Measuring. Part 5. Description
and Use of Instruments in Surveying and Plotting—The Improved Dumpy Level—Troughton's
Level—The Prismatic Compass—Proportional Compass—Box Sextant—Vernier—Panta-
graph—Merrett's Improved Quadrant—Improved Computation Scale—The Diagonal Scale—
Straight Edge and Sector. Part 6. Logarithms of Numbers—Logarithmic Sines and
Co-Sines, Tangents and Co-Tangents—Natural Sines and Co-Sines—Tables for Earthwork,
for Setting out Curves, and for various Calculations, etc., etc., etc.

Mechanical Graphics. A Second Course of Mechanical Drawing. With Preface by Prof. PERRY, B.Sc., F.R.S.
Arranged for use in Technical and Science and Art Institutes, Schools
and Colleges, by GEORGE HALLIDAY, Whitworth Scholar. 8vo,
cloth, 6s.

B 4

The Assayer's Manual: an Abridged Treatise on
the Docimastic Examination of Ores and Furnace and other Artificial
Products. By BRUNO KERL. Translated by W. T. BRANNT. *With* 65
illustrations, 8vo, cloth, 12s. 6d.

Dynamo - Electric Machinery: a Text-Book for
Students of Electro-Technology. By SILVANUS P. THOMPSON, B.A.,
D.Sc., M.S.T.E.   [*New edition in the press.*

The Practice of Hand Turning in Wood, Ivory, Shell,
etc., with Instructions for Turning such Work in Metal as may be required
in the Practice of Turning in Wood, Ivory, etc.; also an Appendix on
Ornamental Turning. (A book for beginners.) By FRANCIS CAMPIN.
Third edition, *with wood engravings,* crown 8vo, cloth, 6s.

CONTENTS :

On Lathes—Turning Tools—Turning Wood—Drilling—Screw Cutting—Miscellaneous
Apparatus and Processes—Turning Particular Forms—Staining—Polishing—Spinning Metals
—Materials—Ornamental Turning, etc.

Treatise on Watchwork, Past and Present. By the
Rev. H. L. NELTHROPP, M.A., F.S.A. *With* 32 *illustrations,* crown
8vo, cloth, 6s. 6d.

CONTENTS :

Definitions of Words and Terms used in Watchwork—Tools—Time—Historical Summary—On Calculations of the Numbers for Wheels and Pinions; their Proportional Sizes,
Trains, etc.—Of Dial Wheels, or Motion Work—Length of Time of Going without Winding
up—The Verge—The Horizontal—The Duplex—The Lever—The Chronometer—Repeating
Watches—Keyless Watches—The Pendulum, or Spiral Spring—Compensation—Jewelling of
Pivot Holes—Clerkenwell—Fallacies of the Trade—Incapacity of Workmen—How to Choose
and Use a Watch, etc.

Algebra Self-Taught. By W. P. HIGGS, M.A.,
D.Sc., LL.D., Assoc. Inst. C.E., Author of ' A Handbook of the Differential Calculus,' etc. Second edition, crown 8vo, cloth, 2s. 6d.

CONTENTS :

Symbols and the Signs of Operation—The Equation and the Unknown Quantity—
Positive and Negative Quantities—Multiplication—Involution—Exponents—Negative Expo-
nents—Roots, and the Use of Exponents as Logarithms—Logarithms—Tables of Logarithms
and Proportionate Parts—Transformation of System of Logarithms—Common Uses of
Common Logarithms—Compound Multiplication and the Binomial Theorem—Division,
Fractions, and Ratio—Continued Proportion—The Series and the Summation of the Series—
Limit of Series—Square and Cube Roots—Equations—List of Formulæ, etc.

Spons' Dictionary of Engineering, Civil, Mechanical,
Military, and Naval; with technical terms in French, German, Italian,
and Spanish, 3100 pp., and *nearly* 8000 *engravings,* in super-royal 8vo,
in 8 divisions, 5l. 8s. Complete in 3 vols., cloth, 5l. 5s. Bound in a
superior manner, half-morocco, top edge gilt, 3 vols., 6l. 12s.

Notes in Mechanical Engineering. Compiled principally for the use of the Students attending the Classes on this subject at the City of London College. By HENRY ADAMS, Mem. Inst. M.E., Mem. Inst. C.E., Mem. Soc. of Engineers. Crown 8vo, cloth, 2s. 6d.

Canoe and Boat Building: a complete Manual for Amateurs, containing plain and comprehensive directions for the construction of Canoes, Rowing and Sailing Boats, and Hunting Craft. By W. P. STEPHENS. *With numerous illustrations and 24 plates of Working Drawings.* Crown 8vo, cloth, 9s.

Proceedings of the National Conference of Electricians, *Philadelphia,* October 8th to 13th, 1884. 18mo, cloth, 3s.

Dynamo - Electricity, its Generation, Application, Transmission, Storage, and Measurement. By G. B. PRESCOTT. *With* 545 *illustrations.* 8vo, cloth, 1l. 1s.

Domestic Electricity for Amateurs. Translated from the French of E. HOSPITALIER, Editor of "L'Electricien," by C. J. WHARTON, Assoc. Soc. Tel. Eng. *Numerous illustrations.* Demy 8vo, cloth, 6s.

CONTENTS:

1. Production of the Electric Current—2. Electric Bells—3. Automatic Alarms—4. Domestic Telephones—5. Electric Clocks—6. Electric Lighters—7. Domestic Electric Lighting—8. Domestic Application of the Electric Light—9. Electric Motors—10. Electrical Locomotion—11. Electrotyping, Plating, and Gilding—12. Electric Recreations—13. Various applications—Workshop of the Electrician.

Wrinkles in Electric Lighting. By VINCENT STEPHEN. *With illustrations.* 18mo, cloth, 2s. 6d.

CONTENTS:

1. The Electric Current and its production by Chemical means—2. Production of Electric Currents by Mechanical means—3. Dynamo-Electric Machines—4. Electric Lamps—5. Lead—6. Ship Lighting.

Foundations and Foundation Walls for all classes of Buildings, Pile Driving, Building Stones and Bricks, Pier and Wall construction, Mortars, Limes, Cements, Concretes, Stuccos, &c. 64 *illustrations.* By G. T. POWELL and F. BAUMAN. 8vo, cloth, 10s. 6d.

Manual for Gas Engineering Students. By D. LEE. 18mo, cloth, 1s.

Telephones, their Construction and Management.
By F. C. ALLSOP. Crown 8vo, cloth, 5s.

Hydraulic Machinery, Past and Present. A Lecture
delivered to the London and Suburban Railway Officials' Association.
By H. ADAMS, Mem. Inst. C.E. *Folding plate.* 8vo, sewed, 1s.

Twenty Years with the Indicator. By THOMAS PRAY,
Jun., C.E., M.E., Member of the American Society of Civil Engineers.
2 vols., royal 8vo, cloth, 12s. 6d.

Annual Statistical Report of the Secretary to the
Members of the Iron and Steel Association on the Home and Foreign Iron
and Steel Industries in 1889. Issued June 1890. 8vo, sewed, 5s.

Bad Drains, and How to Test them; with Notes on
the Ventilation of Sewers, Drains, and Sanitary Fittings, and the Origin
and Transmission of Zymotic Disease. By R. HARRIS REEVES. Crown
8vo, cloth, 3s. 6d.

Well Sinking. The modern practice of Sinking
and Boring Wells, with geological considerations and examples of Wells.
By ERNEST SPON, Assoc. Mem. Inst. C.E., Mem. Soc. Eng., and of the
Franklin Inst., etc. Second edition, revised and enlarged. Crown 8vo,
cloth, 10s. 6d.

The Voltaic Accumulator: an Elementary Treatise.
By ÉMILE REYNIER. Translated by J. A. BERLY, Assoc. Inst. E.E.
With 62 illustrations, 8vo, cloth, 9s.

Ten Years' Experience in Works of Intermittent
Downward Filtration. By J. BAILEY DENTON, Mem. Inst. C.E.
Second edition, with additions. Royal 8vo, cloth, 5s.

Land Surveying on the Meridian and Perpendicular
System. By WILLIAM PENMAN, C.E. 8vo, cloth, 8s. 6d.

The Electromagnet and Electromagnetic Mechanism.
By SILVANUS P. THOMPSON, D.Sc., F.R.S. 8vo, cloth, 15s.

Incandescent Wiring Hand-Book. By F. B. BADT, late 1st Lieut. Royal Prussian Artillery. *With 41 illustrations and 5 tables.* 18mo, cloth, 4s. 6d.

A Pocket-book for Pharmacists, Medical Practitioners, Students, etc., etc. (*British, Colonial, and American*). By THOMAS BAYLEY, Assoc. R. Coll. of Science, Consulting Chemist, Analyst, and Assayer, Author of a 'Pocket-book for Chemists,' 'The Assay and Analysis of Iron and Steel, Iron Ores, and Fuel,' etc., etc. Royal 32mo, boards, gilt edges, 6s.

The Fireman's Guide; a Handbook on the Care of Boilers. By TEKNOLOG, föreningen T. I. Stockholm. Translated from the third edition, and revised by KARL P. DAHLSTROM, M.E. Second edition. Fcap. 8vo, cloth, 2s.

A Treatise on Modern Steam Engines and Boilers, including Land Locomotive, and Marine Engines and Boilers, for the use of Students. By FREDERICK COLYER, M. Inst. C.E., Mem. Inst. M.E. *With 36 plates.* 4to, cloth, 12s. 6d.

CONTENTS:

1. Introduction—2. Original Engines—3. Boilers—4. High-Pressure Beam Engines—5. Cornish Beam Engines—6. Horizontal Engines—7. Oscillating Engines—8. Vertical High-Pressure Engines—9. Special Engines—10. Portable Engines—11. Locomotive Engines—12. Marine Engines.

Steam Engine Management; a Treatise on the Working and Management of Steam Boilers. By F. COLYER, M. Inst. C.E., Mem. Inst. M.E. New edition, 18mo, cloth, 3s. 6d.

A Text-Book of Tanning, embracing the Preparation of all kinds of Leather. By HARRY R. PROCTOR, F.C.S., of Low Lights Tanneries. *With illustrations.* Crown 8vo, cloth, 10s. 6d.

Aid Book to Engineering Enterprise. By EWING MATHESON, M. Inst. C.E. The Inception of Public Works, Parliamentary Procedure for Railways, Concessions for Foreign Works, and means of Providing Money, the Points which determine Success or Failure, Contract and Purchase, Commerce in Coal, Iron, and Steel, &c. Second edition, revised and enlarged, 8vo, cloth, 21s.

Pumps, Historically, Theoretically, and Practically Considered. By P. R. BJÖRLING. *With 156 illustrations.* Crown 8vo, cloth, 7s. 6d.

The Marine Transport of Petroleum. A Book for the use of Shipowners, Shipbuilders, Underwriters, Merchants, Captains and Officers of Petroleum-carrying Vessels. By G. H. LITTLE, Editor of the 'Liverpool Journal of Commerce.' Crown 8vo, cloth, 10s. 6d.

Liquid Fuel for Mechanical and Industrial Purposes. Compiled by E. A. BRAVLEY HODGETTS. *With wood engravings.* 8vo, cloth, 7s. 6d.

Tropical Agriculture: A Treatise on the Culture, Preparation, Commerce and Consumption of the principal Products of the Vegetable Kingdom. By P. L. SIMMONDS, F.L.S., F.R.C.I. New edition, revised and enlarged, 8vo, cloth, 21s.

Health and Comfort in House Building; or, Ventilation with Warm Air by Self-acting Suction Power. With Review of the Mode of Calculating the Draught in Hot-air Flues, and with some Actual Experiments by J. DRYSDALE, M.D., and J. W. HAYWARD, M.D. *With plates and woodcuts.* Third edition, with some New Sections, and the whole carefully Revised, 8vo, cloth, 7s. 6d.

Losses in Gold Amalgamation. With Notes on the Concentration of Gold and Silver Ores. *With six plates.* By W. McDERMOTT and P. W. DUFFIELD. 8vo, cloth, 5s.

A Guide for the Electric Testing of Telegraph Cables. By Col. V. HOSKIÆR, Royal Danish Engineers. Third edition, crown 8vo, cloth, 4s. 6d.

The Hydraulic Gold Miners' Manual. By T. S. G. KIRKPATRICK, M.A. Oxon. *With 6 plates.* Crown 8vo, cloth, 6s.

"We venture to think that this work will become a text-book on the important subject of which it treats. Until comparatively recently hydraulic mines were neglected. This was scarcely to be surprised at, seeing that their working in California was brought to an abrupt termination by the action of the farmers on the *débris* question, whilst their working in other parts of the world had not been attended with the anticipated success."—*The Mining World and Engineering Record.*

The Arithmetic of Electricity. By T. O'CONOR SLOANE. Crown 8vo, cloth, 4s. 6d.

The Turkish Bath: Its Design and Construction for
Public and Commercial Purposes. By R. O. ALLSOP, Architect. *With
plans and sections.* 8vo, cloth, 6s.

Earthwork Slips and Subsidences upon Public Works:
Their Causes, Prevention and Reparation. Especially written to assist
those engaged in the Construction or Maintenance of Railways, Docks,
Canals, Waterworks, River Banks, Reclamation Embankments, Drainage
Works, &c., &c. By JOHN NEWMAN, Assoc. Mem. Inst. C.E., Author
of 'Notes on Concrete,' &c. Crown 8vo, cloth, 7s. 6d.

Gas and Petroleum Engines: A Practical Treatise
on the Internal Combustion Engine. By WM. ROBINSON, M.E., Senior
Demonstrator and Lecturer on Applied Mechanics, Physics, &c., City
and Guilds of London College, Finsbury, Assoc. Mem. Inst. C.E., &c.
Numerous illustrations. 8vo, cloth, 14s.

*Waterways and Water Transport in Different Coun-
tries.* With a description of the Panama, Suez, Manchester, Nicaraguan,
and other Canals. By J. STEPHEN JEANS, Author of 'England's
Supremacy,' 'Railway Problems,' &c. *Numerous illustrations.* 8vo,
cloth, 14s.

A Treatise on the Richards Steam-Engine Indicator
and the Development and Application of Force in the Steam-Engine.
By CHARLES T. PORTER. Fourth Edition, revised and enlarged, 8vo,
cloth, 9s.

CONTENTS.

The Nature and Use of the Indicator:
The several lines on the Diagram.
Examination of Diagram No. 1.
Of Truth in the Diagram.
Description of the Richards Indicator.
Practical Directions for Applying and Taking
Care of the Indicator.
Introductory Remarks.
Units.
Expansion.
Directions for ascertaining from the Diagram
the Power exerted by the Engine.
To Measure from the Diagram the Quantity
of Steam Consumed.
To Measure from the Diagram the Quantity
of Heat Expended.
Of the Real Diagram, and how to Construct it.
Of the Conversion of Heat into Work in the
Steam-engine.
Observations on the several Lines of the
Diagram.

Of the Loss attending the Employment of
Slow-piston Speed, and the Extent to
which this is Shown by the Indicator.
Of other Applications of the Indicator.
Of the use of the Tables of the Properties of
Steam in Calculating the Duty of Boilers.
Introductory.
Of the Pressure on the Crank when the Con-
necting-rod is conceived to be of Infinite
Length.
The Modification of the Acceleration and
Retardation that is occasioned by the
Angular Vibration of the Connecting-rod.
Method of representing the actual pressure
on the crank at every point of its revolu-
tion.
The Rotative Effect of the Pressure exerted
on the Crank.
The Transmitting Parts of an Engine, con-
sidered as an Equaliser of Motion.
A Ride on a Buffer-beam (Appendix).

In demy 4to, handsomely bound in cloth, *illustrated with* **220** *full page plates*,
Price 15s.

ARCHITECTURAL EXAMPLES

IN BRICK, STONE, WOOD, AND IRON.

A COMPLETE WORK ON THE DETAILS AND ARRANGEMENT OF BUILDING CONSTRUCTION AND DESIGN.

By WILLIAM FULLERTON, Architect.

Containing 220 Plates, with numerous Drawings selected from the Architecture
of Former and Present Times.

*The Details and Designs are Drawn to Scale, $\frac{1}{8}$", $\frac{1}{4}$", $\frac{1}{2}$", and Full size
being chiefly used.*

The Plates are arranged in Two Parts. The First Part contains
Details of Work in the four principal Building materials, the following
being a few of the subjects in this Part :—Various forms of Doors and
Windows, Wood and Iron Roofs, Half Timber Work, Porches,
Towers, Spires, Belfries, Flying Buttresses, Groining, Carving, Church
Fittings, Constructive and Ornamental Iron Work, Classic and Gothic
Molds and Ornament, Foliation Natural and Conventional, Stained
Glass, Coloured Decoration, a Section to Scale of the Great Pyramid,
Grecian and Roman Work, Continental and English Gothic, Pile
Foundations, Chimney Shafts according to the regulations of the
London County Council, Board Schools. The Second Part consists
of Drawings of Plans and Elevations of Buildings, arranged under the
following heads :—Workmen's Cottages and Dwellings, Cottage Resi-
dences and Dwelling Houses, Shops, Factories, Warehouses, Schools,
Churches and Chapels, Public Buildings, Hotels and Taverns, and
Buildings of a general character.

All the Plates are accompanied with particulars of the Work, with
Explanatory Notes and Dimensions of the various parts.

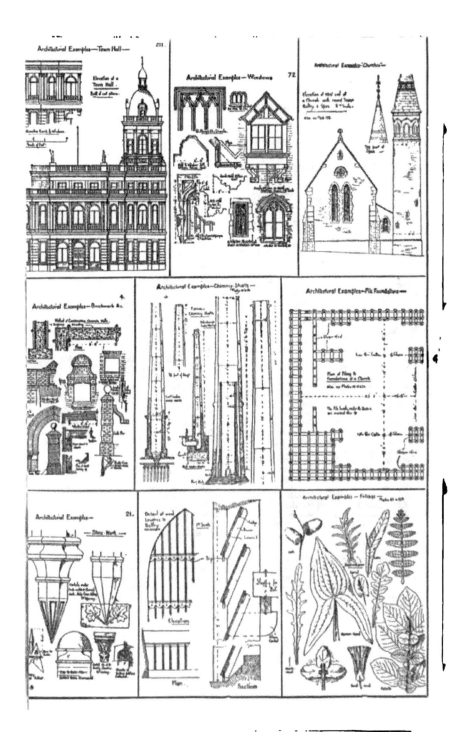

Crown 8vo, cloth, with illustrations, 5s.

WORKSHOP RECEIPTS,

FIRST SERIES.

By ERNEST SPON.

SYNOPSIS OF CONTENTS.

Bookbinding.
Bronzes and Bronzing.
Candles.
Cement.
Cleaning.
Colourwashing.
Concretes.
Dipping Acids.
Drawing Office Details.
Drying Oils.
Dynamite.
Electro - Metallurgy — (Cleaning, Dipping, Scratch-brushing, Batteries, Baths, and Deposits of every description).
Enamels.
Engraving on Wood, Copper, Gold, Silver, Steel, and Stone.
Etching and Aqua Tint.
Firework Making — (Rockets, Stars, Rains, Gerbes, Jets, Tourbillons, Candles, Fires, Lances, Lights, Wheels, Fire-balloons, and minor Fireworks).
Fluxes.
Foundry Mixtures.

Freezing.
Fulminates.
Furniture Creams, Oils, Polishes, Lacquers, and Pastes.
Gilding.
Glass Cutting, Cleaning, Frosting, Drilling, Darkening, Bending, Staining, and Painting.
Glass Making.
Glues.
Gold.
Graining.
Gums.
Gun Cotton.
Gunpowder.
Horn Working.
Indiarubber.
Japans, Japanning, and kindred processes.
Lacquers.
Lathing.
Lubricants.
Marble Working.
Matches.
Mortars.
Nitro-Glycerine.
Oils.

Paper.
Paper Hanging.
Painting in Oils, in Water Colours, as well as Fresco, House, Transparency, Sign, and Carriage Painting.
Photography.
Plastering.
Polishes.
Pottery—(Clays, Bodies, Glazes, Colours, Oils, Stains, Fluxes, Enamels, and Lustres).
Scouring.
Silvering.
Soap.
Solders.
Tanning.
Taxidermy.
Tempering Metals.
Treating Horn, Mother-o'-Pearl, and like substances.
Varnishes, Manufacture and Use of.
Veneering.
Washing.
Waterproofing.
Welding.

Besides Receipts relating to the lesser Technological matters and processes, such as the manufacture and use of Stencil Plates, Blacking, Crayons, Paste, Putty, Wax, Size, Alloys, Catgut, Tunbridge Ware, Picture Frame and Architectural Mouldings, Compos, Cameos, and others too numerous to mention.

Crown 8vo, cloth, 485 pages, with illustrations, 5s.

WORKSHOP RECEIPTS,

SECOND SERIES.

By ROBERT HALDANE.

SYNOPSIS OF CONTENTS.

Acidimetry and Alkalimetry.	Disinfectants.	Iodoform.
Albumen.	Dyeing, Staining, and Colouring.	Isinglass.
Alcohol.	Essences.	Ivory substitutes.
Alkaloids.	Extracts.	Leather.
Baking-powders.	Fireproofing.	Luminous bodies.
Bitters.	Gelatine, Glue, and Size.	Magnesia.
Bleaching.	Glycerine.	Matches.
Boiler Incrustations.	Gut.	Paper.
Cements and Lutes.	Hydrogen peroxide.	Parchment.
Cleansing.	Ink.	Perchloric acid.
Confectionery.	Iodine.	Potassium oxalate.
Copying.		Preserving.

Pigments, Paint, and Painting : embracing the preparation of *Pigments*, including alumina lakes, blacks (animal, bone, Frankfort, ivory, lamp, sight, soot), blues (antimony, Antwerp, cobalt, cæruleum, Egyptian, manganate, Paris, Péligot, Prussian, smalt, ultramarine), browns (bistre, hinau, sepia, sienna, umber, Vandyke), greens (baryta, Brighton, Brunswick, chrome, cobalt, Douglas, emerald, manganese, mitis, mountain, Prussian, sap, Scheele's, Schweinfurth, titanium, verdigris, zinc), reds (Brazilwood lake, carminated lake, carmine, Cassius purple, cobalt pink, cochineal lake, colcothar, Indian red, madder lake, red chalk, red lead, vermilion), whites (alum, baryta, Chinese, lead sulphate, white lead—by American, Dutch, French, German, Kremnitz, and Pattinson processes, precautions in making, and composition of commercial samples—whiting, Wilkinson's white, zinc white), yellows (chrome, gamboge, Naples, orpiment, realgar, yellow lakes) ; *Paint* (vehicles, testing oils, driers, grinding, storing, applying, priming, drying, filling, coats, brushes, surface, water-colours, removing smell, discoloration ; miscellaneous paints—cement paint for carton-pierre, copper paint, gold paint, iron paint, lime paints, silicated paints, steatite paint, transparent paints, tungsten paints, window paint, zinc paints) ; *Painting* (general instructions, proportions of ingredients, measuring paint work ; carriage painting—priming paint, best putty, finishing colour, cause of cracking, mixing the paints, oils, driers, and colours, varnishing, importance of washing vehicles, re-varnishing, how to dry paint ; woodwork painting).

Crown 8vo, cloth, 480 pages, with 183 illustrations, 5s.

WORKSHOP RECEIPTS,

THIRD SERIES.

By C. G. WARNFORD LOCK.

Uniform with the First and Second Series.

SYNOPSIS OF CONTENTS.

Alloys.	Indium.	Rubidium.
Aluminium.	Iridium.	Ruthenium.
Antimony.	Iron and Steel.	Selenium.
Barium.	Lacquers and Lacquering.	Silver.
Beryllium.	Lanthanum.	Slag.
Bismuth.	Lead.	Sodium.
Cadmium.	Lithium.	Strontium.
Cæsium.	Lubricants.	Tantalum.
Calcium.	Magnesium.	Terbium.
Cerium.	Manganese.	Thallium.
Chromium.	Mercury.	Thorium.
Cobalt.	Mica.	Tin.
Copper.	Molybdenum.	Titanium.
Didymium.	Nickel.	Tungsten.
Electrics.	Niobium.	Uranium.
Enamels and Glazes.	Osmium.	Vanadium.
Erbium.	Palladium.	Yttrium.
Gallium.	Platinum.	Zinc.
Glass.	Potassium.	Zirconium.
Gold.	Rhodium.	

WORKSHOP RECEIPTS,

FOURTH SERIES,

DEVOTED MAINLY TO HANDICRAFTS & MECHANICAL SUBJECTS.

By C. G. WARNFORD LOCK.

250 Illustrations, with Complete Index, and a General Index to the Four Series, 5s.

--- -- ————

Waterproofing — rubber goods, cuprammonium processes, miscellaneous preparations.

Packing and Storing articles of delicate odour or colour, of a deliquescent character, liable to ignition, apt to suffer from insects or damp, or easily broken.

Embalming and Preserving anatomical specimens.

Leather Polishes:

Cooling Air and Water, producing low temperatures, making ice, cooling syrups and solutions, and separating salts from liquors by refrigeration.

Pumps and Siphons, embracing every useful contrivance for raising and supplying water on a moderate scale, and moving corrosive, tenacious, and other liquids.

Desiccating—air- and water-ovens, and other appliances for drying natural and artificial products.

Distilling—water, tinctures, extracts, pharmaceutical preparations, essences, perfumes, and alcoholic liquids.

Emulsifying as required by pharmacists and photographers.

Evaporating—saline and other solutions, and liquids demanding special precautions.

Filtering—water, and solutions of various kinds.

Percolating and Macerating.

Electrotyping.

Stereotyping by both plaster and paper processes.

Bookbinding in all its details.

Straw Plaiting and the fabrication of baskets, matting, etc.

Musical Instruments—the preservation, tuning, and repair of pianos, harmoniums, musical boxes, etc.

Clock and Watch Mending—adapted for intelligent amateurs.

Photography—recent development in rapid processes, handy apparatus, numerous recipes for sensitizing and developing solutions, and applications to modern illustrative purposes.

NOW COMPLETE.

With nearly 1500 *illustrations,* in super-royal 8vo, in 5 Divisions, cloth.
Divisions 1 to 4, 13*s.* 6*d.* each ; Division 5, 17*s.* 6*d.* ; or 2 vols., cloth, £3 10*s.*

SPONS' ENCYCLOPÆDIA

OF THE

INDUSTRIAL ARTS, MANUFACTURES, AND COMMERCIAL PRODUCTS.

EDITED BY C. G. WARNFORD LOCK, F.L.S.

Among the more important of the subjects treated of, are the
following :—

Acids, 207 pp. 220 figs.
Alcohol, 23 pp. 16 figs.
Alcoholic Liquors, 13 pp.
Alkalies, 89 pp. 78 figs.
Alloys.   Alum.,
Asphalt.   Assaying.
Beverages, 89 pp. 29 figs.
Blacks.
Bleaching Powder, 15 pp.
Bleaching, 51 pp. 48 figs.
Candles, 18 pp. 9 figs.
Carbon Bisulphide.
Celluloid, 9 pp.
Cements.   Clay.
Coal-tar Products, 44 pp.
    14 figs.
Cocoa, 8 pp.
Coffee, 32 pp. 13 figs.
Cork, 8 pp. 17 figs.
Cotton Manufactures, 62
    pp. 57 figs.
Drugs, 38 pp.
Dyeing   and   Calico
    Printing, 28 pp. 9 figs.
Dyestuffs, 16 pp.
Electro-Metallurgy,   13
    pp.
Explosives, 22 pp. 33 figs.
Feathers.
Fibrous Substances, 92
    pp. 79 figs.
Floor-cloth, 16 pp. 21
    figs.
Food Preservation, 8 pp.
Fruit, 8 pp.

Fur, 5 pp.
Gas, Coal, 8 pp.
Gems.   .
Glass, 45 pp. 77 figs.
Graphite, 7 pp.
Hair, 7 pp.
Hair Manufactures.
Hats, 26 pp. 26 figs.
Honey.   Hops.
Horn.
Ice, 10 pp. 14 figs.
Indiarubber   Manufac-
    tures, 23 pp. 17 figs.
Ink, 17 pp.
Ivory.
Jute Manufactures,   11
    pp., 11 figs.
Knitted   Fabrics —
    Hosiery, 15 pp. 13 figs.
Lace, 13 pp. 9 figs.
Leather, 28 pp. 31 figs.
Linen Manufactures, 16
    pp. 6 figs.
Manures, 21 pp. 30 figs.
Matches, 17 pp. 38 figs.
Mordants, 13 pp.
Narcotics, 47 pp.
Nuts, 10 pp.
Oils and Fatty Sub-
    stances, 125 pp.
Paint.
Paper, 26 pp. 23 figs.
Paraffin, 8 pp. 6 figs.
Pearl and Coral, 8 pp.
Perfumes, 10 pp.

Photography, 13 pp. 20
    figs.
Pigments, 9 pp. 6 figs.
Pottery, 46 pp. 57 figs.
Printing and Engraving,
    20 pp. 8 figs.
Rags.
Resinous and Gummy
    Substances, 75 pp. 16
    figs.
Rope, 16 pp. 17 figs.
Salt, 31 pp. 23 figs.
Silk, 8 pp.
Silk Manufactures, 9 pp.
    11 figs.
Skins, 5 pp.
Small Wares, 4 pp.
Soap and Glycerine, 39
    pp. 45 figs.
Spices, 16 pp.
Sponge, 5 pp.
Starch, 9 pp. 10 figs.
Sugar,   155   pp.   134
    figs.
Sulphur.
Tannin, 18 pp.
Tea, 12 pp.
Timber, 13 pp.
Varnish, 15 pp.
Vinegar, 5 pp.
Wax, 5 pp.
Wool, 2 pp.
Woollen. Manufactures,
    58 pp. 39 figs.

In super-royal 8vo, 1168 pp., *with* 2400 *illustrations*, in 3 Divisions, cloth, price 13*s.* 6*d.* each ; or 1 vol., cloth, 2*l.* ; or half-morocco, 2*l.* 8*s.*

A SUPPLEMENT

TO

SPONS' DICTIONARY OF ENGINEERING.

EDITED BY ERNEST SPON, MEMB. SOC. ENGINEERS.

Abacus, Counters, Speed Indicators, and Slide Rule.

Agricultural Implements and Machinery.

Air Compressors.

Animal Charcoal Machinery.

Antimony.

Axles and Axle-boxes.

Barn Machinery.

Belts and Belting.

Blasting. Boilers.

Brakes.

Brick Machinery.

Bridges.

Cages for Mines.

Calculus, Differential and Integral.

Canals.

Carpentry.

Cast Iron.

Cement, Concrete, Limes, and Mortar.

Chimney Shafts.

Coal Cleansing and Washing.

Coal Mining.

Coal Cutting Machines.

Coke Ovens. Copper.

Docks. Drainage.

Dredging Machinery.

Dynamo - Electric and Magneto-Electric Machines.

Dynamometers.

Electrical Engineering, Telegraphy, Electric Lighting and its practicaldetails,Telephones

Engines, Varieties of.

Explosives. Fans.

Founding, Moulding and the practical work of the Foundry.

Gas, Manufacture of.

Hammers, Steam and other Power.

Heat. Horse Power.

Hydraulics.

Hydro-geology.

Indicators. Iron.

Lifts, Hoists, and Elevators.

Lighthouses, Buoys, and Beacons.

Machine Tools.

Materials of Construction.

Meters.

Ores, Machinery and Processes employed to Dress.

Piers.

Pile Driving.

Pneumatic Transmission.

Pumps.

Pyrometers.

Road Locomotives.

Rock Drills.

Rolling Stock.

Sanitary Engineering.

Shafting.

Steel.

Steam Navvy.

Stone Machinery.

Tramways.

Well Sinking.

JUST PUBLISHED.

In demy 8vo, cloth, 600 pages, and 1420 Illustrations, 6s.

SPONS'
MECHANICS' OWN BOOK;
A MANUAL FOR HANDICRAFTSMEN AND AMATEURS.

CONTENTS.

Mechanical Drawing—Casting and Founding in Iron, Brass, Bronze, and other Alloys—Forging and Finishing Iron—Sheetmetal Working —Soldering, Brazing, and Burning—Carpentry and Joinery, embracing descriptions of some 400 Woods, over 200 Illustrations of Tools and their uses, Explanations (with Diagrams) of 116 joints and hinges, and Details of Construction of Workshop appliances, rough furniture, Garden and Yard Erections, and House Building—Cabinet-Making and Veneering — Carving and Fretcutting — Upholstery — Painting, Graining, and Marbling — Staining Furniture, Woods, Floors, and Fittings—Gilding, dead and bright, on various grounds—Polishing Marble, Metals, and Wood—Varnishing—Mechanical movements, illustrating contrivances for transmitting motion—Turning in Wood and Metals—Masonry, embracing Stonework, Brickwork, Terracotta, and Concrete—Roofing with Thatch, Tiles, Slates, Felt, Zinc, &c.— Glazing with and without putty, and lead glazing—Plastering and Whitewashing— Paper-hanging— Gas-fitting—Bell-hanging, ordinary and electric Systems — Lighting — Warming — Ventilating — Roads, Pavements, and Bridges — Hedges, Ditches, and Drains — Water Supply and Sanitation—Hints on House Construction suited to new countries.

E. & F. N. SPON, 125, Strand, London.
New York : 12, Cortlandt Street.

2787
3